THE MOORLANDS
OF
ENGLAND AND WALES

THE MOORLANDS
OF
ENGLAND AND WALES

An Environmental History 8000 BC to AD 2000

I. G. SIMMONS

EDINBURGH
University Press

© I. G. Simmons, 2003

Edinburgh University Press Ltd
22 George Square, Edinburgh

Typeset in Minion
by Servis Filmsetting Ltd, Manchester, and
printed and bound in Great Britain by
Antony Rowe Ltd, Chippenham

A CIP record for this book is available from the British Library

ISBN 0 7486 1730 2 (hardback)
ISBN 0 7486 1731 0 (paperback)

The right of I. G. Simmons
to be identified as author of this work
has been asserted in accordance with
the Copyright, Designs and Patents Act 1988.

Grateful acknowledgement is made for permission to reproduce
material previously published elsewhere. Every effort has been
made to trace the copyright holders, but if any have been
inadvertently overlooked, the publisher will be pleased to make
the necessary arrangements at the first opportunity.

Unacknowledged plates, figures and tables are © Ian Simmons

Contents

Preface

My childhood was spent in East London and East Lincolnshire and until I was about twelve years old I had never realised that water actually moved in streams and rivers, since the River Thames and fenland dykes were not quite in the babbling brook category. I had certainly never been more than about 300 feet above sea level until I went with my school to the Lake District for a week in the mid-1950s. This was a revelation and it fixed in me a love of rocky streams as well as an interest in upland areas. In my final undergraduate year, we were asked to compose a mock examination paper dealing with the course on Britain and I remember my embarrassment when mine was singled out as having questions almost entirely on upland Britain. So a PhD thesis on the vegetation history of Dartmoor was a pretty obvious development and ever since then I have maintained a research interest in the prehistoric vegetation changes of the areas now moorlands, through my own work and that of post-graduates and Research Assistants. But more recently I have wanted to chron-icle the big picture, which I did for the whole of Britain in my 2001 book* from this same Press. This account of the moorlands' history is thus nested within the previous book. Unlike many books on upland Britain, it avoids the truly mountainous areas like The Lakes and Snowdonia since their physiognomy is often so very different, and likewise Scotland is sufficiently dissimilar to need its own book.

Intellectually, the most obvious debt is to W. H. Pearsall's *Mountains and Moorlands*, a classic of the Collins 'New Naturalist' series. There is another link, too, for Pearsall was the external examiner for my PhD thesis, setting a pattern for conducting vivas which I have often tried to emulate by making sure that the wider issues were not overlooked in the scrutiny of the detail of the diagrams. After that, there are few enough treatments of large areas over long time scales, though I have been strongly influenced by the work of such scholars as Christopher Smout and Donald Hughes. Even so, the flavour of what I choose to include (and that is the main problem of every author working at this type of temporal and spatial scale) is different from them and I expect to be criticised by some people for not adhering to some emerging ideology of how environmental history should be written. But this is a first

---

* *An Environmental History of Great Britain from 10,000 Years Ago to the Present Day*, Edinburgh University Press, 2001.

attempt at this type of account in the UK and I expect – and indeed hope – that others will in due course do better.

The end of the book sets out fairly starkly what I see as the dissonance between the history of the moorlands as seen from an ecological viewpoint and the place they hold in the national consciousness. On the one hand there is a story of diminishing biological productivity and biodiversity, and on the other the enshrining of scenery as National Parks and Areas of Outstanding Natural Beauty. I am inclined to the view that more biodiversity ought now to be the primary aim of moorland management and that I regard moorland scenery as such as something we were taught to like after the eighteenth century and could now understand that a new lesson was needed. Some moors and their valleys are indeed beautiful; others are far from it and both would, for example, benefit by far larger areas of deciduous woodland. So while I ended the 2001 book by 'trumpeting the place' (to adapt Dylan Thomas), my sound here is more that of the cor anglais or even, as in my last paragraph, the mournful hooter of a steam locomotive.

Along the track however, there have been some lovely people. In the Dartmoor phase, I was much encouraged by Palmer Newbould and Frank Oldfield and indeed without their support would probably never have got on with the work properly at all. Both are scientists of renown but both have never been afraid to believe that poetry and music for example were important features of life as well, and maybe something of that is reflected here in one chapter at least. Geoffrey Dimbleby was crucial on the subject of the Mesolithic because at the time the conventional wisdom was that only farming communities could alter woodlands. Since those days, many people have helped draft papers or with fieldwork: Margaret Atherden and Don Spratt come especially to mind. Close associates like Judy Turner and Jim Innes made certain pieces of work possible which otherwise would not have been undertaken. In the case of the present volume, I have gained greatly from a period of residence as a Visiting Scholar at St John's College, Oxford and I am grateful to the President and Fellows for their hospitality; the Forestry Library of that University was also very helpful. I was able to take the final draft away for a few quiet days thanks to the generosity of Chris and Alisoun Roberts whose Lake District house is an ideal place for quiet reflection without distractions other than the scenery. The final stage of preparation of the manuscript and illustrations was aided by Vicki Innes and I am grateful for all her willing work on all the boring bits, as I am to the British Academy for a small grant that made it possible to pay her.

Carol helped in the field on Dartmoor fieldwork in 1960 and has been there for all the other work ever since: no grant would be big enough to repay her.

I. G Simmons

Note to readers

Following criticism that some of the detail in my earlier book on Britain was difficult for overseas readers, I have included a list of the major places mentioned in the text with their National Grid references, so that they can be located on Ordnance Survey maps and on those atlases of Britain that use the National Grid. A list of acronyms and specialist abbreviations has also been included. The glossary is intended to make the book accessible to readers from any background, scientist and historian alike. The sources are documented in the Endnotes but there is also a select Bibliography for anybody who would like to see some core material on the moorlands' history and ecology. Much of the source material uses pre-metric units, and conversions are given where that might help the reader to judge absolute amounts, though envisaging large areas ('153,000 acres of moor were converted to . . .') is always rather difficult for most of us.

CHAPTER ONE

'The huge expanse of the moor'

THE MEANINGS OF 'MOORLAND'

If you drive south from Edinburgh along the A68 then the first English environment that you see is moorland, for at Carter Bar the view across the border is of open upland, with some afforestation and a reservoir.[1] It is a typical moorland landscape, though not many such scenes can be viewed to the accompaniment of a kilted bagpiper.

The linguistic transition is from 'muir' to 'moor' which reminds us that the word 'moor' appears in a number of European languages, but in English as used in upland areas it has come to have a special connotation. It becomes freed from its simple association with 'mire' and with the addition of '-land' comes to describe the particular combination of landform, vegetation, land cover and human usages that characterise many areas over about 300 metres above sea level. The low slopes of the summit plains distinguish such areas from mountains (like Snowdonia and the Lake District) and so help to formulate a landscape type which is instantly recognisable in the field, in photographs and on maps. The photographs which comprise Plates 1.1–1.3 exemplify these features and Figure 1.1 shows the distribution of the feature in England and Wales: clearly the moors form an important environmental element in what has for many years been labelled 'Highland Britain', following the archaeologist Sir Cyril Fox.[2]

The topographic occurrence of moorland is mostly of three types. The first and most obvious is that of the block with clear edges to adjoining lowlands and a single name: Dartmoor, The Yorkshire Dales, and the North York Moors are examples. Some of these may be joined together to form a moorland chain as in the Pennines from Derbyshire to the Scottish border, interrupted only by river valleys. The second type is of a moorland fringe to mountainous areas, where the moors to the east of the Lake District (such as Shap and Crosby Ravensworth fells) join the steeper terrain of the Lakes to the Howgill Fells and thence to the Yorkshire Dales and there are analogous areas to the south-west of the Snowdon massif. The third type is of fragments of moorland among more intensively manipulated environments, which clearly represent a former continuous swathe. The north-east projection of Rothbury Forest in

Plate 1.1 *A wild moor heartland. This is Tavy Cleave on Dartmoor (Grid Ref SX 548827), looking north-east in the direction of the high northern plateau of the Moor, with Black Hill in the middle distance and Dartmoor's highest point (619 m) just visible at eleven o'clock. The landscape is underlain entirely by granite, though none of the characteristic tors of Dartmoor is visible here. The steep sides of The Cleave are not alto- gether typical of this moorland, but the shallow slopes that dominate the rest of the picture certainly are. Combined with the high rainfall, they are often covered with blanket peat, which shows as black areas in the middle distance. The Cleave is not far into the Moor with enclosed land starting about 1 km downstream but is the start of an apparently unpopulated and seemingly natural tract of country. We have to imagine it as once covered with deciduous forest which was then cleared by a succession of human activities. The slopes to the sides of The Cleave and in the foreground have scarcely visible hut circles of presumed Bronze Age date. None of the later millennia of grazing, burning, occasional phases of reclamation and exploitation of minerals are obviously manifested here. Evidence for them has to be sought out, which adds to the illusion that these Moors are areas of 'natural beauty' when in fact the upper surface of the land and its ecosys- tems are largely the result of human activity.*
Source: Photography by Cambridge University Air Photograph Collection.

PLATE 1.2 *The moorland essentials in Mid-Wales. Rhuddallt (SN 788475) looking north towards the dam of Llyn Brianne at 190 m ASL, which was plugged in 1972. The heart of Wales in in fact a set of plateaux which carry moorland, whose concordant summit surfaces are well picked out by the snow cover. The snow can also stand proxy for blanket bog, which likewise covers the interfluves picked out by the sunlit areas. Sheep densities are implicit in the spread of bracken up the foreground slope. The human impress is more obvious than in the picture of Tavy Cleave since the valley is enclosed to improved grassland and has oak woodland on its steeper slopes: the spur in the centre is an RSPB reserve which was the sole breeding ground of red kites in 1899. The reservoir is surrounded by Forestry Commission plantations and is part of a scheme to improve the supply of water to Llanelli and west Glamorgan. The upper zone of the hills is not devoid of human traces for there are prehistoric period settlements on them, and on the large foreground plateau there are walls which mostly date from periods of enclosure in the nineteenth century when landowners were enabled to carve up commons and enclose them, though without necessarily 'improving' the land itself. As with Dartmoor, the impression from the air is of a wilderness, but on the ground, the dissection by valleys with enclosed land infrequently gives that impression.*
Source: © Crown Copyright: RCAHMW.

Northumberland is one of a number of areas of such land in that county; although rather low in altitude for the usual definition of moorland, there are some areas of Cornwall like St Austell Moor and the Goonhilly Downs that feel more like moor than lowland heath. The current book will however concentrate on the main blocks which show up on images derived from satellite data (Fig. 1.2).

Taking all the regions together, certain attributes are shared. The most obvious is the treelessness of much of the land except where obviously recent plantations of coniferous trees spread over the hillsides. Occasionally, the lower slopes exhibit a covering of scrub, usually of hawthorn or mountain ash

PLATE 1.3 'Industrial moorland'. Looking north-west up Old Gang Beck, a tributary of the Swale in the Yorkshire Dales (NY 975006). The geology is dominated by the Carboniferous Limestone series which has produced a series of steps on the valley sides, best seen in the right middle ground. There is some suggestion of agricultural activity in the enclosure walls on the south-west side of the valley but two ecologies dominate visually. The first is the now-dead lead industry which until about 1900 extracted ore from veins in the limestone, crushed it and smelted it into lead pigs which were taken away

and there are a few small woods, often dominated by one tree species, like the ash woods of the Craven Pennines or the oak copses of Dartmoor. In North and Mid-Wales, a continuous cover of oak wood may clothe the lower slopes but in the Valleys of the south, there is none. The lower boundary of the open land may be distinct, with a fence or wall separating obviously improved grassland from the rougher pastures above. By contrast, it may be irregular as if maintenance of the boundary had ceased and the two types of land cover were merging. Above such a boundary the land cover is dominated by a low vegetation in which grasses, sedges, rushes or dwarf ericaceous shrubs (such as heather) are present in varying amounts of cover, and the whole is interrupted by wet green areas where bog-moss indicates the presence of a spring. The valley sides often become steep for a stretch before flattening to summit plains of a very low angle of slope. These plateaux are often covered with a blanket of peat which may be as much as 3 m deep, often eroded into linear channels or into a reticulation of wasting islets among a maze of passages. Any of the areas of low slope may show a linear pattern of modern drains, designed to carry off the water quickly and dry out the peat surface to some extent. Seen from the air, the drier moors are striped into the patterns made by managed moor burning. Animal life is relatively sparse: on eastern moorlands the red grouse is the most frequently seen bird, but on all moors the sound of the curlew is usually heard. Other species like skylarks and pipits are frequently seen, and the conifer plantations attract roe deer; Exmoor has wild red deer and the lower slopes everywhere have rabbits. Predators are present but less often casually seen since they are fewer in number: hen harriers and merlins would fit that category. Scavengers such as carrion crows and ravens are more obvious, and the story of the red kite is detailed later in the book. For some years the Staffordshire moors and adjacent scrub carried a naturalised population of wallabies, but these have now disappeared. Bodmin Moor especially (but others from time to time) provides reports of big **feral** cats of the size of the lynx or the puma.

A key element in the moorland history is the relationship between a core of moorland (once largely created out of **broadleaved woodland**) and the enclosed land below it. The boundary is continually shifting, most often

PLATE 1.3 *(continued)*
by cart and by packhorse, on routes such as the linear track along the middle slope. The remains of the Surrender smelting mill (350 m ASL) are visible in the valley bottom, surrounded by waste material and a flue runs straight up the hill. This carried lead-rich toxic fumes from the smelter to a dispersal chimney, now disappeared. The hillsides in many places are dominated by waste material, which shows pale on the photograph. The open moors are managed for grouse production by the rotation burning of heather in strips and patches and this gives the quiltwork appearance to much of the landscape since there is a mosaic of heather of different ages. A third but less obvious ecology, that of sheep grazing, is probably the most important in terms of today's livelihoods.
Source: Reproduced by permission of R. White, Yorkshire Dales National Park Authority.

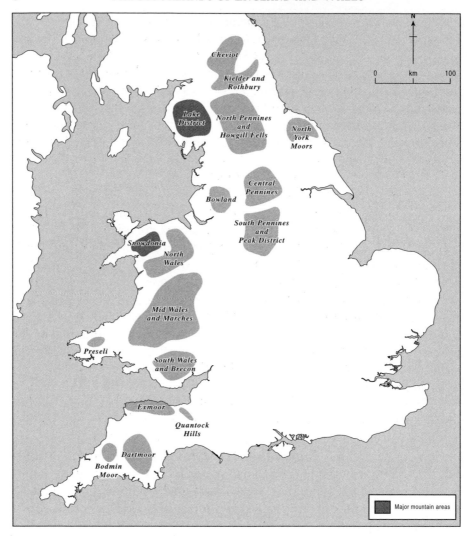

FIGURE 1.1 *The main areas of moorland dealt with in this book. There are some small areas of moor and heath that are only rarely referred to and these are omitted from this map. The two large areas of true mountain are identified but they are not part of the subject matter except in so far as both have peripheries of moorland.*

between open land covered with apparently natural plants, and enclosed land where farmers have sought to manage the pasture for grazing and hay production, by fertilising, draining, supervising grazing times and densities and maybe seeding. In response to climate and to economic factors (and more recently to environmentally-informed or recreational pressures) the interface between the open moorland and the enclosed ground has risen and fallen like the tides. Hence, the debris from a high tide level can be seen in different vegetation patterns, for example, or in rushes colonising former **enclosures** now bounded by broken-down walls. In a very long time-span, we might also

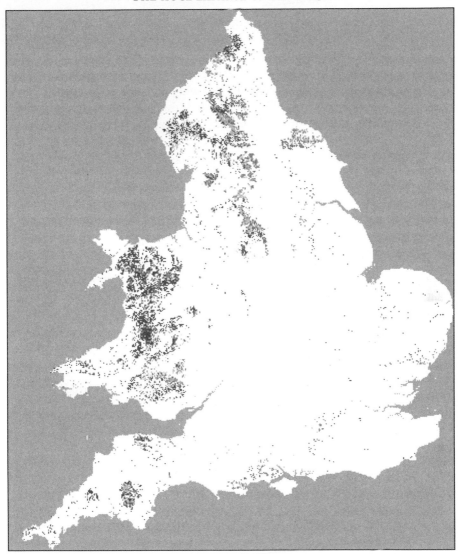

FIGURE 1.2 *Wales and England, with the land classes of upland areas in 2000. The classes include acid grasslands, bracken, dwarf shrub heaths, bog and montane habitats but neither woodland nor coniferous afforestation. These represent the core areas of semi-natural vegetation and are surrounded by and interspersed with more intensively managed regions which also contribute to the perceptions of moorland areas.*
Source: © Natural Environment Research Council, Centre for Ecology and Hydrology, Monks Wood.

visualise it as being like the pulsing of a jellyfish: there is a constant core which seems to expand and contract as we watch. This core was for long called the **waste**, in modern times carrying many of the connotations of that word.

The valleys which intersect the uplands are also part of the characteristic environments. They are usually divided into small fields with some stone somewhere in their boundaries, though this may be surmounted by trees and bushes, especially in the south and west. The land cover is now most often grass rather than **tillage** and small deciduous woods are found, characteristically tracing the steeper slopes of the valley sides. Previous **forestry** practices may have left a legacy of small to medium-sized patches of coniferous plantation. The settlement pattern is above all that of the village and hamlet, nowadays expanded linearly along the valley floor. There are also dwellings in more isolated locations: farms and rows of cottages and 'big houses' are all found away from the village foci. Some of these are clearly associated with industries past or present, which are mostly concentrated upon the extraction of minerals. Quarries into the valley sides are the most obvious, but in the north of England and scattered places in Wales, remnants of the mining and processing of lead are found in remote valleys and on the hilltops. Current energy pricing has peppered some summits and exposed hillsides with windfarms. The whole is generally now much less densely populated than surrounding areas and livings are harder to make: the government designation of 'Less Favoured Areas' (LFAs), has all kinds of landscape, environment and economic resonances.[3]

Thus the landscape, the visible phenomena. In writing an environmental history, however, there has to be a complementary approach, which is that of the dynamics of the systems which have their visual expression as the landscape in front of our eyes. That is to say, we need to have information about the ways in which the **biophysical** systems (what we might term **nature**) have interacted with the cultural systems configured by human societies; we need this information for as long a time span as possible, and we also need to recognise that many of the interactions will in the case of the moorlands produce intermediate states which are no longer solely 'natural' but are not obviously 'man-made' [sic] either. The case of a stand of heather on a moor is an example: the heather is not a domesticated plant (as wheat or pansies are) but its abundance is a result of the grazing and burning practices of the land managers. Since policies are made in the context of an entire culture, and not just its scientific knowledge, we need to appreciate the cultural frameworks through which humans have come to engage with these regions: biology and geology must be juxtaposed with economics, sociology and poetry in order to achieve a fuller delineation of these important places. We start off with a scientific representation, but later in the book we shall consider some of the other ways in which we construct symbolic models of the moorlands and the ways in which they may lead to action.

THE VIEW FROM ABOVE

The aerial photographs of the type in Plates 1.1–1.3 are symbols of the scientific approach, not the least because a great deal of a technology based on science has been used in their production. Their depiction of landforms, of vegetation patterns and of the sky itself remind us of the need to understand the geology and geomorphology of the landscape, the evolution and dynamics of the patterns of plant life, and of the climates which clearly must have exerted some influence on these.[4] All these phenomena have a history as well as a present. Obviously, too, they are part of the human scene and may well have been an influence on the course of the development of resources and of social life. We must beware of thinking though that they are a kind of passive scenery, changing at times of their own choice, against which the human drama moves with its own autonomy. The scenery here has been partially made by the actors, from a very early time. If a 'natural' environment is one in which there is no human influence at all, then the moorlands of England and Wales have not been 'natural' for much of the last 10,000 years. But systems with a strong human input are susceptible to description by the methods of the sciences even though the explanations have to be sought in the cultural sphere.

Geology

In England and Wales, a number of rock types have proved sufficiently resistant to erosion that they form the basis of upland terrains; these are summarised in Table 1.1. The youngest are of Jurassic age (North York Moors) and the oldest Ordovician (parts of Wales). Some of the geological strata are of limestone and thus predispose to soils with a high base status (e.g., the Carboniferous Limestones of the Pennines), others are highly siliceous and easily develop soils with a tendency to acidity. The granites of the south-west peninsula are of this type. The Millstone Grit forms some very characteristic upland features: a line of jagged rock forms the western boundary of the

TABLE 1.1 Rock types of moorland England and Wales

Period	Region	Lithology	Summit heights (m)
Jurassic	North York Moors	Limestones, sandstones, shales,	454
Carboniferous	Dartmoor and Bodmin	Granite	619 (DM) 572 (BM)
	Pennines, South Wales	Whin Sill, Coal Measures, Millstone Grit, Yoredales, Carboniferous Limestones	893 (NP) 724 (CP) 637 (SP)
Devonian	Cheviot	Granite	816
	South Wales	Red Sandstone	886
Silurian	Wales	Shales and volcanics	705 (MW)
Ordovician	Wales	Slates, shales, volcanics	811

Regional abbreviations: BM = Bodmin Moor; DM = Dartmoor; SW = South Wales; MW = Mid-Wales; NW = North Wales; SP = South Pennines; CP = Central Pennines; NP = North Pennines; NYM = North York Moors. For locations see Figure 1.1

Staffordshire moorlands between Leek and Buxton, for example, with features such as The Roaches, Hen Cloud and Ramshaw Rocks. Over it, **blanket bog** developed to the west and south-west of Buxton at Goyt's Moss may well be the closest such mire formation to an urban area in England and Wales.

The geology of any of the upland areas might however have been affected by the glaciations of the **Pleistocene**. Among a number of glaciations of the British Isles, there have been two major episodes of relevance. In the Wolstonian (about 250–165ky),* most of upland Britain was subjected to glaciation, though the highest areas may have been sufficiently upstanding to divert ice-flow round them. Only the south-west peninsula of England was largely outside the ice limit. The Devensian glaciation was at its height at 18ky and the southern limits stretched from East Anglia to South Wales. Parts of the Pennines and the North York Moors projected above this ice sheet and again the south-west uplands were probably not directly affected.[5] Around 10,000 BP there was a further cold period which resulted in some valley glaciations and ice-cap formations in mountain areas of Britain. The landforms of the moorlands were affected in three main ways by such episodes:

1. Firstly, the formation of erosional features such as **corries** (not so spectacular in e.g., the Brecon Beacons as in mountain areas like Snowdonia but present nevertheless) and over-deepened valleys, incised glacial meltwater channels and other sub-glacial drainage features.
2. Secondly, the deposition of material picked up by ice as many kinds of valley deposit, like moraines, kames and outwash fans. Lakes were formed which may have been short-lived or which (as in kettle-holes) persisted through into the **Holocene**, filling up only slowly.
3. Thirdly, there was **periglacial** activity marginal to the ice, resulting in **solifluction** on slopes and the formation of patterned ground. In limestone areas, drainage may have been on the surface at these times since the rocks were subject to permafrost; the characteristic underground drainage of limestone regions is thus seen as a later development.

Hence, any of the rock types in Table 1.1 may be obscured by deposits of glacial or periglacial origin: the more base-rich rocks may be plastered with patches of **till** of a lower nutrient content; equally, the most siliceous areas may be diversified with sheets of **drift** which locally provide more fertile soils. The outcome of these interactions. together with 10,000 years of post-glacial weathering, is a fairly standard type of upland topography. There are flat summits with accordant high points, long interfluves with large areas of low slope are common, as are sudden breaks of slope falling to steep valley sides

---

* For the very far past, 'ky' is used, meaning 'thousand years ago'. Dates given in years BP are radiocarbon years before 1950. Dates in years BC are calibrated radiocarbon dates; those AD are either from calibrated radiocarbon assays or from documentary sources.

with broad dales. Interruptions may be caused by volcanic intrusions like the Whin Sill in Upper Teesdale, and the south-western granites have their special boulder-strewn slopes or 'clitter'. Transitions between moorland regions and the surrounding lowlands are of interest: the North York Moors present both a steep scarp on the north and west, and an almost imperceptible (to the motorist at any rate) slope up from the Vale of Pickering. The Black Mountains seem almost like an island surrounded by lower terrain, as does Mynydd Preseli. Moors are the commonest location of **tors**, which are exposures of rock in situ, standing out on all sides from the surrounding slopes. At some stage in their history, periglacial processes have stripped away blocks and boulders which are piled up below the foot of the solid mass ('clitter' on Dartmoor) and weathering has often produced fantastic shapes that attract folk names. Examples, mainly on granite or gritstone, are found on Dartmoor, Exmoor, The Pennines, North York Moors, Cheviot and Preseli[6] and at Brimham Rocks near Ripon (N. Yorks) there is an outlier of rocks and scrub rising above the agricultural surroundings.

Soils

A diversity of soil types reflects the variety of combinations of substrate, slope, climate and vegetation. The very thickness of the soil profile may vary from the few centimetres of organic material and quartz overlying Carboniferous Limestone, which may easily be washed into the joints of the rock leaving a bare pavement, to a metre or more of **colluvial** material washed down a valley side to become the basis of a deep **brown earth** profile. Under woodland and on gentle to moderate slopes with grassland at the present day, **brown earths** (acid or calcareous according to subsoil type) are the commonest soil type and it may be presumed that all these were once woodland soils. Underneath most types of moorland vegetation there are now **podsolic** soils and these may be **gleyed** where the drainage is slow. Where water accumulated in the past, and on the highest areas of low slope, peat has accumulated in great sheets (often now eroded) over the landscape. Much of this **blanket peat** in water-shedding sites at high altitudes started to grow in the mid-Holocene and its history will be considered in more detail in the next chapter. The relations between the various environmental conditions and the soil type were summarised in one of W. H. Pearsall's diagrams in his pioneering book (Fig. 1.3).[7] Some of the terminology may have changed a little but the basic outlines are still valid. Although the present diversity of soil types is a valuable ecological indicator, soils and vegetation form an interactive system and if the vegetation has been different in the past then the soil too may have possessed different characteristics. The fossil soils beneath Bronze Age barrows on moorlands, which often suggest the former presence of brown earths in a zone now dominated by podsols, are a clue to such dynamic processes. Thus great care is needed when **retrodicting** soil types back to earlier times.

Though the basic lineaments of the physical environment of the moorlands

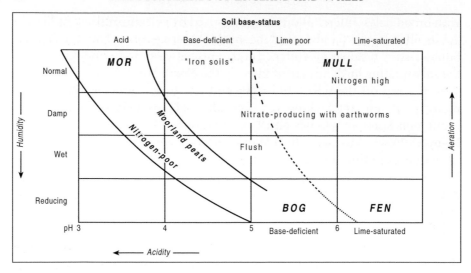

FIGURE 1.3 *W. H. Pearsall's original diagram of soil relations in the British uplands. Later work has confirmed the essentially correct information imparted by it; this book concentrates on the two left-hand columns, though also discussing 'normal' and 'damp' soils with grasslands and woodlands. It is largely those zones marked 'nitrogen-poor' which are undergoing eutrophication from atmospheric sources.*
Source: W. H. Pearsall, Mountains and Moorlands, *London: Collins New Naturalist, 1950, Fig. 16, p. 77.*

show much continuity between past and present in the last 10,000 years, what is now visible is not always an accurate guide to the past. Climate has changed, the **biota** have changed, the soils have undergone some degree of transformation, and many slope profiles have altered quite sharply. Under the woodlands of prehistoric times, the soils were largely dominated by acid brown earths. Now, as many surveys have shown, podzols, gleys and **stagnopodsols** predominate, often with a peaty top even if deep mires have not accumulated.[8] Stream courses on the uplands have altered as peat blankets grew and were eroded, and stream regimes and **alluviation** have altered with the density of vegetation cover. In limestone areas the streams have disappeared underground. Likewise, today's vegetation can only be an imperfect guide to past conditions. Some areas of **ancient woodland** exist in upland valleys but they are probably not very like their prehistoric forebears.[9] Their isolation makes them more open to the wind and its damaging effects on tree form, for example. It seems likely, too, that many if not all of them have been heavily managed during historic times: fuel or bark are obvious uses. This may account for single-species stands such as the Pennine ash-woods or the Dartmoor oak copses in places where it is known that the prehistoric woods consisted of mixed stands. The non-woody vegetation has also been subject to recent use practices which have changed its composition: shifts from cattle to sheep by moorland graziers or the moorburn procedures that accompany grouse moor management have produced configurations of species' dominance which are of quite recent date.

The normally unwanted spread of bracken on so many hillsides is also quite recent.[10] It is clear that the use of all available data is necessary in reconstructing the course of both the natural and the human histories of these regions.

Climate

All the patterns of soils and land cover have developed within the framework of the highly seasonal climate of upland Britain, a regime of low temperatures, severe wind exposure, very high precipitation, cloud and humidity, persistent winter frost and snow cover, a lack of sunshine, continual ground wetness from low evaporation rates, and often poor visibility. Where this varies most from the lowlands is in features which make British **lapse rates** among the steepest in the world: a recent investigation in the North Pennines revealed a rate of 6°C for every kilometre, starting at 171 m ASL.[11] One result is a plant growing season that decreases by about 13 days for every 100 metres gained in elevation: in the North Pennines the growing season lasts only from mid-April to October. There are significant local variations: the effects of topography on mean minimum temperatures are as great as those of altitude, so that at night a summit may be warmer than a valley side at the same height. The gradient of temperature with altitude has other sources of variation as well. It is highest in spring and will also vary according to whether the sun is shining or not across the whole gradient. If there is no sun then the gradient is 7°C per km but in the sun the difference is 10°C per km. In the North Pennines the lower difference is often noticed on the east flank in April–July since the lowlands are covered in sea-fret. There are seasonal differences to the west of the Pennines as well, first noticed by Gordon Manley but not yet satisfactorily explained.[12] Another regional phenomenon is the Helm Wind on the western side of the North Pennines, when the airflow pattern produces a strong cold north-easterly wind into the Eden valley, while all the while a cap of cloud (the 'helm', presumably from 'helmet') sits over the summits of the moors.[13]

The winter gradient is echoed in the rainfall, so that a valley in mid-Wales may average 1,130 mm/yr whereas only the highest part of the North York Moors receive over 1,000 mm/yr. Eastern slopes generally exhibit a rain-shadow effect with even the North Pennines getting 500–1,000 mm less rain than the Lake District fells to the west. Most of the rain is precipitated as air rises over high ground and so totals rise with altitude and precipitation from cloud becomes frequent. In Wales, precipitation rises by about 2.3 mm of rainfall per metre of altitude but as much as 5 mm on exposed windward slopes. Snow is commonest in the north and east of Britain, with the Cheviots topping the moorland league at 30–40 days per yr of snow and sleet; the south-western uplands rarely exceed 10 days. Northerly winds are critical and so the North York Moors may get as much snow as the higher ground of the Pennines; in both places it may fall as late as May.[14] On Dartmoor there were immense snowstorms at the end of the nineteenth century: in January 1881 the snow was 120 cm deep on the moor and in March 1891 near Princetown a train was effectively lost for thirty-six hours.

There are extremes as well as averages. Great Dun Fell in the Northern Pennines has the distinction of having recorded England's highest gust, of 116 knots (215 kph or 133 mph), in 1968. Heavy rainfall may often cause severe flooding in river valleys (Exmoor in 1952 for instance); long periods of drought dry out peat and surface vegetation alike, rendering them vulnerable to fire, as on Glaisdale Moor (North York Moors) in 1976. These events must however be seen in the context of today's land use and cover: neither would have been so severe in a largely forested landscape. But a period of low temperatures might well have affected the ecology by preventing the flowering of a species of tree, or a high one encouraged the immigration of a new parasite. In either case the effect is likely to be lagged, as would have been the case with vegetational response to climatic change. Insects and birds can respond much faster than plants to changing ambient conditions.

The Late Devensian's last two phases lead into the post-glacial or Holocene time. Between 13,000 and 11,000 BP the climate was warm, with temperatures reaching an average of 18°C in the English summer and allowing the development of pine-birch woodland. Towards the end of this warm interlude (termed the Windermere Interstadial) Britain and Ireland became separated. The final kick of the Pleistocene came with a return to colder conditions between 11,000 and 10,000 BP with the Loch Lomond stadial, which allowed the regrowth of an ice-field in the Western Highlands of Scotland, with smaller glaciers in the eastern and northern Highlands, some of the Hebrides, the Southern Uplands, Lake District, the uplands of both North and South Wales and some hills in Ireland. This period was however the last one in which glaciers appeared in the British Isles, and all traces of glacier ice seem to have vanished from these islands by 10,000 BP.

The succeeding period up to the present day is often described as post-glacial and indeed may be *a* post glacial period, though it is not inevitable that it should be *the* Post-glacial. The broad changes have been outlined in many books and monographs.[15] Environmentally, there was a rapid warming rate over the first 1,000 years from 10,000 BP of perhaps 1°C per century and by 9,500 BP temperatures similar to those of the present day were achieved. Evidence from the Greenland ice-cores suggests that some of the major warming took place within decades rather than centuries. By 8,500 BP the first elements of the deciduous forest biome had entered Britain and started the build-up that was complete by 7,000 BP. From 8,800–4,500 BP a **Hypsithermal Interval** or 'climatic optimum' can be recognised in which average annual temperatures were 1–2°C above the present; until about 7,000 BP rainfall was possibly 90 per cent of present but 110 per cent thereafter. Rising sea-levels in the present North Sea basin meant that the Dogger Bank was submerged about 7,500 BP and the southern North Sea formed by 6,500 BP. After then, the rate of rise of sea-level slowed up. After 5,000 BP, some climatic deterioration towards cooler and wetter conditions set in. Short-term variations would no doubt have included very wet years with possibly 50 per cent additional precipitation above average, which would have produced flood hazards even in a forested landscape. As more detailed work on climatic history using subfossil pollen, spores and macro-remains emerges, it seems likely that any picture of smooth changes between one climatic regime and another will be replaced by one in which there are many pauses and small-scale retreats. The Iron Age at May Moss (260 m ASL) on the North York Moors appears to span several fluctuations in surface wetness not accounted for by **successional** changes in the mire surface. These are subsumed into several major fluctuations since the Iron Age.[16] The major transitions are:

1. Iron Age: climatic deterioration
2. Roman period: warm and dry especially AD 200–400
3. 'Dark Ages': climatic deterioration during AD 260–540; drier between AD 650–860 and 690–980 but after the 690–980 span there were wetter conditions but with marked fluctuations.
4. A warm but variable period with a dry phase around AD 1290–1410.
5. Around AD 1500: deterioration into the Little Ice Age (LIA) with wetter and cooler conditions.
6. Even during the LIA there were drier phases: between 1470–1660 and 1520–1800 and possibly 1700–1800.

Even more detailed work, using testate amoebae from upland mires in Northumberland, has allowed a reconstruction of wet periods during the past.[17] The evidence for earlier prehistory is not always congruent from different sites, but there is a good degree of convergence during the first millennium BC. The main changes indicated are:

1. Shifts to wetter conditions at times with mean ages of 1520, 830/900 and 280 BC. At 1900 BC there was a dry shift.
2. A major climatic deterioration about the end of the fourth century AD which continued for 300–400 years.
3. A further peak of wetness at around AD 860, followed by a dry shift.
4. This dry shift lasted perhaps 100 years but there were wetter conditions in the twelfth century, returning to a drier state in the late thirteenth century.
5. Wetter conditions after the late thirteenth century, a brief dry stage in the sixteenth century and then wetter conditions until the mid-twentieth century, with fluctuations in wetness within this general period.

Much interesting research into this level of detail is now appearing but the regional syntheses for the whole of the uplands have yet to appear. There seems little doubt that there will soon be a virtually decadal reconstruction of precipitation in the uplands from the beginnings of peat growth to the start of instrumental records. Given the sparsity of the latter in the uplands until very recently, the subfossil record may well retain its importance even for the last century. In passing, the effect of forest cover upon regional climate should be noted. The interception of rainfall and of insolation would have reduced heat and moisture at ground level; exposure to wind would have been ameliorated, especially in summer. Replacement of woodlands by peat means the emplacement of a habitat with a very low heat conductivity so that even a thin peat cover to a soil means that its growth season is reduced beyond the 'normal' local level and starts later. Any early farmers seeking to use peaty-topped soils would likely experience a higher crop failure rate than those using the remains of forest humus, transported leaf litter or manure from stalled or corralled animals.

For temperatures in the whole of the non-Scottish uplands, Taylor[18] provided a set of corrected curves based on an altitude of 500 m. The calibrations of radiocarbon dates adopted in his work have now been superseded but the overall estimates are still useful. Average temperatures at 10,000 BP are of the order of −2.5°C but by 8,000 BP rise to 4°C, with a slightly later maximum of 8°C. During the period of optimal climate, summer air temperatures for the lowlands were of the order of 17.5°C and in the uplands 15.5°C. Today's figure for the upland summer would be 13.5°C. Lapse rates would ensure that even in periods of better climate, an altitude of 500 m would have a growing season some sixty-five days shorter than at sea-level in the same latitude, from mid- or late April to late November in Wales. In the Principality, the lapse rate is exceptionally high: the mean monthly minimum for Moel Cynedd at 385 m is 3.5°C, whereas at Valley (Anglesey) it is 7.2°C.[19]

Not all climatic change is gradual and even relatively smooth transitions may announce themselves with extreme events. In the case of rainfall, enhanced rates of erosion and more floods are the usual signals. When therefore we find landforms which clearly derive from periods of, for example, gully erosion or heightened alluviation episodes, we need to be aware that climatic change may be implicated. Equally, climate may be permissive but not the immediate cause: the introduction of grazing by sheep (for example) may allow the formation of erosional features because the vegetation cover is loosened, just as the removal of trees permitted the growth of peat because the water relations of the soil had altered. Examples of these kinds of interaction will be found at appropriate places in the following chapters.

Climate remains a powerful factor in upland land use and biota, in spite of all the power of humans to change their environments and to evade some of the perceived difficulties. Thus any future climatic changes of the kind which might result from what is usually called 'global warming' are very likely to have an impact upon the uplands. In the case of wild species, the ability of species to adapt to climatic change is adversely affected by the fragmentary nature of the habitats through which they have to migrate. Many of the species have disjunct distributions, and occur on relatively isolated upland areas and can move upwards only on those blocks of country on which they occur. They may drive out the species of the upper zones and in turn come under pressure from the species accustomed to life at lower altitudes: climatic change seems inevitably to mean extinction at one spatial scale or another.[20]

The present climatic fluctuations (however caused) are only the latest in a million-year history of change. This book continues by going back part of the way into that far past to see how the ecology of the uplands was affected by natural variations and by the roles of human societies. Climatic change, however caused, must be set in such a multi-stranded context, including perhaps the possibility that we are living in an **interglacial** period of unknowable length.

Notes and references

1. Hence the chapter title, which comes from Arthur Conan Doyle's Sherlock Holmes novel of 1912, *The Hound of the Baskervilles*, set mostly on Dartmoor.
2. C. F. Fox, *The Personality of Britain: its Influence on Inhabitant and Invader in Prehistoric and Early Historic Times*, Cardiff: National Museum of Wales, 2nd edn, 1933. It is not a classification that I shall use again in this book.
3. The LFAs derive from an EU Directive (75/268/EEC) of 1975 which provided special measures to assist farming in areas of less fertile land which produces returns appreciably below the national average, and have a low or dwindling population largely dependent on agriculture. A peripheral designation of land not quite so badly affected was made under Directive 84/169/EEC. (There are further amendments in 91/25/EEC.) There are statutory maps showing the 'moorland line' in England; this is usually a smaller area than the LFA. In UK government documents LFA is also used as an acronym for Low Flying Areas, which are often the same places.
4. For a general account of the physical geography of the whole of upland Britain, see M. A. Atherden, *Upland Britain. A Natural History*, Manchester: Manchester University Press, 1992.
5. An argument for the glaciation of Dartmoor is presented by S. Harrison, 'Speculations on the glaciation of Dartmoor', *Quaternary Newsletter* no. 93, 2001, 15–26. There will be a considerable body of contrary opinion to overcome.
6. There was a vigorous argument in the 1955–62 period about whether deep weathering under supposedly tropical conditions was necessary to produce the material later stripped away by periglacial action. See D. L. Linton, 'The problem of tors', *Geographical Journal* 121, 1955, 470–87 and the dissenting J. Palmer and J. Radley, 'Gritstone tors of the English Pennines', *Zeitschrift für Geomorphologie* 5, 37–51. The summary entry for tors in the Second Edition of *The Encyclopedic Dictionary of Physical Geography* (eds A. Goudie et al., Oxford: Blackwell, 1994, 511–12) is helpful.
7. W. H. Pearsall, *Mountains and Moorlands*, London: Collins New Naturalist, 1950, 77.
8. See for example, L. Curtis, *Soils of Exmoor Forest*, Harpenden: Soil Survey of England and Wales, Special Survey no. 5, 1971; D. M. Carroll and V. C. Bendelow, *Soils of the North York Moors*, Harpenden: Soil Survey of England and Wales, Special Survey no. 13, 1981.
9. J. S. Rodwell (ed.), *British Plant Communities. Vol. 1. Woodlands and Scrub*, Cambridge: Cambridge University Press, 1991.

10. There is one particularly good review of vegetation dynamics though quite a lot of it is derived from work in Scotland: R. J. Hobbs and C. H. Gimingham, 'Vegetation, fire and herbivore interactions in heathland', *Advances in Ecological Research* **16**, 1987, 87–173.

11. P. Ineson *et al.* 'Effects of climate change on nitrogen dynamics in upland soils. 1. A transplant approach', *Global Change Biology* **4**, 1998, 143–52.

12. R. J. Harding, 'The variation of the altitudinal gradient of temperature within the British Isles', *Geografiska Annaler* **60** A, 1978, 43–9; R. J. Harding, 'Altitudinal gradients of temperature in the Northern Pennines', *Weather* **34**, 1979, 190–202. The classic papers on the North Pennines by Gordon Manley focus on 'Meteorological observations on Dun Fell, a mountain station in Northern England', *Quarterly Journal of the Royal Meteorological Society* **68**, 1942, 151–65; and 'Further climatological averages for the Northern Pennines with a note of topographical effects', *Quarterly Journal of the Royal Meteorological Society* **69**, 1943, 251–61; also 'The northern Pennines revisited: Moor House 1932–78', *Meteorological Magazine* **109**, 1980, 281–92. A general overview is J. A. Taylor, 'Upland climates', in T. J. Chandler and S. Gregory (eds), *The Climate of the British Isles*, London and New York: Longman, 1976, 264–87.

13. G. Manley, 'The Helm Wind of Cross Fell 1937–1938', *Quarterly Journal of the Royal Meteorological Society* **71**, 1945, 197–215.

14. D. Wheeler, 'North-east England and Yorkshire', in D. Wheeler and J. Mayes (eds), *Regional Climatology of the British Isles*, London and New York: Routledge, 1997, 158–80.

15. M. Bell and M. J. C. Walker, *Late Quaternary Environmental Change. Physical and Human Perspectives*, London: Longmans, 1992; N. Roberts, *The Holocene. An Environmental History*, Oxford: Blackwell, 1998; J. A. Taylor, 'The role of climatic factors in environmental and cultural changes in prehistoric times', in J. G. Evans, S. Limbrey and H. Cleere (eds), *The Effect of Man on the Landscape: the Highland Zone*, London: CBA Research Report no. 11, 1975, 6–19.

16. R. C. Chiverrell and M. A. Atherden, 'Post Iron Age vegetation history and climate change on the North York Moors; a preliminary report', in R. A. Nicholson and T. P. O'Connor (eds), *People as an Agent of Environmental Change*, Oxford: Oxbow Books, Symposia of the AEA **16**, 2000, 45–59; R. C. Chiverrell, 'Proxy records of recent climate change derived from peat sequences on the North York Moors and ongoing environmental research at May Moss' in R. Charles, S. Wightman and M. Hammond (eds), *Moorland Research Review 1995–2000*, Helmsley: North York Moors National Park Authority 2001, 108–13.

17. D. Hendon, D. J. Charman and M. Kent, 'Palaeohydrological records derived from testate amoeba analysis from peatlands in northern England: within-site variability, between-site comparability and palaeoclimatic implications', *The Holocene* **11**, 2001, 127–48. Although not strictly speaking an upland site, work at Walton Moss and Bolton Fell Moss in east Cumbria is highly relevant: P. M. Hughes, D. Mauquoy, K. E. Barber and P. G. Langdon, 'Mire-development pathways and palaeoclimatic records from a full Holocene peat archive at Walton Moss, Cumbria, England', *The Holocene* **10**, 2000, 465–79.

18. J. A. Taylor, op. cit 1975 (Note 15).

19. G. Sumner, 'Wales', in D. Wheeler and J. Mayes (eds), *Regional Climatology of the British Isles*, London and New York: Routledge, 1997, 131–57.

20. See e.g., B. Huntley, 'The dynamic response of plants to environmental change and the resulting risks of extinction', in G. M. Mace, A. Balmford and J. R. Ginsberg (eds), *Conservation in a Changing World*, Cambridge: CUP, 1999, Symposia of the Zoological Society of London no. 72, 69–85.

CHAPTER TWO

The millennia of the hunter-gatherers

HUMANS IN THE HILLS

The last million years have seen very great changes in the land area we now call the British Isles. A number of glaciations occurred, covering the land and the North Sea basin with ice sheets; mountains had valley glaciers. Locking-up of water as ice led to dramatic falls in sea-level, with subsequent rises in warmer periods. Trees immigrated westwards and northwards as warmth allowed and retreated in favour of tundra as the ice surged southwards again. In some warm periods tropical fauna inhabited the river valleys, to be displaced by the wolves of the **Boreal** woodlands and the foxes of high arctic environments. The past 500,000 years can also be seen as fluctuations between being a set of islands and a peninsula, with consequences for both climate and human occupation.[1]

In the intervals between glacial episodes, members of the genus *Homo* occupied Great Britain. *H. erectus* was present in the Hoxnian interglacial, for instance, at the eponymous location and at Marks Tey in Essex; the names of Swanscombe and Boxgrove also evoke remembrance of human presence a long time ago. None of the earliest traces of occupation, however, is found in the areas which were to become the moorlands of England and Wales. Either those high areas were shunned or, more likely perhaps, the evidence has not been preserved since most of it would have been erased by later ice.

It makes more sense therefore to think of the last presence of glacial ice at about 10,000 years ago (10 ky or 10,000 BP) as a sort of new beginning in a number of environmental processes and also the start of evidence for the presence of humans (by then, members of our own sub-species, *Homo sapiens sapiens*) whose societies included the uplands in their world. In the last years of the late-glacial or Late Devensian period, the human cultures present are generally termed (**Upper**) **Palaeolithic;** as the post-glacial or Holocene progressed, the cultures after 10 ky are labelled **Mesolithic** and they in turn overlapped with the **Neolithic** people from perhaps about 6,200 BP and are culturally invisible after about 5,000 BP. The Neolithic cultures were agricultural and so begin a new era (and a new chapter in this book) altogether. The relationships between the cultures, the radiocarbon dates and the calendar

TABLE 2.1 Lithics and chronology of early prehistory in England and Wales

Period	Lithics	Dates BP	cal BC
Upper Palaeolithic	Bone and antler barbed points Points on stone blades	>10,000	9,260, 9,230, 9,160 1σ = 9,785 (9,260, 9,230, 9,160) 9,053
early Mesolithic	Bone and antler barbed points Broad-blade microliths	10,000– 8,600	7,570 1σ = 7,691 (7,570) 7,534
later Mesolithic	Narrow-blade microliths, often geometric	8,600– 5,000	3,782 1σ = 3,905 (3,782) 3,708

The lithics are highly simplified; for more detail see S. Mithen, 'Hunter-gatherers of the Mesolithic' in J. Hunter and I. Ralston (eds), *The Archaeology of Britain*, London and New York: Routledge, 1999, 35–57. Calibration is of a nominal date for each horizon, with an assumed standard error of ±100, 75 and 50 years on the three dates. See also P. A. Mellars, 'A major 'plateau' in the radiocarbon time-scale *c*.9,650 BP: the evidence from Star Carr (North Yorkshire)', *Antiquity* 64, 1990, 836–41; S. P. Dark, 'Revised 'absolute' dating of the Early Mesolithic site of Star Carr, North Yorkshire, in the light of changes in the early Holocene tree-ring chronology' *Antiquity* 74, 2000, 304–7.

years BC are set out in Table 2.1. We shall start with the Mesolithic since the presence of Palaeolithic people in the uplands is largely undocumented. The Mesolithic is conventionally divided into an early Mesolithic which yields a lot of bone and antler tools together with broad-bladed flint implements, and a later Mesolithic with narrow-bladed microlithic flints which true to their name are very small in size, about 20 mm long. The year 8,600 BP (7570 BC) can act as a rough divider between the two phases. It is assumed that these cultures were not sedentary but moved through their environments on some kind of a yearly cycle, tapping different resources at different times of year. The word 'nomadic' is sometimes used but this must not be taken to mean some kind of random or purposeless wandering. With fixed points in a yearly cycle, though, they were perhaps the first New Age travellers in Britain.

It is an obvious fact that hunter-gatherer people must live closer to an unmanipulated nature than any others and so we must consider in some detail the natural environmental changes of this first 5,000 years of the Holocene before moving on to think of the relations between the Mesolithic cultures and their non-human surroundings. Since some of the uplands of England and Wales come down to the sea, the history of the coastlines must be an integral part of the story.

CHANGES IN THE NATURAL ORDER

As the account in Chapter 1 has stressed there was a rapid warming during the first thousand years of the Holocene and by 9,500 BP temperatures similar to those of the present day were achieved. Some of the major thermal increases took place within decades rather than centuries. Earlier ideas of a single climatic 'optimum' or Hypsithermal Interval have been re-thought[2] and replaced with episodes of higher values within a broad zone of summer temperatures of 15.5°C in the uplands and 17.5°C in the lowlands, compared with today's upland figure of 13.5°C.[3] Even in periods of better climate, the growing season at 500 m ASL would have been some sixty-five days shorter than at sea-level in the same latitude.

Vegetation change

Thus the early Mesolithic was set against a period of rapid change of climate and vegetation, especially the change from tundra through open woodland to the establishment of a deciduous forest, at any rate in the lowlands. The later Mesolithic culture developed in an overall environment of greater stability of major vegetation type, the warmest climate so far achieved in the last 20,000 years, and witnessed the final insulation of Britain from the European mainland.

The climatic amelioration of the Holocene period allowed the expansion of deciduous forests into England and Wales. These formed the major ecosystem type of the lowlands and extended far into the uplands as well. The assembly of the forests has two main phases in the years 10,000–5,000 BP. The first of these is the immigration of forest trees tolerant of relatively cold conditions and their colonisation of formerly open-tundra-like habitats. Birch, pine and hazel are key examples. The second is the arrival of more warmth-demanding (thermophilous) tree species such as elm, lime and oak, and the ways in which they compete with the existing species and form a woodland with a different species composition. This forest was in many respects a stable ecosystem where natural disturbances led to further deciduous forest, though not necessarily of exactly the same composition. Further, small changes of climate continued to shift the balance in favour of some species at the expense of others. So the early Holocene is seen in the British Isles as a period of rapid change of vegetation type, the mid-Holocene of a slow change of composition. But in the mid-Holocene, there was deciduous forest all the way from the coast to the upper limit of woody growth.

The forests in the uplands were not all simply thinner variants of their lowland neighbours. A compilation from some of the sources is given in Table 2.2. A number of points can be made about this 5,000 year period. For instance, the idea of a '**wildwood**' which represents some form of mid post-glacial climatic equilibrium is almost certainly mistaken. Adjustment to changes in climate on both glacial/interglacial and secular scales, transformation of soils under differing water and vegetation regimes, and inter-specific competition all brought about a vegetation characterised by 'patch dynamics' rather than predictable successions to a single climax of mixed oak woodland. Another feature is the importance of the hazel (*Corylus avellana*), which features strongly in the transitional woodlands of the early Holocene and in the

TABLE 2.2 Woodlands during the Holocene/Flandrian in the uplands

Upland	Flandrian I	Flandrian II: composition	Flandrian II: physiognomy
Bodmin Moor	Early presence of oak (9,050 BP). Birch & oak first then elm-oak-birch (8,229 BP) to which alder is added (6,451 BP)	<30% Arboreal Pollen at Dozmary Pool; oak dominant but hazel abundant: Dozmary Pool, oak only on hillsides	Scattered woodland at most
Dartmoor	oak wood predates hazel rise (8,250 BP); pine declines	Forest still climbing – oak with some elm Hazel dominant at 490 m	Closed forest to 457 m
South Wales	Late alder rise mid-Glamorgan	Waun-Fignen-Felen: hazel dominant but all trees present by 8,000 BP except alder (7,000 BP)	well wooded at 530 m
Mid-Wales	Prolonged birch phase with low pine at 395 m. Birch taken over by pine and oak	Pine dominant locally, as is oak. Hazel very frequent in early Fl II but oak & alder then more successful; low lime	
Southern Pennines	Birch-hazel-willow maximum by 7,000 BP Mosaic of pine, birch, hazel in valleys	Pine disappears often, oak a replacement along with alder, ash and lime. Birch, oak & hazel in Nidderdale at 350 m	Forest not continuous above 400 m
Craven Pennines (Central Pennines)	Pine-hazel then elm. Tree spread dependent on soils. Alder rises at expense of pine	Alder-oak-elm dominant, some ash, lime. Yew on scars ? Pine on limestone	Open: cliffs, screes, springs and flushes

slower-changing forests of the mid-Holocene. Its role is such that the uplands can be characterised during the mid-Holocene as belonging to a kind of oak-hazel region on a European scale,[4] rather than the oak-lime region of the lowlands. Its ability to grow on a variety of soils and to adapt to being a shaded shrub in closed canopy forest as well as forming the dominant vegetation at the forest edge and beyond, ensured for it the occupance of a variety of terrain. Several studies show that it was the woodland species which formed the upper edge of woody growth, even if in the form of a scrub.

The mix of successful tree species is strongly affected by soil type. This is shown in an extreme form in Upper Teesdale where the crumbly Sugar Limestone (baked by contact with hot basalt) formed a substrate for pine when that species had been competed out from most other habitats;[5] in a roughly similar fashion, limestone scars in Craven probably bore yew trees.[6] Limestone cliffs are one case of a more general category of openings in forest brought about by geomorphic processes. Steep and shifting slopes such as screes and landslides, and undercut river cliffs are for instance unlikely to be stable enough to bear woodland. Mires and lakes are obviously unsuited for tree roots, though at the edge of both there may be trees, and both may dry out at the margins and become colonised by woody species. River channels themselves need to run in the same course without catastrophic floods for long periods if they are not to retain sub-woodland vegetation in which perhaps willows are the largest element.

Thus any image of a more or less inert blanket of forest stretching unbroken from coast to coast must be dissolved. The 'wildwood' was in a constant state of change, though to any two or three human generations in the mid-Holocene these would have seemed relatively small in scale and in overall terms predictable. Note however that the current distinction in the landscape between upland and lowland which is so often defined by the limit of enclosure would have had no equivalent in the forests. Altitudinal zonations were undoubtedly present, culminating in some places in an upper forest limit but the main forest type at say 400 m would have extended downwards to sea-level, so that the ecology of the uplands (though not exempt from physical processes such as the temperature lapse rate and orographic rainfall totals) is less distinct an entity than it now appears to be. The sea was the more obvious and important boundary zone.

Biotic factors are also at work creating gaps in forests. Senescence leads to trees falling over entire or to snapping off above the base of the trunk; wind-throw may produce the same effect. Animal populations, too, can affect forest dynamics: a succession of poor mast years may mean that every acorn is eaten by herbivores and so regeneration to forest containing oaks is delayed; in hard winters ungulates will eat tree bark and so effectively ring-bark individual trees. Any opening where herbivores concentrate is likely to be slow in regenerating woodland because the animals will feed selectively on young shoots. A last biological influence in mid-Holocene forests is that of the beaver, *Castor fiber*. It is a doughty feller of small trees (especially aspen, *Populus* spp) and a dammer of streams, resulting in open areas of water within the forest with a fringe of dead trees where the water-table has been raised.[7]

But were all the uplands covered with forest? During the Mesolithic period the tree-line would be at its highest because that period also encompasses the Hypsithermal interval or Climatic Optimum of the Holocene. Maximum summer warmth in the uplands was probably in the years 9,000–8,000 BP. Given that plant growth tends to lag behind climatic changes, the period 8,000–7,000 BP was likely to have seen the maximum extent of tree cover in the uplands. For England and Wales, the data are summarised as Table 2.2. On Bodmin Moor, there was only birch-hazel scrub at 265 m, and oak woods were confined to sheltered hillsides though soil variability may also have contributed to woodland variation. In the 9,200–8,500 BP period, birch and hazel scrub appear to have been dominant, with scattered stands of oak; limited elm and pine were confined to sheltered localities. By the mid-Holocene, hazel woodland was a dominant vegetation type which formed a closed canopy over many of the hillsides of Bodmin Moor and may even have covered some of the tors. Alder did not expand until after 6,500 BP but then was not confined to valley woodlands but was also present on damper soils at higher altitudes.[8]

On northern Dartmoor, there appears to have been hazel-dominated wood-land as high as 551 m. The summits themselves may have been open species-rich heaths dominated by crowberry (*Empetrum nigrum*) which later underwent acidification and species loss. On southern Dartmoor, oak-hazel woodland was present at 457 m but above that level open ground can be inferred. In South Wales, at Waun-Fignen-Felen, mixed woodland with a rising proportion of hazel is deduced for 530 m; elsewhere, open hazel woodland is suggested for 660 m, with no trees somewhere above 715 m. In the Lake District, birch wood has been found in peat at 518 m and at Shelf Moss (Pennines) at 594 m. Some of the only work specifically aimed at elucidating this problem comes from the South Pennines. Summarised in outline, the upper limit of mixed woodlands containing alder is postulated for 425 m, for continuous forest with pine and oak 500 m, and thereafter scrub with hazel and birch. In the North Pennines, there appears to have been a substantial upland forest cover between 200 and 760 m ASL. Within this forest, there was more hazel (*Corylus*) as alti-tude increased and much less lime, *Tilia* south of the latitude of the Tyne valley. The tree-line may have been higher than 760 m, for analyses from Cross Fell in the North Pennines (893 m) suggest that there was open herb-rich birch, (*Betula*) woodland on the summit slopes during the optimal period for tree growth. Pine (*Pinus* sp)does not appear to have been confined to any particu-lar soil type; elm (probably *Ulmus glabra*) increased with altitude but may have decreased on the Millstone Grit's highly siliceous soils.

The substantially less high region of the North York Moors (highest point 454 m) might therefore be expected to be totally tree-clad in any 'Optimum'. There was however very likely some open land during the mid-Holocene. At 402 m, for example during the mid-Holocene, total tree pollen was generally less than 30 per cent of total pollen, whereas 1.5 km away at 346 m there was closed-canopy forest (evidenced by the trunks of birch, oak and pine) just after 5,000 BP. Intra-regional variation may produce diversity. In the Craven Pennines, there is a difference in mid-Holocene vegetation patterns between the plain limestone areas and those covered with acid drifts. The former, for instance, had higher levels of pine and elm but were also subject to greater intensities of disturbance (and the subsequent increase in *Sphagnum* peat, sedges and bracken) during the Mesolithic. In some places, hazel formed more or less pure stands.[9]

In summary, the most careful work shows that there were variations in the forest composition but that as upper limits were approached, the proportion of hazel increased to the point where it may have formed a scrub higher up the slope than the closed canopy woodland with oak and pine. Since hazel is unlikely to grow on **gley** and **podsol** soil types, then increasing wetness and acidity of soils would likely produce a shift to birch and willow in this upper zone. For the mid-Holocene and the later Mesolithic, however, much of the research makes the point that evidence of fire is often present in the deposits which also contain the micro- and macro-fossil evidence for the tree-lines. It thus raises the question as to whether the tree-lines might have been affected by the occurrence of natural, lightning-set, fire or whether human activity was involved. The evidence for Cross Fell suggests that on climatic grounds there is no reason why most uplands should not have carried open woodland (even if only a birch or birch-hazel scrub) in the mid-Holocene. If they did not, then local explanations must be sought.

Animal resources

As vegetation changes, so do the animal communities available to hunters. The large animals typical of the crowberry, *Empetrum* heaths with tree and shrub thickets of the early Holocene included the wild cattle or aurochs, *Bos primigenius* (Plate 2.1) as well as remnant populations of reindeer and moose. As the deciduous forests became established, the balance swung towards the dominance of red and roe deer, with aurochs still present. Wild pigs are also a feature of the forests, and both the wolf and bear probably grew in numbers. Small mammals like fox, badger, wild cat and otter were present. The presence of beaver has already been noted and its activities would have diversified the

PLATE 2.1 *A* Bos primigenius *bull from Lascaux, one of the wild fauna of the mid-Holocene forests. The animal was 1.5–2.0 m high at the shoulder and so one-third bigger that today's domestic cattle.*

stream regimes and allowed the expansion of the fish populations of slow-moving waters to complement the trout and other salmonids. Inland, relatively few birds are obvious candidates as food resources, except possibly at nesting time, though in many cultures around the world feathers of predators or showy species are valued for personal adornment. No eagle-feather head-dresses have however been found in the British Mesolithic.

Coastlines

In one sense the whole of the British Isles are a set of coastal environments, for the influence of the sea upon climate is pervasive. A number of upland areas abut onto coastal areas and even where the uplands do not as it were fall directly into the sea (which some do spectacularly, in the case of cliffs on Exmoor and the North York Moors), they often come within the kind of distance which is envisaged as being a possible part of the territory of a hunter-gatherer group. Within 50 km of most upland areas in England and Wales, one or more of the following types of coastline ecosystem can be found:

1. Very high cliffs as at Boulby (North York Moors) which fall more or less directly onto a shoreline which is exposed at low tide but which may reach onto the foot of the cliff at high tide.
2. Low and medium-height cliffs with a normal width of rocky, shingly or sandy beach, not all of which is covered at normal high tides.
3. Low coastline with one or more parallel lines of dunes. These may pond back lagoons. They may be fronted by any of the normal kinds of fore-shore: sandy, rocky or salt-marsh, for example.
4. Sheltered bays, estuaries and other inlets with accretive inter-tidal areas such as mud-flats, saltmarshes and sandbanks.
5. Near-shore islands, accessible by small boats.

Given the variety of geomorphological processes within small areas in Britain, more than one of these types is likely to be found within this wholly arbitrary 50 km distance.

The coastline underwent a number of transitions during the early and mid-Holocene. The period 9,200–6,800 BP was the time of the most rapid post-glacial rise of sea-level, with the years 9,000–8,500 and 7,800–7,600 being especially fast.[10] At about 8,500 BP in Lancashire, for instance, sea level was at −15.2 m OD but by about 7,700 BP its level was at −6.5 m OD. The rate of rise then slackened and after 6,980 BP the level fell to about −4.3m OD, with a band about 2 km in width at Downholland Moss being removed from marine influence. But between 6,800–6,300, sea-level rose again by over 1 m to −3.2 m OD, fell at 6,000 BP and then rose again rapidly (at perhaps 0.6 cm/yr) between 6,000–5,800 to −2.2 m, then falling again to −3.4 m by 5,500 BP. Between 5,500–5,000 BP, the level rose again at a rate of 0.5 cm/yr. Comparable patterns are recorded for many other parts of England and Wales, with the details varying according to topography and the degree of warping of the coastline following **isostatic recovery**.

The significance of such changes for Mesolithic populations cannot be dismissed. At the very least, some groups would find their familiar resource-yielding territories being rapidly reduced by fast-rising sea levels. Reconstructions of the coastline of Cornwall during the Mesolithic suggest that the rises of sea level between 8,000 and 5,000 BP would have diminished the land area by 25 per cent. Marine incursions no doubt would have brought unpredictable hazards when for instance storms coincided with very high tides. Examples of habitat change would include the **perimarine** zone of peat mosses where such existed: these would have been flooded and peaty substrates replaced by clays and silts under the influence of rising water-tables at a regional scale. Tidal flats and lagoonal zones are invariably affected quite quickly and their extent oscillated with the advance and withdrawal of the sea, provided there was space for them to develop. Sand dunes, sandbanks and shingle spits develop little during eras of overall sea-level rise, for if they build up during times of falling sea-level, they are soon overtaken by later rises and their materials dispersed. All these types of change are most developed in quiet-water environments where deposition can take place and there is no reason to expect comparable complexity along for example coasts with high cliffs easily reached by high tides. In such places, however, it is reasonable to expect that any inlets would have an additionally important place in Mesolithic resource scheduling and that rapid changes would bring adaptational challenges. The overall story, though, is one of mean tide levels which have risen throughout the Mesolithic period, from about −20 m in 9,000 BP to about −2 m in 5,000 BP. This must mean that much evidence for human occupation of coastal zones is either buried under later deposits, was washed away during periods of sea-level rise, or is beneath several metres of water as well as any other cover. Coastal sites such as those at Eskmeals, Cumbria[11] which relate settlement to sediments (such as shingle ridges, peat-filled basins, and sand dunes) take on an added importance; perhaps even more significant are regions like the coasts of south-west Wales where there appear to be no later Mesolithic **lithic** remains at all except at the present-day coastline.[12] The lesson drawn by many workers is that coastal zones were plentiful sources of food and that they could therefore support sedentary populations and/or 'nomadic' populations of a higher density than inland areas. So any hunter-gatherer population that spent time in the hills would be advantaged by having a stretch of coastline within their access, preferably one which included lagoons or an estuary.

THE HUMAN OCCUPANCE

Compared with any later period, the evidence for the upland mesolithic is sparse. Compared with the same period in for example Scandinavia, we lack detailed remains of bodies, structures and food remains. Much has to be inferred from stone artefacts and from palaeoecological evidence, but nevertheless some reasonably confident statements can be made about the culture and economy of a people who started out the Holocene as members of a pan-European population who were the last people who could walk to France and the Low Countries until the opening of the Channel Tunnel. During their time, the last land bridge with the continent was severed by rising sea-levels, with 7,800 BP as the date after which sea-sickness became an inevitable concomitant of continental travel.[13]

Archaeology

The next few paragraphs briefly put the material evidence (Table 2.1) into a regional chronological pattern and foreshadow some of the aspects of a human-oriented environmental history which will be developed later.

In the **Upper Palaeolithic**, the transition to the Mesolithic is usually placed at 10,000 BP, with some overlap of stone tool types into the later cultural period. The **lithics** of this (Creswellian) culture are dominated by a range of points made on large (40–90 mm long) blades and these have strong affinities with some cultures of mainland Europe, such as the Brommian and Ahrensburgian. Close resemblance is not surprising since mainland Britain was not then separated from continental Europe. The whole of the **Mesolithic** is characterised by the presence of microlithic industries, which are rarely larger than 40 mm in length and often smaller. None of the evidence for the entire Mesolithic suggests anything other than the dominance of a food-collecting (i.e., pre-agricultural) economy until possibly the very end. An early phase is usually separated from the later by the presence of broad-bladed microliths. Examples of the early Mesolithic in England and Wales are dated between 9,700–9,000 BP. The culture is distributed throughout England and Wales with the possible exception of the uplands of Wales. Smaller microliths with narrow blades and the appearance of more geometric shapes (including a small scalene triangle) are diagnostic of the later Mesolithic, about 8,800–9,000 BP. However, separation from the early phase is not always complete: narrow-blade industries show up before the insulation from the continent. Equally, some geometric industries are accompanied by larger, non-geometric forms. Pure geometric assemblages are especially common on the uplands of northern England. There is every appearance of some continuity between early and later Mesolithic periods, as indeed there is often some between the Mesolithic and the succeeding Neolithic.

Lithic material is the foundation for the recognition and delineation of the Mesolithic in the whole of Britain, notwithstanding that palaeoenvironmental evidence and radiocarbon dating has now added further and integral data which form a more complex whole.

In the early Mesolithic, in the period to about 8,800 BP, the non-geometric industries include a restricted range of large microliths and a range of distinctive non-microlithic tools. Some barbed points of antler and bone are found.

The later Mesolithic lacks any antler and bone technology but has a range of smaller geometric types such as scalene triangles and rod-forms. In both periods, the finds are mostly attributed to projectile technology, although wear studies at, for example Thatcham in Berkshire have rejected the use of microliths as projectile points in favour of composite knives for vegetable processing. In the early Mesolithic two or three large armatures on a projectile are postulated; in the later Mesolithic there appear to be a larger number of smaller armatures with a variety of numbers and layout. The raw material also differs: the later Mesolithic uses stones with a broader range of mechanical qualities including an increased use of low-quality chert and even flints from river and beach gravels. The absence of antler and bone projectile points from the later Mesolithic suggests that the microliths form a substitute. An overwhelming proportion of later Mesolithic sites however are unfavourable environments for the preservation of organic materials.

In a close analysis of form, function and distribution of Mesolithic stone technology, the frequency of sites from the later Mesolithic suggests a filling up of the landscape, though with smaller sites.[14] The shift to microlith-dominated tools indicates the need for reliable projectiles, capable of functioning without the need for time-consuming repair. All archaeologists agree that there were seasonal movements of human groups, and that the uplands may not have been permanently settled. Perhaps the whole of a band (about twenty-five people) might move there for the summer and send off sub-groups of males to the higher ground to engage in hunting and the maintenance of the hunting environments.

Settlement pattern and structures

At the scale of the individual site, spreads of charcoal and microliths are mostly below 20m^2 in area but an area on Snilesworth Moor, North York Moors (NYM) totally exposed after a fire showed 5,000 flints within an 8 m radius of a spring-head. At Cock Heads (NYM) an area 30 × 17 m yielded only thirty-three tools. Over 50 m^2 is common for assemblages with a diversity of tool types: at Upleatham (NYM) the main area of flints was 100 × 60 m, thinning towards the edges. Most of the upland spreads of stone tools and charcoal are near water, either in the form of streams, spring-heads or standing water. Examples of proximity to contemporary lakes include Waun-Fignen-Felen in South Wales, the Malham Tarn area of the Craven Pennines, Cefn Gwernffrwd and Rhosgoch Common in mid-Wales; on the North York Moors most of the sites are near the spring-lines. In Swaledale (Yorkshire Dales), scatters of flint artefacts are found above the spring-line and on south-facing slopes. A sheltered aspect not far from the tops of hills and ridges seems also to have been sought: in south-west Yorkshire, most of the sites are found in the 415–500 m zone, on sites facing east through south and off the crests. This is also evident in the hills east of Manchester where the areas above 450 m are mostly devoid of the worked stone that is plentiful between 350–450 m, in a terrain which has been extensively worked over. There is a concentration of sites along the edges of some of the steeper valleys which might reflect peat erosion, though in a wider context it seems to be a valid distribution. This narrow strip of gritstone between Skipton and The Peak has a very high concentration of sites which are regularly re-visited by local archaeologists and so patterns are probably reliable. In the North Pennine valley of Upper Teesdale, flint is most often found on alluvial flats or on limestone shelves overlooking steams; to what extent blanket peats obscure other sites is uncertain. On the limestone uplands between Shap and Kirby Stephen, most sites are in the range 275–350 m but show no obvious relation to the topography; like several other places, they do show some degree of continuity between implements associated with both Early and later Mesolithic.[15]

At a regional scale, Mellars points out that in northern England, a transect from east to west reveals a clustering of sites between 350–480 m.[16] In the Central Pennines, Stonehouse calculated the average altitude of forty-nine narrow-blade lithic sites at 385 m.[17] There is discussion about whether this is an artefact of land use type, peat erosion and the concentrations of amateur archaeologists and of flint collectors. It seems likely that these distributions are reliable where many sites have been found but absence of finds should not be taken axiomatically to mean absence of mesolithic presence. That is, the high densities of charcoal and stone tool sites on the Pennines and North York Moors are real (though both individually and collectively they may represent a considerable chronological depth) but absence of comparable sites in the lowlands may be due primarily to lack of discovery. In Wales one Mesolithic site in six is above the 300 m contour in the uplands which comprise about 40 per cent of the Principality.[18] In the uplands of south-east Wales these have been classified into locations around former small lakes, those on river terraces, those commanding views across valleys and the localities at the heads of valleys near passes in the mountains.[19]

Most sites which have been seriously investigated have evidence of charcoal concentrations which are taken to be the remains of hearths. These are typically at the junction of the mineral soil and overlying peat and so radiocarbon dates lack conviction as to the real age of the organic material. Some places have a concentration of several of these spreads within a small area and are presumably the result of either a number of groups camped near together or repeated visits by smaller groups. A number of Early Mesolithic sites on the central Pennines occur in pairs, as close as 20 m apart, and one site at least (Pule Hill Base, 361 m ASL) was probably re-occupied in the later Mesolithic. On Mynydd Hiraethog in North Wales there is a complex of later Mesolithic bowl-shaped pits with scorched sides in which layers of charcoal are each covered with layers of fresh soil. When the pit was full, a new one was dug. The lithics exhibit a diverse assemblage and the radiocarbon dates suggest the Mesolithic/Neolithic boundary: 5,700, 5,350 and 5,250 BP are quoted.[20] A few places have been deemed to show evidence of structures. At Broomhead 5,

(~8,570 BP, South Pennines) there are five stake holes as if to form a wind-break and the charcoal areas are all within or adjacent to the flint distribution. At Deepcar (mid-tenth millennium, 12 km north-west of Sheffield), stones are arranged in an oval to circular pattern and Dunford Bridge A exhibits an oval area of stone paving.[21] There is a dearth of organic materials at upland sites. Wood is very scarce, bone totally so. Peat is generally thin, burnt and abraded. Organic muds are scarce, with Stump Cross (Craven Pennines) one of the few exceptions. Charcoal spreads at the interface of mineral soil and peaty top are difficult to date and unrewarding to identify to species. Commonest are the shells, often burnt, of hazel nuts. Hints of worked lengths of wood are given by the layout of microliths: these are generally interpreted as arrowheads, though D. L. Clark argued for their consideration as aids to gathering plant material.[22] A very interesting organic upland structure is found on Bodmin Moor. Birch bark strips were laid horizontally over a layer of birch stems and twigs and then the bark layer was covered with roots and branches of willow. It may have been a platform overlooking a marsh and used for wildfowling.[23]

CULTURE INFLUENCES ENVIRONMENT

In this material, we consider whether the environmental systems of Great Britain were changed by human activity during the hunter-gatherer era. We are certain that natural changes took place but here we want to know if human agency produced similar trends or metamorphoses that were different. We will assume that the Upper Palaeolithic groups were too few in number and limited in technology to produce anything but the most local and temporally limited effects.

Mesolithic cultures and vegetation

The analysis of pollen stratified in peats and lake deposits of the early Holocene shows occasional traces of the opening or repression of pine or birch woodland and of the burning of heathy plants (Plate 2.2). None of this can be unequivocally tied to human activity, but neither is it sufficiently syn-chronous to be tied to climatic oscillations. Examples include the recession of juniper-pine-birch woodland in the presence of charcoal on northern Dartmoor before 8,758 BP; at about the same time, channel peats on the North York Moors show inwash stripes of silt which appear to result from soil erosion. In South Wales, an upland birch forest was apparently kept open by human activity, resulting in heathy vegetation.[24] It seems to be a phenomenon which is restricted in scale. But there are sufficient hints of a process in which fire in the landscape happened often enough (and perhaps at critical times) to produce a vegetation different from the pre-existing condition. Mesolithic communities were certainly present on the uplands in the period 8,500–5,500 BP but it does not necessarily mean they were absent from the lowlands which

PLATE 2.2 *A section through Holocene peats on the North Pennines. The mineral soil is visible at the base and a prehistoric tree layer surmounts some peat growth by which it is then overtaken. Trees can colonise peats providing the growth is not too thick nor the surface too wet. In this case however later peat growth has killed the woody species whose remains can be seen.*

formed another sector of a yearly cycle. Evidence for upland forest recession at that time is quite plentiful: the pollen of forest trees is partially replaced by species of open ground; there are many deposits with charcoal in them, and woodland is smothered by bog vegetation. It is often difficult to disentangle natural processes from human actions and much of the discussion revolves around the probabilities of the two processes.

The recession of woodland in upland England and Wales in later Mesolithic times meant its replacement with other kinds of vegetation, in contexts which mean that the possibility of human agency cannot be ignored. No type of woodland was necessarily immune from disturbance at that time. For example, because of the nature of upland climates, pine-birch-hazel woodland is sometimes found in the later Mesolithic at the time when a mixed deciduous forest had developed at lower altitudes. Thus at Dufton Moss in Upper Teesdale at 368 m OD and *c.*8,000 BP, there is evidence for the sudden burning of a woodland of willow, birch and pine, which results in the better establishment of the hazel (*Corylus*) component and a gain for birch (*Betula*). At West House Moss in the North York Moors peaks of herb pollen in the peat, together with bracken, suggest that the birch-pine-hazel woodlands were being disturbed. Pine seems to decline at the expense of birch, and hazel begins its familiar rapid expansion. The main forest type to exhibit the process is the mixed deciduous woodland dominated in the pollen spectra by that of oak. Disturbance and recovery often afford a foothold for species like ash (*Fraxinus*) as well as letting hazel flower more abundantly.

The upper edge of the woodland was the scene of disturbance. Such an **ecotone** is of course always the most susceptible to climatic change, but this zone is also prone to the kinds of disturbance in which fire is implicated. The clearest indications of the processes at work come from central Pennine sites.[25] At 6,000 BP, work at Robinson's Moss proposes that the upper level of the 'lowland forest' was at 425 m and that of the 'upland forest' at 460 m OD. This date, however, coincides with a second temporary retraction of the limit of the upland forest which is associated with evidence of burning. This is seen by Tallis and Switsur[26] as just one of a series of fires in which burning probably prevented the upward spread of the tree species all the time from when upward forest expansion was just commencing in the early Holocene.

The principal woody species at the tree-line at 6,000 BP was probably hazel. In spite of the fires the tree-line moved up slowly between 6,800 BP and 5,500 BP, which probably signifies a continuing response to climatic change. Given that exposure, soil type and slope are all additional variables, it is to be expected that the impacts of burning at the upper edge of woodland are highly variable from place to place. Between 8,000 and 6,000 BP on Dartmoor, hazel is dominant in the woodland community but its quantity appears to have been negatively influenced by fire after 7,500 BP.[27] On neighbouring Bodmin Moor no convincing evidence of Mesolithic management of vegetation was found for the Rough Tor area, raising the possibility that the population density of the region was so low as to make manipulation unnecessary. There may have only been three or four small bands of hunters on the upland at any one time.[28]

In many investigations, estimates of areas involved in disturbance rely simply upon the degree to which certain pollen frequencies alter. The outstanding example of a more reliable data-set comes from Waun-Fignen-Felen in South Wales, where analysis of multiple profiles within a small basin has allowed the construction of a series of diagrams reconstructing the vegetation at 8,000, 7,500, 6,500, 5,700, 4,700 and 3,700 BP. The area contains an 'early mesolithic' flint knapping site and abuts a small shallow lake, all probably set in *Empetrum* heath. This kernel forms a nucleus for the spread of blanket bog to the north-west but on other sides continues to be set in mixed woodland until 4,700 BP when mixed woodland appears to form islands in a sea of blanket bog (the lake being now covered with acid peats), a reverse of the position at 6,500 BP (Plate 2.3). Local topography is clearly important in determining the actual sites of vegetation change but it is clear that fire is implicated in the basal layers of the blanket peat and played a role in its inception. It is not certain, though, that vegetation manipulation was practised in the early Mesolithic.[29]

Even without the accuracy of such studies, some locations of forest recession in the uplands of England and Wales often recur. These include the proximity of lakes. There were a number of small lakes in the uplands during hunter-gatherer times which are now overgrown with blanket peat and near the margins of which fire seems to have been frequent. In the few riverside sites which have been subject to analysis, forest recession is usually seen, as investigated[30] at Buxton in Derbyshire at 300 m OD. Another zone falls between the hill-tops and the spring-line, where the latter is altitudinal and under geological control. The exposed interfluve 'ridge' (perhaps the wrong word for many of the swale-backed hills of upland England and Wales) is a less common site. On such hills, recessions of woodland have often been detected near to the spring-line, and especially just above it altitudinally and then downhill along stream-sides.

Forest recession brings about vegetation change. What had been the ground layer of a woodland might become a dry grassy glade with a variable amount of bracken and/or heather. Herbs which are intolerant of shade but do not require soils of high base status are found, including cow-wheat

PLATE 2.3 *A low-level oblique photograph of Waun-Fignen-Felen in the Black Mountain of South Wales. A former lake has become grown over by acid ombrogenous peats which are now partly eroded. The lake and its surrounding woodlands were the site of Mesolithic occupation and management of the vegetation.*
Source: © *Crown Copyright: RCAHMW*

(*Melampyrum*). Another possibility would have been a dry heathy glade with accumulating *mor* humus and a dominance of heather (*Calluna vulgaris*). If the environment was wetter then there might be a damp grassy glade, with wet-tolerant grasses and herbs. The common nettle (*Urtica dioica*) might be found as a natural component of areas of this kind. Along this spectrum of dampness the next stage might well have been a glade in which there was a carpet of wet-tolerant, peat-forming plants such as *Sphagnum* moss or a sedge like cotton-grass (*Eriophorum*). There might also have been openings with pools of open water in them. These might be brought about by partial deforestation, for example, or wallowing by large mammals. Water plants and acid mire plants are all likely colonists.

The growth of blanket peat in the uplands

On many uplands above about 400 m, the original topography is blanketed with a layer of peat commonly 2–4 m deep, sometimes more. It is found in two main topographic situations, the first of which is water-receiving sites which are the recipients of natural drainage, such as dry basins, basins with small lakes, channels and poorly drained cols between stream-heads. The second set of locations are water-shedding sites, of which the convex slopes of low angle which form the majority of the upland terrain today are the major category. Such terrain underlies much of the blanket bog characteristic of higher ground today but this type of peat has also overgrown sites where there were small lakes, as at Waun-Fignen-Felen in South Wales (Plate 2.3).

These blanket bogs are mires of the type classified as **ombrogenous** since the nutrition sources of the plants are either the mineral content of rainfall or the partly decayed remains of their immediate predecessors. Any time between 9,000 and 1,000 BP might have seen an episode of onset of such growth. If inception happened shortly after a major climatic shift to wetter conditions, as is postulated for the British Isles around 7,500–7,000 BP, then the role of climatic factors must be suspected, as permissive if not necessarily decisive.[31] The evidence for fire is also germane to any study of the origin of ombrogenous peat. There may be a period of nutrient enrichment after fire, detectable in some peat profiles by remains of the moss *Drepanocladus*, a species of **flushed** mires. Another ecological effect of fire is to block the pores in the surface layers of soil with fine charcoal and thus to reduce its permeability to water and hence to increase the amount of water on the soil surface which is dependent upon evaporation for transport.[32]

The accumulation of water is enhanced in the woodland case by the removal (by any means) of trees. Deciduous trees act (a) as a shelter layer, intercepting precipitation and re-evaporating it from the canopy and trunk, and (b) as water-pumps, removing water from the soils via their root systems and transpiration mechanisms. Experiments have shown that runoff increases by as much as 40 per cent after clear felling of deciduous forest. The pathways to peat accumulation are to some extent more complicated than allowed in

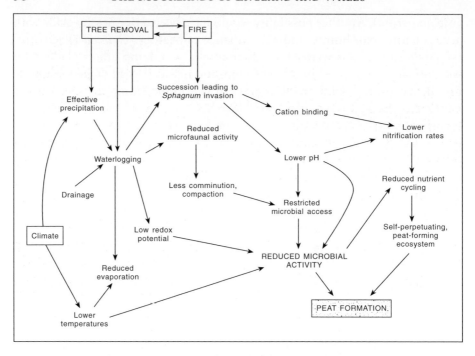

FIGURE 2.1 *Pathways to peat accumulation, with the major forcing functions in boxes. Note that peat accumulation can be reached via more than one pathway. Tree removal and fire may each precede the other but in upland Britain fire most often followed tree removal except at the upper edge of woodland.*
Source: I. G. Simmons, The Environmental Impact of Later Mesolithic Cultures, *Edinburgh: Edinburgh University Press, 1996, Fig. 3.1, p. 116.*

this brief description: Figure 2.1 shows some of the complexities but prior to them all is the removal of woodland and/or the presence of fire leading to waterlogging. Once waterlogging has started then a sequence of biochemical mechanisms is initiated which will usually lead to peat formation. If the peat grows only slowly and in a thin layer then it may be dry enough to permit rec-olonisation by trees: birch and pine form layers in peat profiles at Lady Clough Moor in the south Pennines and at Bonfield Gill Head on the North York Moors. Generally, the re-growth of trees coincides with lower levels of all types of charcoal in the peat, suggesting that fewer fires as well as (or even rather than) any climatic shift was a factor in their re-growth.

Plots of the dates of initiation of blanket peat in England and Wales show a range of dates between *c.*9,500 BP and 500 BP. One inference is that human impact over time may be an important factor, but also that wet periods like 7,500–7,000 BP were influential in the Pennines and that 2,500 BP in the Berwyn Mountains was a similarly critical time. At all events there is now about 214,000 ha of blanket peats in England and about 159,000 in Wales, comprising almost 68 per cent of all peatland types in those two coun-tries.[33]

PLATE 2.4 *Westerdale in the North York Moors. Shallow blanket peat in the foreground overlays several spreads of Mesolithic microliths and charcoal. The hunters of the region were implicated in the initial disappearance of the mixed deciduous forests that covered this whole landscape.*

In every case, to the upland populations of the later Mesolithic, heather moors accumulating *mor* humus and becoming seasonally waterlogged (with underlying soils undergoing gleying) and invaded by wet-tolerant sedges and *Sphagnum*, cotton-sedge mires which were wet year-round, *Sphagnum* bogs, open hazel and birch scrub with a variety of wet-tolerant ground flora species and a high proportion of dead trees,[34] as well as wet mires in water-collecting sites, must have been a familiar part of their environment (Plate 2.4).

Causes of fire: lightning strike?

The ability of fire to run through heath and moorland habitats is not in doubt. Nor is there any uncertainty about the fire-prone nature of largely **coniferous** forests. The fire-characteristics of deciduous forest are more difficult to elucidate, since most analogous woodlands in continental Europe are intensively managed and those in North America are, likewise, carefully guarded against fire or are largely secondary woodland rather than mature communities.

Among the various natural possibilities of forest disturbance, lightning must rank as the most likely. It is known that most ecosystems in the world will at some season burn if struck by lightning. The question becomes, are the mid-Holocene woodlands of upland England and Wales exempt from this generality and if not what are the likely temporal and spatial results? Though the present climate is not especially prone to the convectional storms that are the most frequent sources of lightning strike, it is known that the uplands are subject to the phenomenon. Walkers are hit from time to time, and there is a record of twenty-three lightning-caused fires in the Galloway Hills (SW Scotland) during a two-day period in June 1970. These fires were however in coniferous woodland. One summary[35] asserts that in Europe there are severe fires only in periods of exceptional drought; the normal fire season is bimodal in spring and autumn and the average fire size is 0·97 ha, which is a patch about 200 × 50 meters (0·5 ha in intensively managed forests) with a **return period** of more than 6,000 years.

Observers of today's deciduous woodlands, especially in upland England and Wales, tend to regard them as likely to burst into flame as a Friday-afternoon tutorial on social theory. But a lightning strike which hit after a noticeably dry period in spring or autumn might kill a tree and/or ignite the fine fuels of the forest floor or, for example, a thick but dry tangle of bramble. Any resulting fire would however be confined in area and would at most kill a few saplings and perhaps a few birch or aspens. There would certainly be no widespread **crown fire**.

Other natural factors: senescence and windthrow

Upland forests in England and Wales would have been potentially subject to this form of disturbance and the probabilities of gaps from this source would seem to be higher than those for fire. High winds bring the possibility of large areas being felled as well as single trees. Windstorms have the potential to wreak extreme disturbance if accompanied by intense rainfall: soils can then be lost to erosion and landslides may remove the entire ecosystem over a small-medium sized area.[36]

Beaver

The lifeways of the beaver (*Castor fiber*) are well-known in the sense that they fell trees to make dams which then inundate areas of woodland, killing the stands adjoining the pond which they have created. They eat aspen, birch and willow as well as aquatic plants; one effect might be to create a kind of natural coppice from stump-sprouting. They aggregate in lodges of 5–12 animals per lodge, with an overall density (as judged by North American evidence) of 40–90 per 1,000 ha. The size and duration of beaver clearings is very variable and it is difficult to estimate the area of forest that might be subject to forest disturbance at any one time; filled-in pools would create grassy clearings even after the beaver had left.[37] It would be a different story if beavers behaved otherwise in Europe, e.g., by not building dams.

Grazing

There is no suggestion that domestic animals were part of the Mesolithic culture but there is a question about whether the wild fauna could create and/or maintain openings in forest. If, for example, the aurochs spent the year in groups of at least four to six animals then some impact of vegetation can be imagined. Smith and Cloutman[38] affirmed this kind of possibility on the Black Mountain in South Wales:

> The [flora revealed by pollen analysis] gives the impression of a damp, grassy, somewhat disturbed area such as might have been quite intensively grazed by wild herbivores. In these circumstances it could have been kept open since early Holocene times. Alternatively, a clearing may have been created by prevention of woodland regeneration.

Could natural populations of aurochs, red and roe deer, together with wild boar either keep open a clearing established by some other form of disturbance or actually create the opening in the first place? Of the former, there seems no doubt, provided that some critical level of intensity of grazing and browsing is achieved by the wild creatures. Many studies have shown that regeneration of woodland can nowadays be prevented by a concentration of ungulate herbivores, especially if predator numbers are for some reason low.[39] On Dartmoor, the tors might well have provided foci, acting as shelter and as territorial markers. John Evans suggests that the tors themselves bore trees and that there would have been a halo of open vegetation around them.[40]

THE POSSIBLE ROLES OF HUMANS

It is clear from the archaeological discussion that the later Mesolithic people had no very advanced technology with which to manipulate their environment. The visible tools are not massive, for example, nor are they highly honed as with the subsequent Neolithic polished stone axes which have been shown to be so effective in cutting trees. It has to be assumed that fire could be used and controlled away from the hearth and that devices such as pits and nets could be constructed. So the tool kit in essence consisted of fire, the means to capture and/or kill animals up to the size of *Bos primigenius*, coupled with ways of gathering plant material.

Tree-felling?

It seems unlikely that any stone axe found in later Mesolithic contexts (or indeed any other) would have made much impression on mature forest trees such as oaks, elms and limes, and would not have been very effective against smaller individuals. Where relatively small tools might have been effective was in ring-barking. If a complete circle of bark is removed then it takes with it the tissues responsible for water and nutrient conduction as well as the live tissue adding wood each year to the trunk. Ring-barking a group of oaks would in due course cause their death. The trees would be left standing but there would be no summer canopy to deter shade-intolerant species of the forest floor or shrub layer.

The evidence does not suggest that fire was used in an attempt directly to clear woodland. This is not surprising since it is unlikely that deciduous wood-land in Great Britain, especially in the uplands, would be vulnerable to the type of crown fire that would be needed to kill mature trees. Secondly, there has become embedded the idea that fire was used to increase the quantity of hazel as a food for animals or humans. This may certainly have happened, though there are also areas where it diminishes in quantity as the result of fire. Fire may though have been used in existing openings to burn off grasses that do not recover much in the next year but then have five to six years of better growth, and which also keeps bracken down for about three years before it comes back. It seems unlikely that leaf-litter and cured herbaceous material would be dry enough to burn in spring. Hence, a late summer or autumn inci-dence of fire is most likely. Grasses are quickly killed off by the fire but come back soon and stay high, They will diminish if there is no more fire, partly because this may well be the end of the disturbance phase and shading by woody species takes place. Grasses will under certain circumstances compete successfully with heather: the accumulation of nutrients (especially nitrogen) is one circumstance. Another is trampling and grazing by large mammals, which will lead towards the replacement of heather by grasses such as *Deschampsia flexuosa*. Heather will also disappear under conditions of contin-uous waterlogging, though not of merely seasonal quagginess.

The wild animal resource

The evidence of microliths suggests that wild animals were hunted in the uplands. Was this a simple chase or were there attempts to improve the hunters' chances of making a kill? The management of open areas (and espe-cially those near water) to attract wild animals may have had a wider context. Humans might seek to attract game to sites where killing was relatively easy. Such a site, in this context, would be open and attractive to wild animals by virtue of providing food and/or water and possess good cover for the hunters from more than one side, to take account of varying wind directions.

To provide such conditions, human groups might try to manipulate the vegetation so as to improve the attractiveness of the fodder for game species, in terms of species available and quantity; this might attempt to compensate for seasonal deficiencies or extend a seasonal resource. They might also inter-vene in the flow of water should it be possible to improve its appeal to the animals; or manipulate the surrounding vegetation to provide cover and/or good sight-lines (which sounds contradictory) for the benefit of the hunters. Human-induced manipulation of vegetation to attract the larger species would have been focussed on improving the grass and herb content of the forest floor. (Heather is also eaten by deer, but not bracken.) Three potential practices immediately come to mind:

1. Maintain any existing openings free from the encroachment of woody vegetation or indeed any species not eaten by the game animals;

2. Create openings with the same purpose in view;
3. Attract animals with fodder: leaves are an obvious source and seasons lacking browse resources for chiefly browsing animals (red and roe deer especially) a possible purpose.

In general, the first tactic seems the most likely since it involves the least effort and does not require people's presence at frequent intervals. The second tactic shares the latter characteristic but needs distinctly more effort.

The wild plant resource

Apart from the frequent finds of hazel nuts in Mesolithic contexts, very little evidence of what might have been any kind of resource has accumulated for the British Isles: finds of small-seeded plants such as *Chenopodium* (goose-foot) and the remains of wetland genera like *Menyanthes, Phragmites* (bog-bean and common reed respectively) and *Nymphaea* (water-lily) whose tubers are edible, have added to the list a little and remains of pear have been found in Ireland. No over-stretching of the speculative facility is needed to consider that edible fruits, nuts, seeds and possibly tubers, were in fact eaten and that some tools (e.g., antler mattocks) might relate to such procurement. Oil-rich seeds must have had special attractions.

A SUMMARY OF ENVIRONMENTAL CHANGE INVOLVING HUMANS

A possible progression might start with a 'natural' condition which consisted largely of a mixed oak forest on the uplands, covering almost all the terrain, though broken where there was open water and bog accumulation in water-receiving sites. The upper edge of woody vegetation was marked by a hazel scrub. The forest underwent normal processes of death and renewal, which included gaps of various sizes caused by windthrow. The lower edge near streams underwent a sharp transition to an alder wood with other deciduous species, including elm.

   Human activity might then concentrate upon the creation or maintenance of openings in the forest with a ground cover of grasses and herbs which attracted mammal herbivores differentially on a seasonal basis. This had two consequences. In the first, the concentration of animals made regeneration of the woodland less likely, and in the second, humans noticed the concentration of animals and wished to maintain and/or enhance it. Hence, humans took over some of the existing openings and maintained them, using fire to keep a grassy sward rather than allow bracken to cover the ground in the early years. Heather grew as well and was burned, though it too attracted grazing animals (Plate 2.5). Attempts to extend the virtues of a grass-herb sward to the alder woodland were also made. Increasing human populations (or more resource-hungry groups) wanted to try to create extra clear patches in the woodlands or at their ecotones. They did this by killing trees using ring-barking and by opening the canopy by breaking off leafy branches, which were also useful in attracting animals to feed, especially in winter; the merits of such openings did not entirely depend upon human presence and they could be left behind as a general encouragement, for instance, rather than be immediate bait. In spite of the human involvement, none of the processes induced by the Mesolithic populations totally replaced natural events: natural openings continued to be formed.

During this time, climatic change and the natural Holocene processes of soil maturation brought about the accumulation of *mor* humus over podsolic and gleyed soils. Some of the openings underwent rises in water-table and became invaded by rushes; thereafter peat accumulation began to get under way even on ground which was water-shedding rather than water-collecting in a micro-topographic sense.

HUNTER-GATHERERS AND ENVIRONMENTAL CHANGE: OVERVIEW

Although the processes of nature were dominant in the uplands of England and Wales during the Mesolithic period, human activity also produced some significant changes. The scientific data have shown that humans probably prevented the woodlands from advancing as far and as fast up the slopes as the climate would have allowed. Additionally, openings in woods were maintained in order to attract wild animals and maybe leafy branches were torn

PLATE 2.5 *One of the Broxa plots on the North York Moors which approximates to a maintained opening in the woods during Mesolithic time. The open area has an acid soil and is vegetated with heather (*Calluna vulgaris*). The birches represent the succession of secondary woodland which hunter-gatherer management kept at bay.*

down to enhance the attraction of particular locations surrounded by trees and shrubs that gave cover to the hunters. The value of shrubs like hazel in providing browse for deer, cover for hunters and nuts for snacks led to its encouragement on the edges of denser forest. The prevention of tree growth allowed peat to accumulate in these wet and cool climates. This process was encouraged by the effects of fire in clogging soil pores with fine ash.

So one main characteristic of later times on the moors was established in the mid-Holocene: their openness. Though most of them were largely covered with woodland (or at the very least a woody scrub) at one time, the combined effects of climate and human presence (with fire as the chief tool of manipulation) produced a landscape with unwooded areas, some of which can best be labelled 'mires'. The upper edge of the woodlands showed the effects most clearly, followed by the upward expansion of open patches. Since the woodlands were by medieval times confined mostly to strips along the steepest slopes of the valley-sides, and remembering the importance of the monasteries in colonising the uplands, we might call this the 'tonsure model' of forest recession.

By the establishment of agriculture about 5,500 years ago, therefore, the environments of upland England and Wales were no longer entirely 'natural environments'. The unmanipulated processes of the **biophysical** systems were still the dominant element but had satellite photography existed, the results would show some obvious traces of human alteration of the pre-existing ecosystems; there would be some less obvious changes as well, of which the growth of blanket peat would be the most widespread. So whoever first used cultivated cereals and domesticated animals did so in a landscape which was showing distinct signs of human impress. Ted Hughes opens his poem 'Moors' with the dramatic image that the open uplands

> Are a stage for the performance of heaven
> Any audience is incidental.[41]

But we now know that the audience have in fact been helping to build the scenery from a very early stage of the production.

NOTES AND REFERENCES

1. M. J. White and D. C. Schreve, 'Island Britain – Peninsula Britain: palaeogeography, colonisation, and the Lower Palaeolithic settlement of the British Isles', *Proceedings of the Prehistoric Society* 66, 2000, 1–28.
2. K. Briffa and T. Atkinson, 'Reconstructing Late-Glacial and Holocene climates', in M. Hulme and E. Barrow (eds), *Climates of the British Isles. Present, Past and Future*, London and New York: Routledge, 1997, 84–111.
3. J. A. Taylor, 'Upland climates', in T. J. Chandler and S. Gregory (eds), *The Climate of the British Isles*, London and New York: Longman, 1976, 264–87.
4. B. Huntley, 'European vegetation history: palaeovegetation maps from pollen data 13,000 yr BP to present', *Journal of Quaternary Science* 5, 1990, 103–22.

5. J. Turner, V. Hewetson, F. A. Hibbert, K. H. Lowry and C. Chambers, 'The history of the vegetation and flora of Widdybank Fell and the Cow Green reservoir basin, Upper Teesdale', *Philosophical Transactions of the Royal Society of London* B **265**, 1973, 327–408.

6. C. D. Piggott and M. E. Piggott, 'Late-glacial and Post-glacial deposits at Malham, Yorkshire', *New Phytologist* **62**, 1963, 317–34.

7. J. M. Coles and B. Orme, '*Homo sapiens* or *Castor fiber*?', *Antiquity* **57**, 1983, 95–102; B. Coles, 'Further thoughts on the impact of beaver on temperate landscapes', in S. Needham and M. C. Macklin (eds), *Alluvial Archaeology in Britain*, Oxford: Oxbow Monographs **27**, 1992, 93–9; B. Coles, 'Beaver territories: the resource potential for humans', in G. Bailey, R. Charles and N. Winder (eds), *Human Ecodynamics*, Oxford: Oxbow Books, 2000, 80–9.

8. B. R. Gearey, D. J. Charman and M. Kent, 'Palaeoecological evidence for the prehistoric settlement of Bodmin Moor, Cornwall, southwest England. Part I: the status of woodland and early human impacts', *Journal of Archaeological Science* **27**, 2000, 423–38.

9. A fully referenced account is in I. G. Simmons, *The Environmental Impact of Later Mesolithic Cultures. The Creation of Moorland Landscape in England and Wales*, Edinburgh: Edinburgh University Press, 1996.

10. M. J. Tooley, *Sea-Level Changes in North-West England during the Flandrian Stage*, Oxford: Clarendon Press, 1978; K. Lambeck, 'Late Devensian and Holocene shorelines of the British Isles and North Sea from models of glacio-hydroisostatic rebound', *Journal of the Geological Society of London* **152**, 1995, 437–48.

11. C. Bonsall, D. Sutherland, R. Tipping and J. Cherry, 'The Eskmeals project: late Mesolithic settlement and environment in North-West England', in C. Bonsall (ed.), *The Mesolithic in Europe*, Edinburgh: Edinburgh University Press, 1989, 175–205.

12. N. David, 'Some aspects of the human presence in West Wales during the Mesolithic', in C. Bonsall (ed.), *The Mesolithic in Europe*, Edinburgh: Edinburgh University Press, 1989, 241–53.

13. The best synthesis is C. Smith, *Late Stone Age Hunters of the British Isles*, London and New York: Routledge, 1992.

14. A. Myers, 'Reliable and maintainable technological strategies in the Mesolithic of mainland Britain' in R. Torrence (ed.), *Time, Energy and Stone Tools*, Cambridge University Press New Directions in Archaeology series, 1989, 78–91.

15. Simmons 1996 (Note 9).

16. P. A. Mellars, 'Settlement patterns and industrial variability in the British Mesolithic' in G. de G. Sieveking, I. H. Longworth and K. E. Wilson (eds), *Problems in Economic and Social Archaeology*, London: Duckworth, 1976, 375–99.

17. P. B. Stonehouse, 'Some Mesolithic sites in the central Pennines: comments on 23 years of fieldwork', *Manchester Archaeological Bulletin* **5**, 1990, 58–64.

18. R. M. Jacobi, 'The early Holocene settlement of Wales', in J. A. Taylor (ed.), *Culture and Environment in Prehistoric Wales*, Oxford: British Archaeological Reports British Series **76**, 1980, 131–206.

19. C. Y. Tilley, *A Phenomenology of Landscape*, Oxford and Providence, RI: Berg Press, 1994.

20. P. B. Stonehouse, 'Two early Mesolithic sites in the central Pennines', *Yorkshire Archaeological Journal* **64**, 1992, 1–15; F. Lynch, *Excavations in the Brenig Valley. A Mesolithic and Bronze Age Landscape in North Wales*, Bangor: Cambrian Archaeological Society Monographs no. 5, 1993.

21. J. Radley and P. A. Mellars, 'A Mesolithic structure at Deepcar, Yorkshire, England, and the affinities of its associated flint industries', *Proceedings of the Prehistoric Society* **30**, 1964, 1–24.

22. D. L. Clarke, 'Mesolithic Europe: the economic basis', in G. de G. Sieveking *et al.*, 1976 (Note 16), 449–81.

23. M. J. C. Walker and D. Austin, 'Redhill Marsh: a site of possible Mesolithic activity on Bodmin Moor, Cornwall', *Cornish Archaeology* 24, 1985, 15–21.

24. C. Caseldine and D. Maguire, 'Lateglacial and early Flandrian vegetation changes on northern Dartmoor, south-west England', *Journal of Biogeography* 13, 1986, 255–64; R. L. Jones, 'The activities of Mesolithic man: further palaeobotanical evidence from north-east Yorkshire', in D. A. Davidson and M. L. Shackley (eds), *Geoarchaeology. Earth Science and the Past*, London: Duckworth, 1976, 355–67; M. B. Bush, 'Early Mesolithic disturbance: a force on the landscape', *Journal of Archaeological Science* 15, 1988, 453–62; A. G. Smith and E. W. Cloutman, 'Reconstruction of Holocene vegetation history in three dimensions at Waun-Fignen-Felen, an upland site in South Wales', *Philosophical Transactions of the Royal Society* series B, **322**, 1988, 159–219.

25. J. H. Tallis, 'Forest and moorland in the South Pennine uplands in the mid-Flandrian period. III. The spread of moorland – local, regional and national', *Journal of Ecology* **79**, 1991, 401–15.

26. J. H. Tallis and V. R. Switsur, 'Forest and moorland in the south Pennine uplands in the mid-Flandrian period. II. The hillslope forests', *Journal of Ecology* **78**, 1990, 857–83.

27. Caseldine and Maguire, 1986 (Note 24).

28. D. Charman, B. Gearey and S. West, 'New perspectives on prehistoric human impact on the uplands of Devon and Cornwall', in M. Blacksell, J. Matthews and P. Sims (eds), *Environmental Management and Change in Plymouth and the South West*, Plymouth: University of Plymouth, 1998, 1–19; B. R. Geary *et al.*, op. cit. 2000 (Note 8), p. 437.

29. Smith and Cloutman, 1988 (Note 24); R. N. E. Barton, P. J. Berridge, M. J. C. Walker and R. E. Bevins, 'Persistent places in the Mesolithic landscape: an example from the Black Mountain uplands of South Wales' *Proceedings of the Prehistoric Society* **61**, 1995, 81–116.

30. P. E. J. Wiltshire and K. J. Edwards, 'Mesolithic, Neolithic and later prehistoric impacts on vegetation at a riverine site in Derbyshire, England', in F. M. Chambers (ed.), *Climate Change and Human Impact on the Landscape*, London: Chapman and Hall, 1993, 157–88.

31. P. D. Moore, 'The origin of blanket mires, revisited', in F. M. Chambers (ed.), *Climate Change and Human Impact on the Landscape*, London: Chapman and Hall, 1993, 217–36.

32. A. U. Mallik, C. H. Gimingham and A. A. Rahman, 'Ecological effects of heather burning. I. Water infiltration, moisture retention and porosity of surface soil', *Journal of Ecology* **72**, 1984, 767–76.

33. There is a good general account of blanket peat initiation in the British Isles in D. Charman, *Peatlands and Environmental Change*, Chichester: Wiley, 2002, 80–4.

34. This need not have been a very lengthy process. There is an eighteenth century description of part of Scotland (Lochbrun) in which the author tells of the transformation of a 'Firr' wood into a bog:

    In the year 1651 . . . This little plain was at that time covered over with a firm standing wood; which was so very Old that the trees had no green leaves but the Bark was totally thrown off . . . the outside of these standing white Trees, and for the space of one Inch inward, was dead white Timber; but what was within that, was good solid timber, even to the very Pith, and as full of Rozin as it could stand in the wood. Some Fifteen years after . . . there was not so much as a Tree . . . but in place whereof . . . was all over a plain green ground, covered with a plain green Moss. . . . the green Moss (there in the *British* language called Fog) had overgrown the whole timber; . . . and they said none could pass over it, because the scurf of the Fog would not support them. I would needs try it; and accordingly I fell in to the Arm-Pits . . .

George, Earl of Cromertie, FRS, 'An account of the mosses in Scotland', *Philosophical Transactions of the Royal Society of London*, **XXVII**, 1711, 296–301. Most of the spelling has been modernised.

35. C. Chandler, P. Cheney, P. Thomas, I. Traband and D. Williams, 'Fire as natural process in forests', in (eds), C. Chandler *et al.*, *Fire in Forestry. Vol I. Forest Fire Behavior and Effects*, New York: Wiley, 1983, 293–393.
36. In major storms of the twentieth century, most native species fared much better than exotic imports.
37. J. M. Coles and B. Orme, 1983; B. Coles, 1992 (Note 7).
38. Smith and Cloutman op. cit. 1988 (Note 24), p. 168.
39. P. C. Buckland and K. J. Edwards, 'The longevity of pastoral episodes of clearance activity in pollen diagrams: the role of post-occupation grazing', *Journal of Biogeography* **11**, 1984, 243–9.
40. J. G. Evans, *Land and Archaeology. Histories of Human Environments in the British Isles*, Stroud: Tempus Publishing, 1999, ch 3.
41. In *Remains of Elmet*, London: Faber and Faber, 1979, 19.

CHAPTER THREE

The millennia of an agriculturalist economy

Though it may seem curious to group together the far reaches of Neolithic time with AD 1750, two properties link them both. Those are firstly the focus on domesticated plants and animals as sources of food and clothing, rather than wild species as in the hunter-gatherer time and secondly the fundamental dependence upon the flows of solar energy, unsubsidised by fossil fuels as in our time. Other common characteristics can be detected like vulnerabilities to disease and periods of bad weather. There may be even more for which there is poor evidence in pre-literate times: in many agricultural societies a major cause of famine, for example, is a civil breakdown. Everything takes place within systems of trade and exchange that originate in prehistoric periods and which gain in volume as time progresses. An overview of agricultural production of all kinds from medieval times to the near-present suggests that there are periods of near-Malthusian stress of population on food supply and contrasting times of relaxation;[1] we would expect these to apply to the uplands as well as more intensive production systems though perhaps more as the outer ripples of a perturbation whose centre was lower down the hill. For present purposes we can subdivide this long stretch of time into an earlier phase (ending about AD 1500) and a later one when population growth and intensification of land use were beginning to take off.

FROM COLLECTION TO PRODUCTION 3500 BC–AD 1500

One contextual factor is important in discussing the environment of the new economic system. The contrast between today's bare and indeed bleak moorlands and the enclosed, tree-studded lands below 300 m would have been much less if large portions of the uplands were themselves wooded. Temperature lapse rates were still important and it was still very wet, but the perception of the environmental potential might well have been favourably skewed or at any rate reckoned to be no worse than anywhere else at the time.

The earliest agriculture in the uplands

A major consequence of the reduced difference between upland and lowland mentioned above is that the introduction of agriculture does not appear to

47

have been dissimilar. It does not appear any later in time, for example, nor were the crops any different. The immediate context of soils and vegetation, too, had similarities in the sense that either naturally or by human action there had come to be a mosaic of woodland with openings. The small clearings in the woods in both upland and lowland areas were, we presume, seen as good sites for the new economic practice (Plate 3.1).

What is without doubt different above 300 m is the absence of the full panoply of Neolithic monuments and settlement types associated with, for example, the chalklands of southern England. Nevertheless, where palaeoecological data are available there is evidence in the uplands of a period of intensified environmental impact just before the decline of elm pollen that is present in most of western Europe either side of 5,000 BP. In some sites, this period is accompanied by cereal pollen, usually that of wheat (*Triticum* spp.); at others there is no such evidence. Absence of cereal pollen in the organic deposits does not necessarily mean there were no cereals, for cereal pollen is not widely dispersed like tree-pollen and it may only get its limited dispersal when the tree canopy is opened up. The result of all the detailed discussion suggests that wheat was indeed being grown in small openings in upland woods from about 5,000 BP.

An upland example of this phenomenon can be seen in detailed pollen

PLATE 3.1 *Pentre Ifan, a Neolithic burial structure in Dyfed, with the Preseli mountains in the background. It reminds us that while Neolithic structures are not common on moorlands, the uplands themselves would have been part of the environment of early agriculturalists. Yet the difference between upland and lowland was less marked, with both being more forested than in more recent times.*

analyses at North Gill in the North York Moors[2] at 354 m ASL. Pollen, macro-fossil and charcoal analyses at 1 cm intervals display a typical sequence. A mixed deciduous forest is complemented by high levels of shrubs such as *Corylus*, and there are varying but consistently present levels of grasses, heather (*Calluna vulgaris*) and intermittent levels of plants of open habitats. Microscopic charcoal (of size <180 μm) occurs throughout, though in highly variable quantities with peaks at the base and in the zone where the analysis showed the presence of verified cereal pollen. The count shows a typical mid-Holocene elm (*Ulmus*) decline at 60 cm. This is immediately preceded by a radiocarbon date of 4,640 ± 50 BP. The overall conclusion, therefore, is of a largely wooded environment in which alder is gaining ground but in which there is sufficient open ground, some of it damp, for herbs to show in the spectra. If there were a great deal of unshaded area, however, more grasses and heather would be expected.

The cereal-containing levels can be flanked by phases with no cereals, thus:

0. Initial state woodland
1. Disturbance with *Melampyrum* (cow-wheat)
2. Disturbance with cereals and *Plantago* (plantain)
3. Some recovery of woodland

The increase of alder in phase 0 suggests that either climate or locally damp areas were providing favourable conditions, which is confirmed by the presence of two grains of *Thalictrum*, (meadow-rue) probably *T. flavum*, now a plant of wet meadows and streamsides. The pollen spectra are therefore recording open grassy areas by the stream, a fringe of shrubs and alder, and the oak-dominated woodland beyond, all probably within 25 m of the sampled site.

The rapid changes of phase 1 clearly show an opening in the woodland, with perhaps space made for a growing fringe of alder. Sudden increases in birch and hazel pollen suggest improved flowering caused by increased light levels. The opened area seems to have allowed some increase in heather, bracken and other ferns and a noticeable surge in *Melampyrum*, usually taken as an indicator of the opening of woodland canopies on acid soils. After two samples (=approximately 12 years), these light-indicators are temporarily overcome by about 30 years' growth of more heather (and possibly birch) before the renewed opening reaches apparently greater dimensions in the levels with high *Melampyrum*. The charcoal curves suggest that fire was implicated in the maintenance of this opening.

Given the poor dispersal of cereal pollen, it is tempting to assume that cereals are grown in this phase but are absent for **taphonomic** reasons. But in phase 2 there is a distinct shift, with the disappearance of *Melampyrum* and the appearance of *Plantago lanceolata*. This and the other **ruderal** herbs with similar patterns suggest an assemblage of a cultivated 'field' and its weed flora, among which *Anagallis arvensis* (scarlet pimpernel) and the *Matricaria* (mayweed) type are likely components, the latter being especially found on arable land with light soils. The other plant types are also found on all kinds of rough and broken ground. *Genista* (whin) on the other hand, belongs with heather and bracken in the flora of nascent heathland which was presumably on the edges of the cultivated patch. So here is a 50-year episode of wheat cultivation (in which the gaps are very possibly taphonomic artifacts rather than actual) in an arable patch away from the stream on the drier soils, surrounded by other broken ground and a fringe of heathy vegetation, beyond which is the shrubby ecotone of the oak-elm-birch woodland. The years of wheat growing saw only two fires: the first was six years after the first cereal pollen, there was then a gap of nineteen years before the next fire and the next was six years after the end of the arable phase.

There is then a largely fire-free period of seventy years (phase 3) in which much of the opening seems to be colonised by shrubs like hazel and secondary trees like birch and alder. Oak almost doubles its frequency from the pre-cereal horizons and the overall picture is one of woodland rapidly gaining ground. The grasses, however, seem to continue to expand and *Calluna* maintains a presence: both possibly in woodland whose canopy was not closed.

The dating suggests that the wheat was grown in the decades before 5,430 BP (approx. 4,300 calendar years BC) and before an elm decline at 4,640 BP. It seems to have been planted in the opening after the main period of use of the clearing, perhaps either as an afterthought ('here is this novel stuff, it won't do any harm to try it now') or as a total novelty ('where shall we put this stuff we've just got hold of?').

There is evidence as well of another kind of purposeful management at North Gill, this time at 5,315 BP but in a different sample profile, about 100 m away.[3] Here, the vegetational sequence is dominated by six years of the selective opening of the canopy. There is no fire but some shrubs flower rather better than before and there is a somewhat patternless response of the ground flora. The trees affected (oak, elm, alder) grow back quickly so it seems unlikely that trees were being killed. Ash and lime seem to benefit and it is possible that their pollen is spread more widely as the canopy is opened. This phase is interpreted as an early Neolithic gathering of branches for animal fodder: most likely for domesticated beasts such as cattle. In continental Europe this was done in late summer and early spring and there was a hierarchy of species' value for this purpose in which oak and elm stood high.[4]

Given that the introduction of agriculture is such a radical change in human life-ways, it is a pity that our evidence gives no clue about why it took place. If it were brought in by established farmers from elsewhere, then the mosaic of woodland, open areas, scrub and mire would very likely have been familiar to them from continental Europe at lower altitudes. If it were adopted by an indigenous population then the open areas of their environment would have been the obvious place to grow a grass (even if it was different from their normal wild flora), even supposing that there were no orally transmitted instructions. The sporadic nature of the cereal finds suggest that for some time in the uplands, the new plants were not staples. They might have been an interesting but an exotic item grown in the corner of an opening that had fulfilled its main purpose of attracting deer. The two purposes would not have been compatible, given the liking of deer for grasses. The new cereal might have been a crisis food or the very opposite: a special-occasion food. Knowing that cereals can be fermented with the production of alcohol may be a very old piece of knowledge indeed. A culture used to attracting deer by encouraging grass and perhaps providing leaf-fodder would have taken well to cattle-rearing. Overall, the coming of the Neolithic cultural style need not have been at all traumatic.[5] Eventually, the simple *availability* of cereals and cattle progresses into their *substitution* for hunting and gathering as the foundation of the economy and then to a *consolidation* phase in which there is a dependence upon farming. We can expect both the extension and consolidation of this practice.[6]

The ecology of pre-medieval agriculture in the uplands

Though 'prehistory' is hard to define, we shall take it here as terminating with the effective end of Roman occupation *c.* AD 400. This means that some written evidence may be available as well as that from archaeology and its subsidiary techniques, but it is in general small in quantity for Britain and not very detailed. 'Prehistory' includes, therefore, the overlap of the Mediterranean power's influence and the impress of the indigenous Iron Age cultures, though they themselves were the result of immigration only a few centuries previously.

The feature which all the agriculture of Britain had in common in the millennia from about 3500 BC to AD 400 was that it was mostly mixed farming. As most of its production was consumed locally, then the system needed to provide both plant and animal materials. In the climate and soil context of Britain it had to be a system that kept its nutrient cycles (especially that of nitrogen) free from long-term loss, and its soil structure sufficiently well bound to resist some of the downward pull of gravity towards the runoff. Hence, systems developed which had some fields near the settlement that were manured by domesticated beasts that fed off an area of peripheral 'wild' vegetation such as scrub, woodland or grassland or even moorland and heath. These might not only sustain their local human inhabitants but have something to spare for population growth or to support a non-productive stratum of priests, bards, kings or conquerors. The immediate question is whether the upland systems were different in any significant way.

The evidence for the possession of the land is plentiful in those areas now in upland grassland or moorland cover. Indeed, it was once thought that these terrains were occupied rather than the lowlands but decades of field-walking, aerial photography and excavation have shown something of an equality of spatial tenure. So any review[7] of developed prehistoric agriculture in the uplands will note that there is evidence of settlement sites, of field systems and quite possibly also of large 'estates' stretching from watershed to watershed which, it is thought, contain land types of different productive potential. Clear systems of land allotment have been mapped on several uplands: one system of 'co-axial' stone and earth banks ('reaves') on Dartmoor covered at least 1,500 ha., and analogous systems appear in the Yorkshire Dales. Stone cairns are also found in such systems and together with the layout of the 'fields' suggest that the economy around 1400–1300 BC (i.e., Middle Bronze Age) was arable as well as pastoral. Similar systems are found throughout the period from 2000 BC to the first centuries AD, with variations of shape, wall type and associated living quarters; many of the well-known 'Celtic fields' are examples. On Dartmoor, one area occupied in 1400–1000 BC shows a complete lack of the orderly layout found in the reave systems: there seems to be gradual encroachments on different alignments.[8] One disadvantage of using upland areas for crop growing is the presence of quantities of stony material in the soils, often of cobble size and sometimes larger. So cairns are sometimes found which seem to have doubled as burial places and as collection points for cleared material. Earth-fast boulders may have served as nuclei for such piles of stone. At Crawley Edge in Weardale (at 323 m), a southerly facing site with a saddle quern was underlain by a spread of oak charcoal; downslope was a layer of silt which possibly came from soil creep. Radiocarbon dating gave a range of 1,490–1,310 ^{14}C years BC (uncalibrated) which suggests the Bronze Age and which range is common in upland Co Durham. A cereal-growing agriculture seems to have been firmly established at 300–400 m ASL in the North Pennines during Bronze Age times.[9] All these enclosures and field systems may be contained within linear ditches that traverse hilltops (and which sometimes use barrows as way-marks) and which are reckoned to be territorial delimiters within which can be found all the resources needed to support a particular group. Thus on the North York Moors, for example, one such 'estate' has its hut-circles and fields down by the Wheat Beck (Plate 3.2) and then what was presumably the grazing and woodland stretching up towards the skylines; a Bronze Age date is usually accepted. In some places they are closely reflected in modern township boundaries. Even more spectacular are the later (c.1000 BC) multiple dykes found on the Tabular Hills region of the North York Moors. The Scamridge dykes are six abreast and most likely reflect a 'tribal' boundary rather than that of a single group of farmers and pastoralists.[10] Later political divisions may also be marked by linear features: Offa's Dyke is well-known in both its lowland and upland phases but there are cross-valley examples in Swaledale which may mark the division between Anglian and native lands.[11]

The later prehistoric period also saw the construction of hilltop enclosures of varying sizes usually designated 'hill-forts'. In Wales for example, they are frequent through the uplands and there is an especially dense band in the

PLATE 3.2 *The valley of the Wheat Beck on the North York Moors. In the foreground there is a Bronze Age field system and its boundaries carry on into the enclosed fields beyond. It is also thought that 'estate' boundaries continue on to what is now moorland, so that the land holding of the prehistoric group contained both lowland and upland resources. Just how different these were has to be established by closely spaced pollen diagrams.*

Marches. At Breiddin (300 m ASL), the enclosure contained a pond and cistern which yielded abundant environmental evidence, the tenor of which was to show that in the Late Bronze Age and Iron Age, the pond lay in an area of open ground with a few trees (oaks, birches) and scrub (gorse or broom) nearby. Some of the open ground was bare and the fauna and flora suggest that it was well-grazed but not heavily so. The environment, in short, was much like that of the present day but the big surprise was a lack of evidence for the presence of people or their structures, since Iron Age hill-forts have usually been given a major role in the socio-economic-political systems of the time.[12]

The example of early cultivation on the North York Moors has a parallel on Bodmin Moor where settlement and 'major landscape disturbance' in the earliest Neolithic are inferred from palaeoecological evidence. There was then some limited woodland regeneration in the later Neolithic before Bronze Age activity led to intensive pressure on the environment. The Dartmoor and North York Moors' examples seem to be confirmed here. On Bodmin Moor the Iron Age was also a period of intensive use of the upland, a situation which continued through the Roman occupation into medieval times. One outcome, unsurprisingly, was the replacement of woodland by grassland, whose pollen record suggests that hay meadow management and seasonal grazing was practised at least in the limited areas of study. The settlement of medieval times and its associated grazing pressures seem to have diminished the diversity of the moorland vegetation.[13]

Even though the uplands were clearly more popular for food production than nowadays, the environment still exerted strong constraints. Only long periods of major climatic amelioration would have lengthened the growing season by more than a few days and there is strong evidence for fluctuations of climate over the whole country (increased continentality from about 2500–1900 BC onwards; a pronounced decline around 1500–1400 BC and then far cooler and wetter from about 900 BC onwards), with variations in the uplands.[14] Lapse rates may have been gentler in the centuries around 2000 BC for example, but the trend to cooler and wetter conditions in the first millennium BC (with an average drop in temperature of 2°C for Britain, which would have lopped five weeks off the growing season) seem to have caused an abandonment of settlement at higher altitudes.[15] One result of a wetter climate and the recession of woodland (from whatever cause) was the growth of blanket peat over areas of high altitude (above c.400 m) and low slope (<3°); this took place at any time from the Mesolithic onwards and must have diminished the areas of good grazing as well as potentially cultivable land. The discovery of shards of volcanic **tephra** in upland (410–510 m ASL) west-central Wales dated c.3900 BC[16] reminds us of the potential of volcanic eruptions in Iceland to interrupt normal weather patterns; in the case of these examples the concentration of particles was perhaps rather low to suggest much meteorological disturbance.

We are sure that the area of forest was decreasing. No doubt some of it was managed carefully, but the absolute area available for hunting wild animals must have shrunk. The extinction of *Bos primigenius* took place sometime in the Neolithic-Bronze Age continuum and the number of red deer (then a largely woodland animal) also fell, though roe deer might have benefited from increased quantities of secondary woodland and scrub. Though venison was not entirely off the menu, it must have become a status and special-occasion food rather than a more everyday item as in the Mesolithic. Moorland vegetation contains fewer soil invertebrates to attract and maintain wild swine. Evidence for the growing of oats, rye and even barley in upland zones during the Iron Age and Roman periods is firm, though most interpreters place more stress on the pastoral component of the economy.[17] In ninth century Wales, south-east facing slopes above 100 m were cultivated for cereals[18] and in Upper Teesdale at about 457 m on Weelhead Moss, pollen analysis shows the presence of cereals as early as 3370 BC and large conversions of open woodland to grass and blanket bog between 1000–600 BC. Such openings in the Neolithic and later prehistoric times seem to be a continuation of the recession of woodlands sporadically found in the Mesolithic in the same region of the Upper Tees.[19] Exmoor seems to have been less affected by the climatic deterioration after 900 BC as there is pollen evidence of agriculture (at sites in the

430–460 m ASL range) from the terminal Bronze Age through the Iron Age to the early Romano-British period. After about AD 500 **pastoralism** dominates in an era of rising water tables and soil acidification. The West Saxons declared Exmoor to be a royal hunting ground and this probably preserved much of the marginal woodland. Continued grazing pressures resulted in the disappearance of woodland and scrub from the Iron Age onwards and thus allowed the formation of blanket peat. Below this zone, nevertheless, woodland came and went with phases of varying intensity of pastoral and arable farming: the Romano-British centuries seem to have allowed some recolonisation of uplands which were formerly grassland or cultivated fields.[20] In Wales, cattle were very important products during the Roman occupation though in its latter phases, sheep appear to have gained more favour.

We cannot conceive of the prehistoric occupation of the uplands for any purpose without being aware of the environmental changes at human hands that resulted. One of the processes found in the uplands just as almost everywhere else in western Europe is the decline of elm (*Ulmus*) pollen in lake and bog deposits at around 5,300–5,000 BP, though not entirely synchronously. At one time, this was thought to be the first evidence of neolithic cultivation, resulting either from cattle eating the leaves or from leaf-foddering. Later, the involvement of elm disease (carried then as recently by the elm bark beetle *Scotylus scotylus*) has been regarded as central, with the possible amplificatory effects of soil changes after tree loss making regeneration of this genus less likely. The decline of elm has to be put into the context of a widespread loss of woodland generally. In the uplands, this is usually sporadic in the Neolithic: indeed often on no greater a scale than forest recession in the preceding Mesolithic. The Bronze Age was often a period of considerable deforestation, which is confirmed by the archaeological evidence and the Iron Age likewise a major period of destruction, though the level of archaeological finds may not be a good indication of the intensity of agriculture or forest recession.[21] The first major impacts in different parts of the Cheviots have been chronicled for 3700–2350 BC and for 550–50 BC, which emphasises the lack of regional, let alone national, synchronicity of the first major forest losses.[22] In Cumbria and Northumberland, the area near to Hadrian's Wall experienced sparse impact during the Bronze Age, but massive indicators of clearance are found during the time of the occupation and construction of the Wall. Possibly the Roman army cleared land not for subsistence but for military purposes, such as construction of forts and improving the field of view. On the other hand, they may have come into a well-used set of Iron Age environments and intensified food and materials production by their presence.[23] When they ran a road over Stainmore in the North Pennines, they made it in an environment that had been more or less totally treeless since the end of the first millennium BC and they themselves made very little impact.[24]

If the woods vanish, the scene is set for soil changes as well (Fig. 3.1). One probability is the conversion of the acid brown-earths commonly found

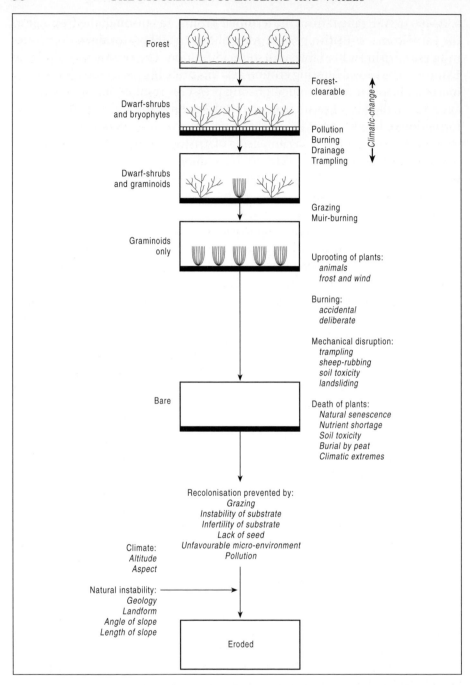

FIGURE 3.1 *A scheme for the development of eroded soils undergoing impact and management after deforestation. The dwarf shrub phase is likely to be dominated by* Calluna *(heather) and at each stage thereafter the possible influences are detailed. Not all of these will necessarily happen at one site: the diagram emphasises the variety of process that may impinge and also that long-term climatic and physiognomic influences are never eradicated.*

beneath upland deciduous woods into soils with greater evidence of separate horizons and of waterlogging. Thus podsol and gley soils are common on the uplands, as is the virtual absence of any profile at all beneath some areas of blanket peat. The soils may indeed erode away to some extent, as the binding effect of the roots is lost. Erosion may happen as sheet-wash or gully formation. The material thus lost to the upland will be deposited as **colluvium** on lower slopes (modifying their profile, especially that of the steeper valley sides where fans are sometimes formed) and as **alluvium** in the valleys. Where alluvial transport is concerned then changes to water quality must have been noticed, especially after storms. Normally clear streams would then have been swollen with silt and faster runoff with higher and sooner flood peaks. Some of this might have been damped down by comparison with today if the beaver populations had been intact and not vanished into fatty stews and Davy Crockett-style hats. It would have been exacerbated by the moves to winter-sown crops suggested for the period after 650 BC. Another major influence in a few valleys would have been the digging-over of gravels for metal ores. Cornish tin is often reckoned to have been a major source of the tin used in the bronze of that Age but few traceable remains are found since the streams were more thoroughly worked and re-worked during the medieval period. A peat profile from Tor Royal shows increased values for copper, zinc and lead at about 2500 BC (tin is not detectable with current technology) which could result from biomass burning or even from metal mining.[25] Overall, iron mining and working may have been an environmental influence wherever ore was found near to settlements. On Exmoor, there are two large slag heaps and smelting platforms of Romano-British age as well as smaller charcoal-burning platforms, all of which add up to a noticeable environment impact in their locality.[26] In Wales, it seems possible even that the advent of the Romans meant a hiatus in production (presumably while control was established over this important activity) before demand seems to have increased.[27]

From the Neolithic transitions to the Roman withdrawal, changes of the flora and fauna resulted. In part these were changes in abundance, as we have surmised for deer species. Yet there were extinctions as well: regionally the beaver almost certainly disappeared and nationally, *Bos* (the wild ox or aurochs) was lost. Introductions were more common, however. Most of them were associated with domestication as with cattle, sheep and goats. Pigs might have been imported as well as tamed from the local wild populations. Wheat and barley, oats and rye, are not native cereals and they no doubt (as with the animals) brought their own free-riding populations of parasites. Management meant, among other things, that sheep, goats and cattle had to be kept out of

FIGURE 3.1 *(continued)*
Source: Original diagram by J. H. Tallis. Reproduced from Blanket Mire Degradation *(1997) by permission of the editors*

growing crops and young coppices. The weed species so common in pollen diagrams of the period were mostly plants of edge habitats and unstable slopes. They had been present in the pre-forest times of the late-glacial and had hung on in suitable areas; now they could expand into the broken soils, opened canopies and trodden greenways of an agricultural and pastoral landscape. Some seem to have been virtually ubiquitous, as with plantains (*Plantago major* and *P. lanceolata*) and the wormwood (*Artemisia vulgaris*); others are noted as cornfield weeds and soils with enhanced nitrogen levels, as with nettles (*Urtica dioica*). Plants such as goosefoot (*Chenopodium album*) have an ambiguous status: long considered weeds, they may have been gathered or even cultivated for their fatty seeds. In the extensive woodlands of the Neolithic, the mollusc *Spermodea lamellata* was common but it gradually became restricted to a few relict patches.

The evidence for environmental impact, though, does not clarify beyond doubt some key questions about the uplands. It is unable to say for the Bronze Age what mixture of arable and pastoral economies were present, though detailed work at a site on the East Moors of Derbyshire shows that field systems were indeed used mainly for arable rather than as stock corrals though some pastoralism was present.[28] For the whole agricultural period up to early medieval times, palaeoenvironmental findings do not track the development of **transhumance**. These key elements in the environmental manipulations of their times are very little known.[29] There seems a consensus that the Iron Age was an important marker. The land was certainly fully allocated by then to a stratified social system, with highly organised major tribes. The lowlands and upland valleys probably presented an aspect of being totally managed.[30] The upland landscape was largely open and looked less managed but was almost certainly earmarked, with dykes-and-banks impressed across the land. Peripheral woodland came under pressure when climatic deterioration caused the abandonment of settlement (and perhaps of some seasonal grazing as well) during the first millennium BC. So the major lineaments of later times, the control of the land but its abandonment for permanent settlement, were in place during the Iron Age. (As many folk then occupied the lowlands, they established some patterns of field, wood and settlement which were to persist through into medieval and early modern times, though that is peripheral to the moorland story.) For the uplands, though, it was the achievement of prehistory and especially of the Iron Age, as Fowler puts it, '. . . inadvertently to create the now much-prized wildernesses of west and north . . .'.[31] The earliest Arthurian tale in Welsh, probably from the period soon after the withdrawal of the Romans, records slash-and-burn agriculture and it seems probable that this was a continuation of Iron Age practice. The Chief Giant sets a task for the hero Culhwch: 'Dost see the great thicket yonder? . . . I must have it uprooted out of the earth and burnt on the face of the ground so that the cinders and ashes thereof be its manure; and that it be ploughed and sown so that it be ripe in the morning against the drying of the dew' Later, the ancient Owl of Cwm Cawlwyd reports that the wood to be seen in a great valley is the third wood to grow there, men having laid the previous growths to waste.[32]

Our view of upland agriculture is highly coloured by the fact that '. . . areas once farmed [in prehistoric times] have tended not to be cultivated again, and consequently the monumental remains of the exploited, relict landscape, fossilized in stone, have been well preserved to the present day'.[33] We must remember that the lowlands too were intensively used but that the evidence is more difficult to find in those areas; but then when found, it tends to be more complete in its preservation of bone, wood and seeds, for instance. Sensitivity to the nature of the evidence need not preclude the quite robust conclusion, though, that the upland zone of England and Wales underwent a distinctly thorough make-over between the coming of agriculture and the departure of the Romans. Further, the main cause was agriculture itself in both its cultivation and pastoralist modes. Climatic change was also implicated as was working for metal ores in a few places, but the primary instrumental pathway was through the removal of woodland, the cultivation of soils and the deployment in the landscape of domesticated animals. Behind the instruments of transformation, of course, lay the mental maps of cultural practice: then as before and after. Some archaeologists have postulated the importance of the symbolic and the **numinous** in environmental change: various combinations of ancient thorn trees, standing stones, mistletoe, pools and even polished stones and engraved stones might all have added a non-material dimension to a locality that might have enhanced its potential for change or the entire lack of it.[34]

If the Roman soldiers had been able to take photographs of the moors as they retreated along upland roads soon after AD 400, the landscape thus captured would have been familiar to us in the sense that its basic open quality was well established. It is likely that there were more people around and that the boundary between 'enclosed' land and the open moor was marked by a zone of scrub and woodland (especially where there were steep slopes) rather than a wall and a cattle grid. The vegetation of the flat swales above 300 m was grassy rather than heathery though the ericoid shrub crowberry (*Empetrum*) was a useful source of berries in late summer. In the flattest and wettest places, much mire vegetation dominated the scene albeit mostly of bog mosses such as *Sphagnum* rather than sedges. To the culturally attuned eye of the commanders, however, one feature would have been certain: the marks of the division and 'ownership' of the land were everywhere. In no sense was this an unclaimed waste: it was part of the **ecumene** of the people they had conquered and had in its time yielded tribute to Rome, including, said Strabo, cattle, gold and silver as well as lead and hunting dogs. No question that some or perhaps most of these came from the upland zone that included the present-day moorlands. The mines at Dolaucothi in Powys were worked for gold in Roman times, and probably before then.[35]

Land use and cover in the medieval period

Questioning visitors to our uplands often stop short when the relics of a field system are traced out on a hillside at perhaps 300 m above sea level. The evidence is in different patterns of grass, bracken and squat baulks thrown into relief by a low sun or in documents showing farms to have been founded in the thirteenth century. 'They grew cereals *here*?' is not so much disbelieved for prehistory – when everything was supposed to be different – but for medieval times, when more continuity with today is assumed.

Plant and animal production

One continuity is clear. A dominant pattern of resource use in the uplands throughout the period from the departure of the Romans until the eighteenth century was of settlements in valleys and on valley sides which relied on the local production of cereals for part of their nutrition, however much that might be supplemented by the consumption of dairy products and by the export of sheep and cattle products or minerals.[36]

Given that cereal production was liable to fluctuation and failure in even the most favourable parts of the British Isles, its fortunes in the uplands were even more precarious. Small wonder, therefore, that wheat, the most desirable crop of the lowlands, was replaced by hardier genera. Barley is more likely to ripen successfully than wheat but above all it is oats that can tolerate the shorter growing season, the wetter conditions and the absence of a period of sunny weather to ripen the crop, being grown even in basins surrounded by uplands far from the influence of the sea as in North Wales.[37] In exceptionally wet years like 1315 and 1316, though, even the oats failed. Oats had become markedly more common during Roman times and remained central thereafter. In the upland areas of northern England, for example, the preponderance of oats was almost complete during the high middle ages. Oat straw is also very useful in manure production since it absorbs a high proportion of the urine of stalled cattle and so the nitrogen level of the bedding is high when it comes to be spread on an arable field. It occupies a pivotal role between the necessary cereals and the animal husbandry which is the obvious way of using upland vegetation and soils.

But why should populations wish to cultivate in such environments? From time to time, the population of Britain seems to have come close to a Malthusian relationship with the land so that a burgeoning population was pressing hard upon its resource base. The century before the Black Death of 1348–9 was one such period and this certainly looks like a time when farms were established further up-slope and higher up the valleys than had been the case since the Iron Age. Other factors seem to have interacted at other times: the Norman conquest of Wales probably replaced partible inheritance (all the sons) with single inheritance (the eldest son) so a demand for new holdings was created rather than the subdivision of the old. Such a conquest could emplace new lords who might wish to push the areas of their visible power into environments that might otherwise be centres of rebellion and dissidence. The use of the uplands for purposes such as mineral production often brought agriculture in their wake, into such lands as were described for Castle Bolton in Wensleydale as *terra inculta et nemorosa*.[38] In sum, there seems to have been a medieval resurgence of interest in the uplands for settlement and agricultural production, of which one result was the further reduction of both tree cover and the floral diversity of pastures.

If woodland and moor were to be reclaimed for agriculture then the zone between 200–350 m was often the limit of feasibility, for above that few crops of any cereal would be reliably successful. This leads to regional differences, since such a band is wide on, for example, the eastern flanks of the Pennines, where the fingers of high land reach down some 25 km towards the basin of the middle Wear near Durham City;[39] the commons of two of the City's boroughs (Crossgate and Elvet) were not enclosed until the 1760s and Gilesgate ('Gilligate') was still undivided albeit 'very improveable' in Bailey's 1810 account.[40] On the northern and western sides of the North York Moors, however, the escarpment means that the 100–250 m zone is covered in a kilometre or so of steep slope. If this reclaimable belt is analogous to an intertidal zone then there is a contrast between a shallow-shelving beach and a narrow drop-off into deep water.

Medieval communities in the uplands had a core area of cereal production, usually on valley sides of low slope but above any flood zones. Woodlands and **wood-pasture** were adjacent, and above them the open moorlands (Plate 3.3). At any one time, a dyke separated the open moor from the enclosed pastures and croplands. At times of expansion, a settlement might expand its juxtaposed enclosed grasslands and its arable area, or it might colonise a 'waste' area within its jurisdiction that was spatially separated higher up the hill or the valley. The effort needed to turn moorland into cropland, or to clear away secondary woodland[41] (or even perhaps the remnants of primary forest) meant that such labours were scarcely to be undertaken lightly and the role of a landlord able to enforce his wishes must often have been central. Elsewhere, the moorland edge might therefore result from piecemeal encroachment by tenants or even squatters and it would likely be irregular in plan. If the manorial lord exercised strong control a 'harder' boundary might result. Procedures for the control of commons seem to have broken down in Wales in the 16th century, so adding to the first type of boundary.

If twelfth- and thirteenth-century crops were grown high on hillsides, even at 350–440 m ASL in the Berwyn Mountains of North Wales[42] (in 1300, the population of Wales was about 300,000 and had pushed to the margins of cultivable soil), then the key constraint of soil fertility assumes even greater proportions than in lowland regions. The central substance was manure. Hence the presence of enclosed pastures near the farmstead where the animals' dung could be collected, the ability to fold the animals upon the ploughed fields and the right sort of winter housing that would preserve the bedding for later

PLATE 3.3 *Medieval clearance in the woodlands along the flanks of upland terrain, north of Eastgate in Co Durham. The steep slopes and valley bottoms were not attractive for cereal growing, unlike the valley-side benches, once settlement had pushed its way into side valleys. Initially we assume that farms like this were islands of open land within a woodland framework, most of the upper parts of which have disappeared. The field pattern is relatively modern, having been consolidated at some later time.*

dispersal on the fields, were all critical. Other materials were also used to improve soil fertility, of which marl, dug from calcareous rocks of various ages and lithologies, was the most important. The whole range of additives accessible to the lowland farmer (soot, night soil from towns, pigeon guano, seaweed) were not often available in the fastnesses of the uplands (Plate 3.4). After the fulling process, however, the grease from sheep fleeces was a prized fertiliser on Dartmoor.

Longhouse settlement and cereal cultivation on Dartmoor seems to have culminated before 1200–50, and the pollen of cereals and weeds confirm the environmental changes that accompanied the necessary enclosure.[43] Some settlements were large enough to have common open fields and were enlarged by the custom (already old by 1345) of an incoming tenant making a 'new take' of up to eight acres (3.2 ha). Smaller examples tended to have oblong fields with no signs of strips and so **lynchets** were often formed. These new farms had rights of grazing on central Dartmoor, as did the whole population of Devon except for the inhabitants of Barnstaple and Totnes as indeed did the people of Cornwall until late Saxon times. The presence of lynchets (as at

PLATE 3.4 *A monastic dovecot's lower storeys at Llanthony in the Black Mountains of South Wales. Apart from the birds and their eggs as a sources of food, the guano would be a valuable addition to sources of nitrogen for arable land.*

perhaps 350 m above Grassington in West Yorkshire) may not always mean cultivation: some of them may have been stock enclosures. One different pattern of colonisation is downslope. In the Calderdale area of the southern Pennines of Yorkshire after 1250, people living at heights of 215–175 m began to clear woodlands below them, moving into the wetter zone of the valley floor by about 1300. Holmfirth was a colony of an older and higher settlement of Holme; now it is a colony of TV fame.[44] The process of converting wood or moor to enclosed land is called **assarting**, and it was usually organised by leading figures in local society rather than being an action by a single peasant household. This was particularly the case in the interiors of large blocks of moorland, such as the North York Moors; however, peasant-initiated assarting was more common around the fringes.[45] On the Pennine border of Yorkshire and Lancashire, assarting of open woodland was popular in the 1330s and 1340s, sometimes for cultivation and sometimes for pasture only. Scammonden, near Huddersfield, was exploited by its manorial lord for raising cattle, for example. Such processes of clearing woodland were usually halted by the Black Death.[46]

Both new intakes and established farms might engage in temporary reclamation of moor by practising infield-outfield cultivation. Near to the farm, one field is kept in excellent condition by frequent applications of manure and it is supplemented by an outfield whose location varies and which is abandoned when its fertility falls, which might happen after a single crop, although a longer period of yield would have been hoped for. The outfield was usually ploughed or dug over so as to produce a ridge and furrow sequence to aid drainage, and the peaty top of the soil profile burned and then dug in so as to maximise the aeration and the nutrient content of the soil before planting. Any additional nutrients came from rainfall and from folded animals. This system was practised in Scotland and Cornwall until the eighteenth century and it must have been locally common on all uplands in the medieval period.

The enclosed pastures were also the source of the hay that would see the animals through the winter, since out-wintering beasts were uncommon until the later eighteenth century. The grass crop was central, especially as wood-lands declined: every blade of grass that could be cropped between October and April was worth its weight in wool. Both lay and monastic establishments had to be prepared to bring in the flock to a sheep-house in bad weather, and feed them from the hay of the enclosed pastures.[47] When the sheep were also the primary dairy animals then management was more intensive than today and so a greater density of sheep-folds was found on most moorlands, inter-rupting the pre-existing vegetation: trampling it with one set of organs and adding nitrogen with another.

The possession of the right domestic animals was essential. The choice lay between cattle, sheep and goats. In Roman times, there was an active demand for cattle products such as leather and, crucially, beef production for soldiers' rations. Thereafter, sheep seem to have become dominant: in medieval Wales the ratio of sheep:cattle:goats of 200:20:1 has been estimated. So sheep were at some time pastured everywhere except the very highest moors. The main product was wool, though meat and milk were also taken; the utility of their dung is obvious. In medieval times, their breeding ratios were poor, with 60–80 per cent of ewes successfully raising a lamb. Post-natal mortality was also high and doubtless local populations of wolves, buzzards and eagles and, especially, foxes and wolves were killed whenever possible. For cattle, a mor-tality rate of 15 per cent of calves was common but the attraction of the pas-tures meant that, for example, the Dartmoor commons in 1296 contained 5,000 cattle, 487 horses and 131 folds of sheep. The perceived potential of cattle is seen in the scheme of the de Laceys in Lancashire in the thirteenth century to turn the moorland edges of Rossendale, Trawdon, Pendle and Accrington forests into specialised **vaccaries** for the breeding and rearing of cows and oxen. By 1296, twenty-eight of these establishments had been set on the moorland edges of the Irwell and its tributaries, with lodges higher up the hills for summer transhumance. The venture was not however a success (in 1295–6 alone seven cattle were lost to wolves) and these proto-ranches were eventually let out as farms after about twenty-five years.[48] These examples show us the way in which large areas of moorland terrain can be given over to a cattle economy: a land economy in which the moors themselves can become the key element. The central Pennines, for example, had whole valleys which were 'cattle country'. In the Cheviots sheep were dominant from the twelfth and thirteenth centuries onwards, with ratios of 10–20:1 in their favour against cattle. At about the same time, there was peat inception caused by the deliberate burning of grassland and the establishment of *Calluna* heath. Unlike later periods, low intensity sheep grazing might have maintained heather stands, especially if the sheep were taken off the hills in winter.

The distances involved in cropping the high lands and poor grazings above 200 m made seasonal movements a good way of converting the moorland vegetation to animal products. Thus evidence of medieval **transhumance** is often found. This meant a move of whole or part of the community to a set of huts and outbuildings (a sheiling) higher into the hills and remaining there for, usually, the May to August or September period. In the Brecon Beacons, traces of one such summer settlement (in Welsh, a *hafod*) are found at 500 m, facing north-east. The period spent in the summer house might be long enough for some **lazy-beds** to be dug and attempts made at a summer oat crop. Encouragement of summer pasturage was probably given by the prolonged period of drier climate from 1250–1450. The predominance in upland peat columns of the remains of the moss *Racomitrium lanuginosum* between those dates confirms the documented presence of part of the Early Medieval Warm Period, usually dated to 1150–1300. In the southern Pennines, stratigraphic studies of the pollen of crowberry (*Empetrum nigrum*) confirms the 1100–1250 warm period in the uplands.[49] Where the *hafod* or its English equivalent proved to be successful in crop-growing as well as animal fattening, then it might become a permanent settlement at times of pressure upon the land. In Allendale (North Pennines) there were valley-head summer settlements in 1421–2 which had become permanent farms by 1547. In North Tynedale though there were still such places in the side valleys ('hopes') in the mid-sixteenth century. From Yorkshire northwards, the place-name elements – *shield* and – *sett* remind us of that economy.[50]

Beyond the intake walls lay the common grazing, which might be regulated by a landlord or subject to common rights handed down from older times. We lack direct evidence of the manipulative effects of the various socially imposed grazing regimes. The variables must have included the ratio of cattle to sheep, for example, and the exact season of summer pasturing. The assumption is usually made too that animals were not outwintered. But the exact outcomes of these variables on a matrix of acid and basic grasslands, heather moor, blanket bog and even sub-alpine heath are not known. Microfossil analysis is restricted to those areas that now carry peat, though it can produce interesting suggestions. For example, on Bodmin Moor, the Romano-British period had seen the establishment of a species-rich grassland at the high altitudes around Rough Tor (265–355 m) which resulted from light grazing and perhaps hay production. Some parts of the same region produce evidence of more intensive grazing, with more pollen of the Poaceae (wild grasses), *Plantago lanceolata* and the *Potentilla* type. These seem to have been maintained for about

1,000 years.[51] Retrodiction has to carry us past some recent familiarities to vegetation communities in which there was probably more heather but never at
the monocultural levels produced in the nineteenth century, in which cotton
grass (*Eriophorum* spp) was much rarer in the wetter places since there was
much more *Sphagnum* mire (probably with *Andromeda polifolia*), and grassland with more bents and fescues and distinctly less *Nardus* and *Deschampsia*.
If burning was a regular management tool then perhaps *Molinia* still gave rise
to 'white moor'. Bracken was much less common under the medieval grazing
regimes but in any case was cut for fuel and bedding. There is a major piece of
research to be done in reconstructing a set of moorland communities in layers
going back from about 1850.

Even at bad times there were still livings to be made by manorial lords and
their tenants, though the possibility of specialisation on animal production
and its export to other regions (so that corn might be bought in) must have
been attractive to the more entrepreneurially-minded. One overall impression
of land use and land cover in the medieval period was that it was the product
of a well-organised social system. There were few patches of land without an
overlord of some kind, and few peasants might farm or pasture animals
entirely at their own whim. A set of social mechanisms for regulating the land
was clearly in operation by the time of Domesday Book and there is speculation about its origins. One interest here, though, is the way in which environmental alterations were channelled through the power structures of the
societies and hence entrained whatever worldviews and dominant metaphors
were held by those who gave the orders.

Ancillary resources

Peasants and their masters living around the moorland core during the middle
ages had access to a number of materials necessary to living in those places.
Small-scale quarries provided marl where there were soft calcareous deposits
and also yielded stone for the foundations of the better class of buildings and
eventually for some of the wallings as well. Field boundaries might also be
made of dressed stone, as distinct from roughly shaped stones cleared from
fields. The lower moors and the heathy areas into which they graded were the
source of useful materials like furze (mostly gorse bushes) and parings (the
tops of peaty soils) for use as firewood and fertiliser respectively. Heather from
almost everywhere was a common roofing material (since neither wheat straw
nor reeds were easily available) and turves might also function as roofing
material. On the high moors and in water-accumulating areas such as the
swangs of the North York Moors there were deep peat deposits and the right
of **turbary** was frequently granted so that the peat might be dug out and dried
and then used for fuel. In the thirteenth century, when Henry III chartered the
commoners of Dartmoor to use peat he called it 'coal' probably because it was
burnt in that region to form a charcoal for use in smelting tin as well as on the
common hearth.

But the description from Wharfedale of *terra . . . nemorosa* reminds us of the importance of woodland.

The assarting of the twelfth and thirteenth centuries assured the disappearance of many areas of tree cover, so that the primary woodland that had occupied the same site since prehistoric times was now quite rare (Plate 3.5). The description of part of the journey of the Green Knight from the fourteenth century alliterative poem *Sir Gawain and the Green Knight* sounds like a last fragment of primary woodland:

Bi a mounte on the morne merily he rydes	Merrily in the morning by a mountain he rode
Into a forest ful dep, that ferly was wylde,	Into a wondrously wild wood in a valley,
Highe hillds on uche a halve and holtwoodes under	With high hills on each side overpeering a forest
Of hore okes ful hoge, a hundred togeder.	Of huge heavy oaks, a hundred together.
The hascl and the hawthorne were harled al samen,	The hazel and the hawthorn were intertwined,
With roghe raged mosse rayled aywhere	And all was overgrown with hoar-frosted moss[52]

PLATE 3.5 *Some uplands have isolated patches of deciduous woodland, whose history is little known. One such is Black Tor Copse in the West Okement valley of Dartmoor. It is comprised largely of one species of oak and there is speculation about whether this is the result of natural environmental conditions or the wood's management history.*

Given Sir Gawain's journey to the Green Chapel in the Staffordshire moorlands,[53] this may reasonably be claimed as upland forest remnants. The huge and heavy oaks were probably unbranched until about 10 m from the ground but there was clearly enough light at the edge of the wood to allow a thicket of hazel and hawthorn; the moss points to a western, damp climate. The lines read as if the woods occupied steep slopes which were topped with moorland but were too steep to cultivate.

Large trees would yield timber resources for construction and the thicket, if properly managed, was valuable for underwood products such as firewood, tool handles, fencing materials and the woven panels which formed the foundation of wattle-and-daub panels. The undershrubs might be pulled down to feed to stalled animals and holly was an especially valuable winter food. The wood itself might form a sheltered grazing area in the winter for cattle and pigs were often depastured there all year, though manorial lords usually forbade any introduction of goats.[54] A great deal of what is described in medieval documents as *silva pastilis* is probably wood-pasture, which ecologically was grass or heather thickly scattered with bushes and trees.[55] In upland valleys like Swaledale, extensive woodlands were probably in retreat by the end of the thirteenth century, but wood-pasture maintained quite a well-wooded aspect, helped by planting and pollarding of, for example, elm in field boundaries (Fig. 3.2). The trees helped to feed deer so that they might form the nuclei of late-medieval deer parks. A wood-pasture also provided charcoal which could be used in smelting metal ores if any were present; again this seems to have been true of Swaledale.[56]

Even if assarting were absent, the effect of grazing in woodland is to prevent its regeneration unless severe measures are taken. Pigs might be eaters of acorns but they were also aerators of soil and destroyers of competing plants; they might also tread in acorns so that their chances of germination were improved. Cattle, on the other hand, would eat young shoots and saplings preferentially; sheep in a wood (as can be seen in unfenced upland woods today) eat almost everything on the ground layer, bracken always excepted. Where woods were managed for a long-term yield of useful underwood products then domesticated animals had to be kept out of coppice for its first few years; likewise deer, if that was possible given their leaping capabilities and the preference of the nobility to have them available for sport. High fences built on banks and ditches might be needed. So unless it was possible to trade some other product for wood then every community with a need for some degree of self-sufficiency had to keep this resource in good condition. That meant most communities; total clearance was something which only an enemy or extreme land hunger could bring about. There is for example an account of Scottish farmers crossing the border to seek fuel and timber from the far north of England, where there was scrubby birch-hazel-alder woodland that was not cleared until after 1450–1600.[57] Nonetheless, there was almost certainly much less woodland in the upland valleys in the sixteenth century than in the twelfth.

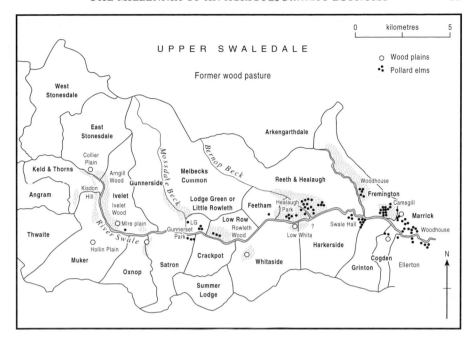

FIGURE 3.2 *Upper Swaledale in North Yorkshire, showing the boundaries of the town-ships and zones of former wood pasture. 'Wood plains' and pollarded elms are also indicators of the former management of valley sides to yield both wood products and pasture. Source: C. W. J. Withers, 'Conceptions of cultural landscape change in upland North Wales: a case study of Lanbedr-y-Cennin and Caerhun parishes c.1560–c.1891', Landscape History 17, 1995, 35–47, Fig. 2 at p. 40. Reproduced with the permission of Andrew Fleming.*

Animals: driving and hunting

Peasants and others who were largely subsistence producers were always interested in increasing the crop, be it of sheep or oats, even though they only had rights of usage. Those who owned and controlled extensive tracts might have other priorities, such as producing the right quality of crop for a market or a particular species of animal for pleasure. We see an example of this in the large-scale movements of cattle from regions like central Wales, the Yorkshire Dales (and Scotland) to lowland areas and especially to growing cities like London. If cattle were to be sold in distant markets rather than used for subsistence, a period of fattening on lowland pastures was often necessary and so cattle were moved in this way as well as up and down the local hills: as early as 1250 there are records of Welsh cattle being driven to Gloucester for fattening, and to Leicester in the 1560s. The visible remnants of this trade are in the drove roads which today often form wide greenways between fields, and some of the narrower bridges on country roads in the uplands. Inns along the routes still sometimes bear appropriate names and the overnight cattle pounds must have yielded a steady and saleable crop of cow pats. The knowledge of other places and different ways that the drovers spread is a reminder to us not to romanticise the moorlands into areas of total isolation and 'underdevelopment'.

An apparently different outlook underlies the establishment and maintenance of hunting grounds in the uplands. The sequestration of land for the conservation of particular species of mammal may well be ancient in England and Wales: hunting was popular in the Celtic and Roman worlds though there is no evidence for active management of game populations. The first active reserves are usually attributed to the Saxon kings (sometimes as far back as Alfred) and certainly established by AD 1000; Cnut (1016–35) is often credited with the first deer preservation areas. Rackham is sceptical of any pre-Norman allocations of territory, claiming that the documents referred to were medieval forgeries;[58] yet the nucleus of an administration for hunting may have existed by the time of Edward the Confessor.[59] More certitude comes from the actions of the Norman conquerors in carrying out King William's wishes that gained him the obituary comments that

> The king [William] set up great protection for deer and legislated to that effect, that whosoever should slay hart or hind should be blinded . . . he loved the high deer as if he were their father

This does not say whether he was fondest of them alive or dead, but is usually interpreted to mean that the provision of a ready supply of animals for great hunts was the priority. The wild beasts and fowls were to be in the safe protection of the king, for his princely delight and pleasure. Rackham, on the other hand, suggests that the actual royal presence was rare and that the supply of venison to the king or his friends was the important function of these preserves, a task entrusted to professional huntsmen. The king might grant similar privileges to lesser nobles, who would set up imitation **Forest** administrations.

Whatever the main motive, the Norman set up numerous hunting areas, which were generally labelled as 'Forests', a term which derives from the word *foris* meaning 'outside', that is, outside the common law and thus inside the Forest* law. It does not mean that trees were present in any quantity and indeed the largest Royal Forests in England, such as Dartmoor, the Forest of Pickering (North York Moors) and the Peak Forest were largely moorland. At their height, the forests covered about one-third of England, and there were considerable numbers in Wales, mostly run by the Marcher Lords. About one-third of these were moorland. Their creation and maintenance was very much a Norman and Plantagenet affair, for by the fourteenth century deer hunting had become a more stylised ritual and the Forests transmuted to the rather less macho 'parks' (Fig. 3.3). In the upland context, we note that many Pennine

---

* Following Rackham 1986 (Note 30), a capital 'F' will be used for areas under Forest law, a lower-case 'f' for wooded terrain.

areas were hunting Forests belonging to the great landowners such as barons and churchmen: Arkengarthdale, Wensleydale, Bowland, Pendle, Trawden and Rossendale do not exhaust the list.

There is no strong evidence to suggest there were special management procedures for upland and moorland Forests as distinct from those of the lowlands. In both, the deer belonged to the king (or other noble in the case of a Chase) and the other resources to the commoners. In the cause of deer management, the king would exact many services from the commoners, and some settlements, such as Bainbridge in Wensleydale were founded in the twelfth century to house the forest servants. Their responsibilities (some were paid, some unpaid) were to enforce the Forest laws (which were generally a constant source of income from relatively minor infringements such as illegal grazing or assarting), to ensure the deer population was kept in good order and to assist in the preparations for an authorised hunt. Red deer were the greatest prize, thereafter roe and then the fallow deer (*Dama dama*) imported by the Normans. The roe was taken out of the protected list in 1339; but hares might be declared honorary deer. The fallow deer were however not very resistant to wet and cold and were mostly confined to parks. Where a Forest covered a large area then the forest servants might be sent to concentrate the deer in a particular area when the king came to hunt.[60] Yet even in a Forest, there might be areas of safe haven for deer: these were known as 'friths' and there is a parish in Upper Teesdale called 'Forest and Frith'.

The revisionist view of the Forests encouraged by Rackham, that they were more a source of income than an aristocratic obsession with slaughter, finds strength in the way in which the Forests were sold off (in 1204 King John disafforested the whole of Devon and Cornwall except Dartmoor Forest, for the large sum of 5,000 marks), and were subject to conversion to cattle ranching, sheep grazing and assarting for cultivation. The income from Dartmoor Forest in 1403–4 derived from horses, cattle and sheep, and the issue of licences to cut peat to make charcoal, rather than from fines or mutilations levied on would-be deer poachers.

Within the Forest there might be enclosed areas called 'hays' where the king's deer were specially protected and from which the king made gifts of timber, underwood or venison. These might be a nucleus if a Forest were transformed into a deer park. Unlike the Forests, the parks were enclosed so that the deer could be kept in. Some had 'deer leaps' which allowed deer outside to jump in but not the reverse process. The increased confinement of the animals permitted a greater degree of management: in *Sir Gawain and the Green Knight*, we hear of a winter hunt with a purpose in a landscape which seems to be that of wooded valleys with moors and mountains above:

> They let the harts with high-branching heads have their freedom,
> And the brave bucks, too, with their broad antlers,
> . . .

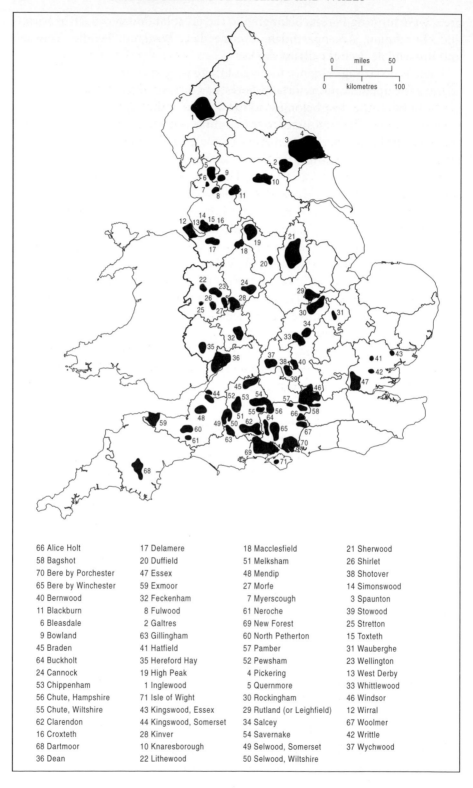

66 Alice Holt	17 Delamere	18 Macclesfield	21 Sherwood
58 Bagshot	20 Duffield	51 Melksham	26 Shirlet
70 Bere by Porchester	47 Essex	48 Mendip	38 Shotover
65 Bere by Winchester	59 Exmoor	27 Morfe	14 Simonswood
40 Bernwood	32 Feckenham	7 Myerscough	3 Spaunton
11 Blackburn	8 Fulwood	61 Neroche	39 Stowood
6 Bleasdale	2 Galtres	69 New Forest	25 Stretton
9 Bowland	63 Gillingham	60 North Petherton	15 Toxteth
45 Braden	41 Hatfield	57 Pamber	31 Wauberghe
64 Buckholt	35 Hereford Hay	52 Pewsham	23 Wellington
24 Cannock	19 High Peak	4 Pickering	13 West Derby
53 Chippenham	1 Inglewood	5 Quernmore	33 Whittlewood
56 Chute, Hampshire	71 Isle of Wight	30 Rockingham	46 Windsor
55 Chute, Wiltshire	43 Kingswood, Essex	29 Rutland (or Leighfield)	12 Wirral
62 Clarendon	44 Kingswood, Somerset	34 Salcey	67 Woolmer
16 Croxteth	28 Kinver	54 Savernake	42 Writtle
68 Dartmoor	10 Knaresborough	49 Selwood, Somerset	37 Wychwood
36 Dean	22 Lithewood	50 Selwood, Wiltshire	

But the hinds were held back with a 'Hey' and a 'Whoa'
And does driven with much din to the deep valleys.
Lo! The shimmering of the shafts as they were shot from bows!

This describes a selective hunt of red and roe deer: the females were being culled so as to keep the population healthy and within the carrying capacity of the area available for their support.[61] (The opposite, regrettably, of what has happened in the Scottish Highlands since 1914.) The detailed land use and management of Okehampton Park on the northern edge of Dartmoor (275–400 m ASL) has been examined and suggests that the park was partly created out of former arable land, by about 1306. The work also emphasises the enclosure of the park by watercourses and walls. It is possible that there was arable agriculture within the park as well, suggesting that landlords were flexible in their attitudes to the mixture of sport and necessity possible on their holdings[62] (Plate 3.6).

After 1080, the Bishop of Durham had rights comparable to those of the king and he established a number of parks, including one in Weardale above Stanhope. It was enclosed with a stone wall some 12–14 miles (19–22 km) in circumference, bounded in the valley by an Eastgate and a Westgate, with a square stone tower as headquarters for the park officials and the Forest courts. Like his other parks in the region, it was stocked with fallow deer. Tenants of the Bishop here, in Stanhope and beyond were bound to render service to the Bishop for his annual Great Hunt,

All the villeins of Aucklandshire, that is North Auckland and West Auckland and Escomb and Newton, provide 1 rope at the Great Chases of the Bishop for each bovate and make the hall of the Bishop in the forest 60 feet in length and in breadth within the posts 16 feet, with a butchery and a store house and chamber and a privy. Moreover they make a chapel 40 feet in length and 15 feet in breadth, and they have 2s as a favour and they make their part of the enclosure around the lodges and on the Bishop's departure a full barrel of ale or half if he should remain away. And they look after the hawk eyries in the bailiwick of Ralph the Crafty and they make 18 booths at St Cuthbert's fair. More overall the villeins and leaseholders go on the roe hunt on the summons of the Bishop.[63]

FIGURE 3.3 (opposite) *The Royal Forests 1327–36. Some of these were indeed wooded, like Dean and the New Forest, but others, like Spaunton and Pickering, Dartmoor and Bowland, are known to have been largely moorland, though possibly with woodland fringes. Many of the upland Forests were larger blocks of country than those in lowlands and presumably were managed on a large scale and with fewer common rights to interfere with hunting management.*
Source: L. Cantor, The English Medieval Landscape, *London and Canberra: Croom Helm, 1982, Philadelpia: University of Pennsylvania Press, Fig. 3.4 at p. 68. Reprinted with permission of the University of Pennsylvania Press.*

PLATE 3.6 *On the flanks of uplands, enclosed areas of wood-pasture might function as grazing land, timber reserves and hunting parks. Near Abergavenny, this park has scattered trees in a grassy matrix and here a drove-way leads from the valley to the moorland of the Sugar Loaf above.*

Here as elsewhere in the uplands, Forest Law (exercised in the Palatinate of Durham by the Bishop as if he were the King), conflicts persisted between the normal processes of an agricultural economy and the leisure tastes of an aristocracy, be they lay or clerical. Vaccaries and grazing rights in Weardale were mostly part of the Bishop's demesne until the fourteenth century, after which grants to lay tenants were made and eventually control over numbers of animals to be grazed was relaxed. The Bishop's powers under the Palatinate are emphasised by the fact that his Chief Forester also regulated the iron and lead mines in the Dale, allowing for example, smelting in Stanhope Park by way of a multiple-use strategy; the remaining records suggest that the Forester's main concern was not with woodlands so much as open country. This is confirmed by the action of the Bishop in hunting with greyhounds (*leporarii*), which rely on sight rather than smell.[64] Presumably the sport was interrupted in 1327 when the Scots camped in Stanhope park and left a great number of deer and cattle slaughtered behind them.

Given that many parks were much smaller than those enjoyed by the Bishops of Durham, it is easy to see that some landowners would opt for an almost domestic approach to their relatively fragile fallow deer and that eventually the parks became more valuable for their timber resources than as sporting arenas. Long distance exchanges of deer took place in the thirteenth century, suggesting that the supply of animals to, for example, the sixty-seven parks created in the North Riding of Yorkshire in the thirteenth century was a flourishing business. The Clifford estates in Wharfedale and the Mauleys of Mulgrave (North York Moors) had deer parks as integral parts of their medieval estates, but they gradually succumbed to more immediate needs. Some might have been hastened on their way by rival nobles engaging in illicit hunting or 'park-breaking', though this was commoner after 1500.[65] (In 1413, the clerks and scholars of the University of Oxford illegally slew game in neighbouring Forests, setting undesirable precedents for their material comforts.)

The restrictions imposed on the livelihoods of common people was never compensated by all the many regulations and court decisions that were supposed to reconcile them and their overlords. Here is an early example of conflict in the uplands (as elsewhere) between large resource owners and the ordinary folk. The main advantages for the latter came only when the rich needed even more money and bargains could be struck. (That possibility seems to have largely vanished.) Lastly, the Forest and Chases were demarcated: **perambulations** were conducted and published and so the division of the land into parcels for specialised use was carried forward. In the broadest perspective, Forests produced environmental change according to their social settings. If large landowners ran demesne stock farms or granted land to religious houses then there was a landscape dominated by a kind of ranching economy of sheep or cattle. Alternatively, if they allowed the peasants to colonise the Forests or leased summer grazing to them, then another set of ecologies was created. We do not know at present what the differences resulted in the upland vegetation composition, for example.

If some uplands satisfied demands for meat and for pleasure in hunting, then others produced another renewable resource, this time for an international market, that of wool. In the medieval period and up to their dissolution in 1536, the monasteries were important actors on that stage. They were not the only players nor were the uplands their only backdrop, but their role in colonising upland areas and harvesting an immense profit is central to moorland history in England and Wales.

The monasteries

The attraction of remote places for the monastic and conventual life had been strong in the Celtic tradition and the Roman church had a similar-thinking strand. The examples of the Desert Fathers echoed the forty days in the wilderness, and several medieval orders left the towns and cities because of the corrupting influence of such places. Of none is this more true than the Benedictines, whose founder rhapsodised over the less favoured environments:

... for in these days the mountains distil sweetness and the hills flow with milk and honey, the valleys are covered over with corn, honey is sucked out of the rock and oil out of the flinty stone . . .

All of which suggests that St Benedict had not visited the Pennine valleys on a rainy day in February; nevertheless, the revival of the Benedictines under St Bernard of Clairvaux in 1097 soon led to the founding of Cistercian houses in Britain: in 1132, for example, both Fountains and Rievaulx in Yorkshire were established. Far from flowing with milk and honey, the first settlers at Fountains thought of it as '. . . a place of horror and vast solitude . . . uninhabited for all the centuries back, thick set with thorns, and fit rather to be the lair of wild beasts than the home of human beings'.[66] In spite of such reactions, Yorkshire had seventy monasteries by about 1200 (having had none at the Conquest), and some of which were large; some 27 per cent of all freehold land in the county belonged to monasteries. After thirty years, for instance, Rievaulx had 140 monks and 500–600 lay brethren. In Wales, at least two great monasteries (Strata Florida and Cwmhir) are among moorlands. Exmoor, curiously enough, had a very low density of monastic establishments: not a sufficiently challenging place, perhaps.

An edict of 1134 required the Cistercian order to reside

In civitatibus, castellis, villis, nulla nostra construenda sunt coenobia sed in locis a conversatione hominum semiotis	Shunning the cities, castles and towns and building our houses far away from the talk of men

So that while the association of the Cistercians and the 'waste' is undoubted in Wales and northern England, the presence of their monasteries elsewhere and in more relaxed spots (the first house in England was at Waverley in Surrey) brings the suspicion that they were regarded by the Norman rulers of Britain and Ireland as part of the Establishment, so to speak.

The acquisition of land by monasteries came in various ways. In north-east Yorkshire the most important single factor was donation by lay people. Much of this appears to be 'waste' which could not be managed by themselves and so could be converted to spiritual capital. The infamous destruction of crops and property by King William's troops created the kind of lands the Cistercians wanted, as later did the Black Death's depredations. To create the right kind of solitude, the monasteries themselves destroyed hamlets and villages, usually moving the inhabitants down the valley away from the moorland cores. The main house, its home farm and outlying **granges** would be buffered in this way (Fig. 3.4). Once acquired, the monasteries moved to manage their holdings in a standard way: the economic activities were carried out in lodges and granges which were supposed to be not more than one day's journey from the mother house, (so that the lay brethren might hear mass

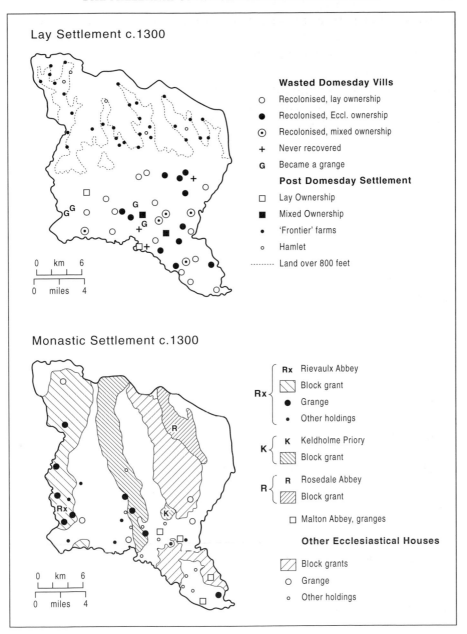

FIGURE 3.4 *Maps of Ryedale (North York Moors) showing the types of lay settlement and then monastic settlements, both in about 1300. The northern half of each map is moorland terrain and emphasised the recolonisation of 'wasted' vills by ecclesiastical ('Eccl') owners and the way in which monasteries acquired blocks of land focused on the valleys. They also had rights in the open land above, thus making unitary management of grazing possible.*

Source: L. Cantor, The English Medieval Landscape, *London and Canberra: Croom Helm, 1982, Philadelphia: University of Pennsylvania Press, Fig. 4.7 at p. 117. Reprinted with permission of the University of Pennsylvania Press.*

regularly), though this rule was often ignored. These were staffed by lay brethren and might be devoted to any one or more of a number of productive activities. Although wool production from sheep is the most commonly remembered activity, there was also cattle raising, arable farming, horse breeding, and metal smelting and working.

Their highest profile activity, and one which depended upon upland pastures, was the wool trade. The records of Florentine merchants tell us that in one year, 3,291 sacks of English wool were bought by them, and of those one half came from Yorkshire monasteries and one half of that from the Cistercians (Plate 3.7). The sheep farming took place mostly on the fringes of large blocks of moorland, so that access to the higher pastures was relatively easy. Seasonal migration might occur and some flocks which were wintered in Nidderdale in the Pennines were taken up to the heights of Wharfedale in the summer. Sheep were dominant in most abbeys' economies after the middle of the twelfth century and typically a sheep grange ('bercary') might have 200–300 animals. The monastery in total would own between 2,000 and 12,000, though numbers have often to be calculated from bags of wool sold, and some monasteries bought in wool and sold it on. The nuns of Rosedale (North York Moors) had 2,000 sheep in one parish in 1308 and Whitby had a total of 4,000 sheep in 1356. Estimates for the end of the thirteenth century

PLATE 3.7 *The vegetation of upland pastures converted into stone at Fountains Abbey, North Yorkshire. Though it had other sources of wealth, sheep were a mainstay of the abbey's economy. Its site was also the scene of environmental manipulation: the River Skell was diverted within the valley, whose sides show the marks of stone quarrying for construction.*

put Rievaulx at the head of the North York Moors monasteries with about 12,000 animals, out of a total of not less than 50–60,000 on the North York Moors and the northern edge of the Yorkshire Wolds. The richest areas were the parishes running up from the Vale of Pickering and on the coastal plateau.

It may indeed have been one of these areas from which we have the latest direct record of wolves in England. In 1395–6, the Abbot of Whitby paid the tanner for dressing thirteen wolf pelts. The wolf was officially hunted from December 25 to March 25 because its skin was then in prime condition but Foresters and shepherds alike were expected to harry the predator all year round: there was no question of preserving it. The monks of Rievalux, for example, had rights of summer pasture in Westerdale and although they were in the middle of the hunting grounds of the Brus family of Skelton, they were allowed to cut timber, to take dogs with them, and to set pit traps for wolves.[67] Wolves were probably plentiful in the uplands even later: under Henry VII (r 1485–1509), a ranger of Teesdale Forest was known as Roast Wolf Ambrose Barnes. The last wolves in Wales were probably killed in the reign of James I (1603–1625). Examination of place-names suggests that there are over 200 places in England named after wolves but only twenty with an element deriving from beavers. One interpretation of such counts is that the wolf was still numerous when the countryside was named (mostly by Anglo-Saxons in this instance) but that the beaver was already scarce. Wolf names are most frequent in upland counties; open country attracted the greatest number of Anglo-Saxon golden eagle names but interestingly the sea-eagle then featured in woodland environments.[68]

Cattle run from vaccaries were also central to the monastic economy; a vaccary would typically have 20–80 cattle and while they were more reliant on enclosed pasture than the sheep they were also grazed on open moors. Together, the moors and wood pasture made excellent cattle country and in the twelfth century most of the valley of the Nidd and its uplands were given over to monastic cattle farming. It has been suggested that the Cistercians 'opened up' this valley above Pateley Bridge.[69] Specialised activities like horse studs were dependent on improved pasture and arable production of oats: Rievaulx in 1131 had pasture rights for mares and foals in much of Upper Teesdale, centering on a grange called Kaveset just downstream from High Force and commemorated today by East and West Friar Houses.[70] Together with assarting of woodland and moorland for arable, these various activities affected the position of the moorland edge and the intensity with which intakes were used. The level of stock densities may however have been an element in the disputes between the grazing tenants of Fountains, Sallay and Bolton Abbeys in the late thirteenth century which led to great lengths of demarcating walls across the moors above Kilnsey and Malham.[71] Environmental management by the monasteries therefore comes from the depasturing of domestic stock, from the assarting of moorland and woodland, and from the complex of activities associated with metals. **Bloomeries** and lead working are especially associated with monastic establishments. Excavations near Rievaulx in 1998 suggest that an early form of blast furnace had been pioneered there, with its potential for much higher iron production than a bloomery. Water courses were diverted in order to supply the mother house, of course, but they were also made to power mills that drove bellows for ore crushers; woodlands were managed as coppice in order to provide charcoal for smelting the ores. It seems that wood and water supplies were considered very important factors in the founding of an abbey. Outside the core area, tracts of scrub (like the *spineta condensa* of the Aire in 1152) could be assarted to improve the grazing or be converted to arable and in a similar manner, woodland could be assarted to provide cultivable land and a single crop of timber. Some foundations exploited their connections in order to be granted timber from princely sources, including the Royal Forests (Fig. 3.5).

Environmental impact here reflects the way in which the population density is kept low ('*nulla nostra construenda sunt coenobia*') but is offset by the controlled intensity of resource extraction of several kinds. In Wales, for instance, the Cistercians notably ran sheep at densities at three to four times that of lay farmers.[72] But the Black Death saw a major step in the breakdown of the 'ideal' monastic pattern, with much more lay leasing of outlying granges. At the end of the thirteenth century, two sets of monastic vaccaries in Upper Swaledale in the Yorkshire Dales which were leased to laymen were at or above the 300-metre contour. Leased-out granges became, in some instances, small hamlets. If these developed into villages, then the former **foldyard** of the grange might become the village green: Upper Swaledale supplies a number of examples.[73] After the Dissolution in 1536, the conversion of many granges into farms and hamlets reinforced the occupation of land which otherwise might have been evaluated as marginal and thus useful only for very extensive land uses.

Inorganic resources

From later prehistoric times onwards, the digging of mineral ores and their processing has been a feature of the British environment. Most of the main minerals of the medieval and early modern periods were found and exploited in the uplands. The uplands are often highly suitable for the former since under moor and heath it is easier to find and get at mineral veins and **adits** can be driven into hillsides; with luck they might be self-draining as well. Further, smelting was aided by hillside locations where winds funnelled up valleys made it easier to heat the fuels to the high temperatures required to bring about the chemical changes in the ore rocks. At 460 m ASL just east of Aberystwyth, for example, copper mining is found within the time-span 2365–900 BC in an environment of open woodland, grassland, heather moor and the beginnings of blanket peat growth.[74] Many similar sites for iron mining and smelting span the Iron Age and early Roman periods in several uplands. In due course the Yorkshire monasteries had a number of holdings

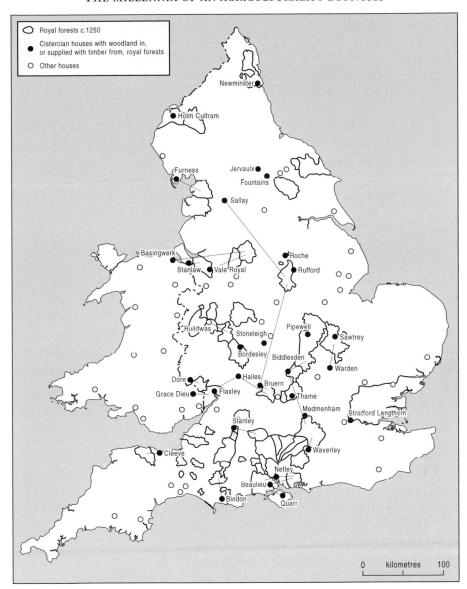

FIGURE 3.5 *Cistercian houses with woodland in, or supplied with timber from, Royal Forests c.1250. Various connections are shown but the uplands might on this basis be seen as much a source of wood as denuded areas dependent on lower areas for their timber supplies. It is assumed that it is timber that is being mapped and not underwood products.*

Source: R. A. Donkin, The Cistercians: Studies in the Geography of Medieval England and Wales *Toronto: Pontifical Institute of Mediaeval Studies, 1978, Fig. 15 at p. 125.*

devoted to mineral working: Guisborough had iron, foreshadowing the great developments on Teesside, Fountains and Byland both worked iron in Nidderdale and Byland produced iron in Rosedale as well; Jervaulx and Bolton had coal 'mines' and Rievaulx seems to have had the advanced technology of the blast furnace for some of its iron production. The North York Moors were an especially good region for iron since there were seven seams which out-cropped over a wide area: 'Smiddale' is probably 'smiths' dale'. The jet which came to high fashion in Victorian times was collected from the beaches and some upland outcrops around Whitby as early as 1394.

Pride of place must nevertheless go to Dartmoor which was the richest source of tin in medieval Europe. The granite dome has in its time yielded copper, iron, lead and arsenic as well as tin but the first mention of the star mineral is in 1156 and it was thereafter a major industry whenever tin prices were high. Since control of output was centralised by the Crown, there are some records from the twelfth century onwards, suggesting that production of refined tin in 1156 was about 60 tons per year and in 1171–89 about 343 tons/year, though by 1300 the working-out of the gravels led to a drop to 33 tons/year. The tin ore was extracted from alluvial gravels and then crushed and smelted locally and the licensed tinners were privileged to dig peat for fuel or to dig over the ground for tin almost anywhere. Even so, their disturbance was not such as to prevent William Camden saying in 1586,

> And this dartemore yeldeth iiij speciall comodities – pasture, corne, tynne and turff cole . . .

Lead was another widely-sought and produced mineral and nowhere more so than in the Pennines, where lead pigs of Roman date have been found; thereafter, evidence of its use in churches is found for York in 690 and for Lindisfarne in 650–700. In Swaledale the documentary evidence reaches back to 1219 but mentions the time of Henry II in the previous century. On Alston Moor (Cumbria) there was silver as well, which led to Royal oversight, not least because the rent was 100 shillings per year in 1130; one result was a sep-arate mining community with its own laws. Lead ore (which contains the silver in low concentrations) has to be followed wherever the veins lead and so its detection at the surface was essential; low moorland vegetation was always a help rather than a hindrance, though we do not know if the Dartmoor saying

> Plough heather, find copper
> Plough furze, find silver
> Plough fern, find gold

is characteristic of other regions, let alone whether it is meant to be literal or figurative, since bracken tends to grow on the deeper and more base-rich soils. Again, most of the northern monasteries engaged in lead production wherever possible and many were roofed with their 'own' lead, with surpluses being sold to secular outlets. Edward III brought 168 pigs of lead from Nidderdale for his alterations at Windsor Castle, using water transport (via Hull) as much as possible. King Stephen (r. 1135–54) granted the use of lead from Weardale to the Bishop of Durham for his cathedral; there was even enough lead to roof a parish church just outside the city walls. Though King Stephen's time was not very happy in Durham, it has given us an account of the separation of silver for the Bishop's mint, found in the Latin verse of the monk and singer Lawrence of Durham,[75] set in the spring of 1143:

> These 'untilled lands' give many wealthy boons;
>   These so-called deserts thus enrich our lord
> Vast weights of silver every year he gets
>   From what you wrongly scorn as barren ground.
> The pallid silver-miner worms through earth's
>   Insides, and sweats to bring the rough mass out.
> He washes it and pounds it in the forge
>   Applying fire, with bellows keenly roused.
> Then to the fire he adds damp elm-tree ash:
>   These split apart the metals here combined.
> The lead sinks down, the silver separates
>   And rises up, now purified of dross.
> To Durham's treasury this money goes:
>   Such is the tribute from the 'empty' vale.

These verses and the ecclesiastical concentration of the records confirm that lead was a high-status commodity (unlike iron), with a peak demand in the thirteenth century for roofing monastic buildings. Landowners might often give away their 'waste' lands with the rights to dig the ores of iron and lead, though not to hunt. Roger de Mowbray gave such land to Fountains before 1145 (reserving the wild boar, deer and birds of prey to himself) and Byland acquired the rights in much of Nidderdale through such gifts. Jervaulx was given the right to dig lead and iron in Wensleydale in 1145, though Count Alan as donor kept the mines of Swaledale and Arkengarthdale, which suggests an early Yorkshire trait.[76]

Everywhere and at all times, and mineral processing exerts strong environmental influence. There is spoil from the extraction and from refining, for example. In the case of tin from stream gravels, whole valleys were turned over leaving parallel heaps of stones flanked by skeins of heavy sand which had been washed through for their ore content. A preliminary smelting with peat charcoal left heaps of slag.[77] Lead veins were exposed by '**hushing**', when a stream was temporarily dammed and its head of water released over the ore-bearing rocks so as to expose the vein, a practice that persisted until the end of the eighteenth century in Weardale at least. Ores had to be crushed and so the diversion of streams to power stamping mills became common at the entrances to mines as well as at refining sites; smelting itself produced massive toxic pollution of land, air and water. Miners might live near the mines and so be given a lease on a small patch of land for their subsistence or to improve their health. Either way, there was a piecemeal enclosure of open ground (an 'intake') and so the lead valleys of the Pennines have terrace housing well above the limit of normal agriculture. In Nidderdale, the hamlet of Greenhow Hill was founded after a Chancery decision said 'that there may be cottages erected for the miners and mynerall workmen upon the said waste and some competant quantity of ground to be improved of the said waste to be laide to them and also for keeping of draught oxen and horses for mayntenance of the mynes always leaving to the tenants sufficiency of common'.[78] The colonisa-tion of upland Dartmoor for agriculture may have accompanied tin-streaming in some places: considerable reductions in the amount of alder are seen in pollen diagrams since the wet-tolerant *Alnus glutinosa* was a common former of streamside woodland. Fuel is another need for both industry and domesticity: in iron-smelting zones there seems to have been an attempt to manage coppice for a renewable output but the ubiquitous peat was more easily obtained in many places and even very thin seams of coal were opened up. Lead must also have been a very great consumer of both peat and wood in areas without ample coal until the nineteenth century.

The demand for fuel for smelting constituted a major environmental impact. Mining leases and bargains often included payments to miners for extracting the wood from old mines. In spite of such recycling, medieval dis-putes between miners and their landlords often stress the rights of the miners to cut trees for smelting as well as mine timber: a dispute at Alston in 1290 over their privileges in taking wood for burning, smelting, building and enclosing seems to have been a sharper version of an 1152 indictment of them having ravaged the local forest.[79] In 1306 there was a complaint that the Bishop of Durham was destroying the church's woodlands in Weardale with forges for iron and lead and the making of charcoal. However as late as the 1420s some of the wood for lead processing came from Wolsingham Park itself where the Bishop used his own land for a smelting **bole** and produced charcoal from his own woods.[80] In Rossendale (Lancs), in 1303–4, a batch of lead ore (91 loads, 6½ dishes) cost £8–8–1½; the cost of cutting down and cutting up the wood for 'burning the said ore' was 7s 5d, and for carrying it to the furnace or 'bole' was 8s 8½d, which may suggest that some distance was involved.[81] In the second half of the seventeenth century, lead smelting at Marrick in Swaledale needed some of its wood to be bought at Braithwaite, south of Middleham, a journey of some 20 km at least. In 1540, Leland spoke of wood for Swaledale 'that they brenne their lead' as coming partly from Co Durham. In 1695, Camden said that the valley had 'grass enough but wants wood'. Unsurprisingly, a lead operator in middle Swaledale was instructing his land agent in 1677 to manage his woodlands as coppice. But by 1715, the same operator had an inventory of

2,000 cartloads of peat and only 200 sacks of 'chopwood'. Smelting had now moved nearer the mines and thus nearer the upland peats at the same time as wood was becoming more expensive. A survey of the Bolton Castle estate in 1765–6 suggests that one output was mine timber, grown on a longer rotation than **chopwood** production.[82] One purchaser of eighteenth century enclosed common near Lanchester in Co Durham found it profitable to plant 300 acres of woodland to larch, pine and mixed deciduous trees. An early return was got by selling thinnings to make 'corf rods', which were woven into the baskets used to haul coal out of the mines of that period; birch wood was popular for the soles of clogs.[83]

Drainage patterns were subject to some manipulation as operators learned to lead water down mine levels to come out lower down the hill. As early as 1304–5 in Rossendale (Lancs) there is a record of 'a certain trench underground to draw off the water from the other trenches' but adits were not in general use until the seventeenth and eighteenth centuries.[84] An important tie between the mineral and the organic economies at this stage was the fact that livestock production was often only a part-time occupation for a family and that meant surplus time could be spent in occupations connecting with mining. Thus many miners also had smallholdings or farms and a dual economy developed.

In upland England, the mining of lead and silver together largely for the silver was over by about 1225 and in some areas, there was little re-opening of earlier mines until new technologies and good prices made the lead trade profitable,[85] even in a context where all the important lead veins in Weardale, for instance, had been discovered by 1660. The Bishops of Durham were often innovators in both the technological and economic spheres and these to some extent overcame the disadvantages of poor ores and high fuel costs. So in 1460, there was production in Weardale and Teesdale but none in Nenthead and Blanchland. The Weardale workings (Fig. 3.6) were very restricted compared with later times and only around Rookhope and Stanhope were they then re-openings of earlier extractive sites. Transport was by ox-cart to Swalwell and thence to the Tyne but by 1530, most of the lead mining was defunct and the Bishop was more interested in leasing the right to smelt iron, not lead, at Stanhope. The effect of this period on woodlands, peat diggings and silting of streams with fine wastes however must have remained. The Yorkshire Ouse, by analogy, retains high levels of lead in its sediments from 1250–1500 and these seem to be the washings from upstream lead mining. From 1500–1750, these levels rise and so probably also relate to metal mining in the Pennines. After 1750, there are additional inputs of lead and other heavy metals from the urbanisation and industrialisation of the River Aire at Leeds and Bradford.[86]

Reading about medieval colonisation of the uplands, and seeing some of the sites, tends to leave us with an impression of the precariousness of some of the developments. It is not surprising therefore to learn of abandonment and retreat. After the population declines associated with the Black Death, for

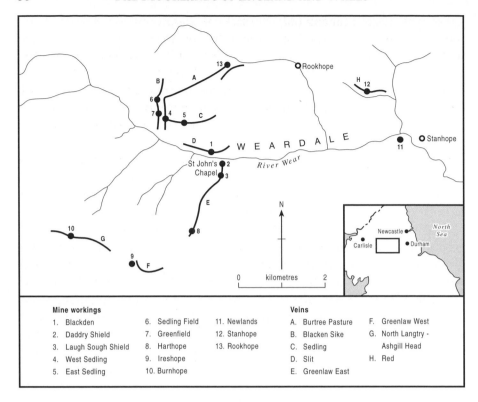

FIGURE 3.6 *The late medieval and early modern lead mines and veins in Weardale,*
North Pennines. These workings were pioneered by the Bishop of Durham and were not
flanked by the same industry to west and north as happened later under lay control: only
Teesdale locally also had lead working.
Source: I. Blanchard, International Lead Production and Trade in the 'Age of the
Saigerprozess' 1460–1560, *Stuttgart: Franz Steiner Verlag Zeitschrift für Unter-*
*nehmensgeschichte Beiheft **85**, 1995, ed. H. Pohl, Map A2–4 at p. 268.*
Reprinted with permission of Ian Blanchard.

example, many tenants found it possible to acquire the leases on lowland farms
which promised an easier life. The harrying of the north by King William's
troops similarly left many upland vills to be described in Domesday Book as
wasta est. The worst of it in Yorkshire at least seems to have been in the dales
and in the Pennine foothills.[87] The Revolt of 1400–10 in Wales was put down
with ferocity (even the monastery of Strata Florida was despoiled) and land
ownership concentrated in fewer hands, resulting in some withdrawal from
marginal holdings. Social reasons might produce environmental change: in
the sixteenth and seventeenth centuries in the Yorkshire Dales the extension of
gavelkind and tenants' rights created a class of smallholders and cottagers
driven off the land to find work in the opening of local lead and coal mines.

    Between 1250–1300, the climatic limit to cultivation in the Lammermuir
Hills of southern Scotland fell from 450 m to 400 m, and another 75–90 m in

the years before the Black Death. This doubled the uncultivable core of the uplands. It seems unlikely that something similar did not afflict the north of England at least, and possibly Wales as well. Certainly, there are records of the blighting of sheep flocks in Wharfedale and of the rotting of the crops in the fields in 1314; oat prices were three times their normal level. Raids as far south as Yorkshire by the Scots often depopulated upland areas and 1314 was a particularly bad year following the Scottish victory at Bannockburn. Nevertheless, since cereals were grown at high altitudes in the northern Cheviots even during the Little Ice Age, earlier cultivation of small plots on **shielings** might still have been possible. Indeed it can be argued that in The Cheviot at least neither of the well-documented periods of climatic deterioration relevant to this chapter (the late Bronze Age and the Little Ice Age) were sufficiently bad to cause farmers to abandon the upland environments.[88]

Consolidation in Early Modern times

In all the accounts of the medieval use of the uplands, the role of the lords of the manor is stressed and by the early modern period, this becomes even clearer, possibly as the records improve, in turn fuelled by the greater propensity to challenge hierarchical orders. In northern England there seem to have been two major categories of land owners: those who kept their forests and chases 'closed' and exploited them by means of cattle farms ('vaccaries'), and those who were 'open' landlords and encouraged their peasants to settle and enclose land, taking a rent therefrom. The central Pennines for example were vaccary country even after the Dissolution. One other major divide which was manifest by the late sixteenth–early seventeenth century was between those uplands that developed some industry, such as textiles (as in Lancashire) or lead mining (as in the Yorkshire Dales), and those which stayed entirely 'rural'. If industry developed then cottage settlement grew, along with a demand for enclosure of moorland; if not, then there was a drift of population away from the hills. Some areas nevertheless shared in the increasing prosperity of uplands where by about 1700 a strong cattle export trade had developed. In border hills such as the Cheviots this reached virtually commercial proportions when Anglo-Scottish wars ceased: shielings were often transformed into stock rearing enterprises. So although minerals might provide an element of environmental determinism, much of the ecology of other land covers was strongly influenced by social controls.[89]

The pastoral economy of Late Medieval and Early Modern times (*c.*1400–1700) was centred upon cattle, and sheep flocks were modest in size. The typical black longhorns were bred to be sold off to the lowlands and dairying was only an adjunct activity. The males were sold off young and oxen kept only if there was significant arable land in the valleys. Sheep were kept by most upland farms before 1650 but they lost popularity after then and any extensive flocks were likely to have been part of large-scale enterprises by big landowners. Sheep were not very important on Dartmoor until the seventeenth century. Their numbers hovered around 7,000–7,500 in the AD 1500–1600 period but by 1700 were more like 100,000.[90] The ecology of animal production was not in essence different from earlier times but was subject to refinement as manors stabilised and intensified their systems. Though most open arable and meadow had disappeared in the uplands, the maintenance of cereal lands (mostly oats and barley) and of hay meadow for winter feed, was still central to the survival of all but the rich. The boundary between the enclosed land and the rough grazing was therefore a powerful ecological marker (for only the former would receive nutrient input from the farm and be intensively managed) as well as a social boundary between land appropriated to individuals and the common grazings, less intensively managed though by no means promiscuously used. At all events this boundary (the **head-dyke**) had to be stock-proof so as to keep animals off the crops and the hay between (in northern England) the beginning of May and early November though some animals were allowed down-hill to graze the post-harvest grasslands after the hay crop had been taken: the 'foggage'. Winter fodder was always critical: the hay crop was basic but tree branches were used if possible though often forbidden because of conflicting uses. (In Weardale there was a scale of fines that reflected value: it ran from ash and holly through elm and alder to hazel.) Sheep were kept on the lower fells and the enclosed land just before lambing in mid-March; if on the fell they might eat some of the 'moss crop' of the cotton sedge *Eriophorum* which greens very early, especially if dead grasses around it have been burned off. The purple moor grass *Molinia caerulea* was sometimes gathered for winter fodder for cattle.

The layout of the lands was complicated in the sixteenth and seventeenth centuries, in northern England at least, by the enclosure of blocks of land that were tacked onto the head-dyke. In Upper Teesdale some 200 acres were enclosed at Cronkley Pasture in 1590; in East Bowland some 4,000 ha (10,000 acres) of 'cow close' were enclosed in 1587–1621. These were often treated as communal cow pasture, with grazing rights or stints apportioned to tenants by the manorial court, but they might (as in Swaledale in the mid-sixteenth century) be subdivided and apportioned to those with existing grazing rights. They were not usually grazed between March and May to allow the grasses to get their early growth. In an analogous fashion, the open fells were often divided into communally accepted but not necessarily physically separated sectors. The lower fellsides closest to the dwellings were especially useful for the milk cattle in summer and for sheltered winter grazing for sheep; they might have become enclosed and assigned by the end of the seventeenth century. A swathe of moorland beyond was summer grazing for bullocks, heifers and horses. The high fells were reserved for summer pastures for sheep. Each owner's flock kept to a specific area of the moor (the 'heaf') and went to and from the farm along routeways ('drifts') that were carefully controlled by the community. Any areas which were common to the manor were controlled: gorse growth and use was one instance; pigs were ringed to stop them grubbing up the soil and goats were also held in check, especially if there was woodland. Goats were never very popular: Durham Priory had had a herd at Muggleswick in the 1290s which disappeared by the 1380s and most such herds survive only as feral animals in, for example, The Cheviot. Not much is known about fox control though in the Lake District it was at one time compulsory to go fox-hunting on pain of a fine of 12 pence.

The early modern period was the last in which wholesale movement to a summer pasture and dwelling ('shieling' in northern England) was important in England and Wales. The decline of the tradition in northern England was marked during late medieval times and although a handful survived into the early seventeenth century in the Borders and Upper Teesdale, this century saw the demise of transhumance in England. In Wales certainly, and probably elsewhere as well, transhumance was by 1800 a feature of memory (usually fond, forgetting the wet and cold days) rather than usage. The practice had been controlled by manor courts, with the summer dwellings generally occupied from late May until mid-August. Some animals might be allowed to stay on the summer pastures for another month. The court was concerned to see that the laws of 'levancy and couchancy' were maintained: that is, that nobody depastured animals in summer that could not be supported in winter. It was a kind of recognition of carrying capacity. When shielings were no longer used in the old way, there was the temptation to let out the grazing to individuals from another manor and here the court was also anxious to see that such animals were confined in number to those who paid the right fee, usually collected by the lord. This practice ('agistment') was especially common where vaccaries had been dominant and happened in the side valleys off Weardale, for instance.

In the seventeenth century, therefore, grazing systems in the English uplands were not uniform. What they had in common, however, was a fair measure of social control. Ecological processes were tied into social processes, and lords of manors still had a strong voice in what was pastured where and on what herbage. Grazing pressure, though mediated by a manorial court, was to some extent a reflection of aristocratic choice and remained so until the eighteenth century.

Controlling the commons

Much of the unenclosed land of England and Wales is designated as common land and some explanation of this term is necessary. Common land is a legal designation which applies to many kinds of land in which several users have an interest: though moors and heaths are common examples, many village greens are also commons in the legal sense. It does not mean however that there is no owner. The lord of the manor owns the land but thereafter different 'common rights' may be assigned to different groups of people. The most frequently found of these rights is that of grazing, which often goes with residence in a particular parish or even county, or perhaps with tenancy of a holding adjacent to the common land. Rights to dig peat or take turves, to gather fallen wood or furze, or to shoot game, are also assignable. Control was usually by a manorial court which allocated for example the number of beasts that might be depastured and the seasons of such use. Such controlled regulation of grazing is called stinting. Public access to common land is confined to rights of way such as footpaths and bridleways.

Commons as legal entities were probably established in the eighth–ninth centuries AD, were in progress of consolidation at the time of Domesday Book and completed by the early thirteenth century. In most senses they appear to be a residue of something more extensive: there appears to have been a time when most resources were held in common but they were gradually appropriated by a smaller number of people, to the point where in this case the lord of the manor actually owns the land. Even as it was becoming consolidated in the landscape, the right to enclose common was granted to the landlord. The Statute of Merton in 1236 gave the lord the right to enclose or improve the waste but only if he left enough common pasture for the free tenants. (The unfree were not mentioned.) Even when an area was designated as Royal Forest, the common rights were usually kept up: the Forest Charter of 1224 for Dartmoor makes it clear that there were pre-existing common rights. We know that Exmoor's Royal Forest area was uninhabited between the thirteenth-seventeenth centuries but also that at the end of the sixteenth century 40,000 sheep, 1,000 cattle and 400 ponies were pastured on it and that the sheep were not outwintered.[91]

Examples of commons in the period of interest here include those of Dartmoor where in the early thirteenth century it was established that there was grazing for all the inhabitants of Devon except those of the boroughs of Totnes and Barnstaple. Exmoor had two Royal manors on its southern flank which were used for summer pasture by all the neighbouring farmers. There are regional differences: upland Derbyshire has no common land, whereas about half the moorland of upland Co Durham is common; many of the Welsh moorlands appear not to be common. The common was also a regulated store of ancillary resources. Peat, wood, turves, furze, rush and heather, bracken, as well as stone were all useful but their gathering was usually subject to control by amount and season. In the case of peat, the depth to be dug might be set at one or two spits depth, accumulated water in the pits might have to be drawn off and the sod replaced. If the common was fired to encourage new growth, then the season of burning was set by the manorial court. Likewise, small valleys with dense and long heather ('thatch cleughs' in northern England) might be carefully regulated and protected from over-gathering and from fire; the thatch might be used within the manor only, not sold off. The same degree of control might also be applied to the common rush *Juncus squarrosus*, unlikely as that now seems. The history of the commons of England and Wales has not been comprehensively chronicled but all commons are now registered, following the work of a Royal Commission which reported in 1958,[92] but which was never followed up by the comprehensive legislation that was promised.

Population and environment

During the middle ages, seasonal settlement had retreated northwards in upland England, and hunting Forests were progressively put to more intensive uses. Wales followed at perhaps a century's interval. By 1550 in England, the northern uplands were well-populated and so the environmental changes wrought by the upland economies could never be negligible. Taking all the changes together, the growing and permanent importance of stockraising, with concomitant effects on, *inter alia,* enclosure patterns, woodland use and survival, and the species composition of swards, becomes the environmental 'driver' to watch through subsequent centuries.

Intensification and fragmentation ad 1500–1750

Although some of the uplands of England and Wales look like islands which rise out of lower ground, they are not isolated from many of the changes of the rest of the country. Though the portrayal of them today may be romantically antiqued, the pressures for environmental change are often as quick to become felt in Central Wales as in Suffolk. So in the period after the dissolution of the monasteries but before the Industrial Revolution, major changes in politics, economics and technology were not necessarily absent from the uplands. Because of the limitations which they imposed upon use, nevertheless, not every innovation nor every shift in power relations was immediately obvious in the landscape.

If we look at the whole period from the Neolithic to the eighteenth century, one of the greatest growths is that of the human population. Though there were set-backs like those of the Black Death and then some periods of slow growth, from an early Neolithic population of perhaps 20,000 people, the population of England in 1801 at the first official census was 9.2 million. In 1430, it was probably about 2 million, so that within the three centuries of interest here, it grew by 4.5 times; though the absolute numbers for Wales are much lower, the rates of growth are much the same.[93] Most demographic research does not separate out the uplands, but it seems likely that growth in prosperous lowlands and above all in the towns and cities accounted for much of the increase in numbers. Building patterns in upland settlements show that there was some expansion of absolute numbers but not on the scale of a city like Bristol, for instance. But because population is increasing elsewhere then any region which is part of the same economic system may share in some of the effects of the growth and clearly in this instance there may well be increased demand for wool and meat which the upland economy is well fitted to supply. Conceivably, wood and water may also be seen to be available in upland regions, though less of the former than the latter, and some mineral ores can only be found in, for example, the rocks of the Carboniferous Limestone series, which are entirely beneath higher land although not all of this is now moorland, as with the Mendip Hills of Somerset for example.

The agricultural economy

Though not isolated from the rest of the country, the agricultural and settlement patterns developed in the uplands had a distinctive quality.[94] Once thought to be exclusively pastoral, it is now accepted that the 'traditional' economy was in fact mixed. The lower ground was cultivated with the aim of producing as much corn crop as possible, in which oats were the favoured species but barley, rye and wheat were grown as well. Pulses might also be grown for fodder and the land might also be in competition with the need to grow hay as a winter keep for livestock. In some areas of both countries, the basic system was an infield-outfield system, with even small infields being held in common (there are examples of fields of less than 20 acres, about 8 ha). The infield received the manure whereas an outfield was rotated round suitable spots and abandoned when its fertility fell off.

This revised view of the economy notwithstanding, animals were the crucial components of the production cycle, for they could probably be sold off for corn but there would never be a surplus of saleable grain. So the breed, stocking, reproduction rates, nutrition and disease patterns of cattle, sheep, goats, horses and ponies all become important to upland communities, as well as the numbers they might be allowed to keep by a landlord or by common custom. In the case of cattle in northern England before 1603, the number that could be stolen from or raided by the Scots was also a factor in stocking rates. The general pattern before the eighteenth century was that the animals were taken in summer to pasture on the moors which were at a distance from the main settlement, in a form of transhumance. The infield would be ploughed in March and April, outfields dug over (wheeled implements could not be used on them) and then the cattle, goats, horses, pigs and geese might all decamp to the summer grazings, variously called the **shieling** in northern England and the *hafod* or *lluest* in Wales. Here there might be a half-way stage called the *meifod*. In September, people would return to harvest the corn. Sheep and ponies do not require any daily tending and so stayed on the moor all year except for round-ups when lambs and foals were taken off for sale. The modification of this pattern most obvious in this period is the disappearance of transhumance. It had certainly gone from Wales by 1790 but was practised in western Northumberland in 1604, to quote two travellers' accounts. So somewhere between 1600 and 1750 the majority of it was no longer found to be useful. In part, this seems related to the development of a vigorous cattle trade between the uplands and the growing markets of the lowlands of England, and of South Wales, and in part also to the enclosure of the upland commons. Detailed use of documents and the landscape has enabled

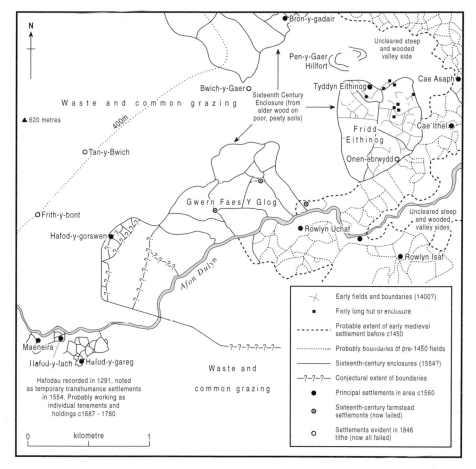

FIGURE 3.7 *An unusually detailed map of the relationship of moorland and enclosed land in upland North Wales c.1560. The gradual expansion of enclosed land from the lower end of the valley is recorded in the dates of the boundaries and of special interest is the fact that some woodland remained intact (north-east corner of the map), whereas some alder wood was cleared in the course of the sixteenth century (west of Fridd Eithinog). The transition from transhumance sites (hafod, pl. hafodau) to permanently occupied settlements is denoted in the south-west corner. The retreat from occupation of the higher areas is seen in the number of 'failed' settlements.*
Source: A. Fleming, 'Towards a history of wood pasture in Swaledale (North Yorkshire)', Landscape History **19**, 1997, 57–73, Fig. 1 at p. 59. Reprinted with permission of Charles W. J. Withers.

Withers[95] to reconstruct one valley's land use dynamics at about 1560, just before transhumance began to disappear (Fig. 3.7).

The upsurge of a long-distance cattle trade between upland areas of Great Britain and the towns and cities is late medieval in origin but grew greatly from the eighteenth century onwards. After 1603, Scotland was a major supplier but because the 'drove roads' along which the cattle were driven to markets ran through the North Pennines, the English uplands were affected as well. Along their routes there was a demand for improved pasture and so the manure provided by their overnight halts was recycled into enclosed grassland whose productivity was thus raised. In the later eighteenth century, limekilns were built even in the high country and small local coal-seams opened up to burn the lime to improve these fields. Many inns and cottages still remain in isolated places as reminders of these developments.

Apart from cattle and sheep, ponies were useful in all kinds of local ways and for sale, but horses are another matter and were in some places another animal product of the uplands. Notably, it was the horses from the uplands of North Yorkshire which bore the brunt of wars in France and Scotland in the 1540s and by 1544 all the horses from Northumberland supplied to the King for service in Scotland had vanished. Such was the concern that in 1540, King Henry ordered that all horses kept on open commons such as heaths and moors had to be 15 hands high*, so that their offspring did not render them too small for military use. In the four northern counties of England, this edict was relaxed to 14 hands.

An even more noticeable change in the uplands was the process of enclosure. In these areas this took two basic forms though with many regional variations. The first was the extension of the enclosed area by means of additions to the existing improved land: a kind of encroaching tide of better grassland replacing the heather and coarse grasses of the moor (Plate 3.8). The second was the parcelling-out of upland commons with fences and (above all) walls so that grazing could be assigned to individuals or at any rate small groups rather than whole communities. Enclosure might happen through a number of routes. There was for example a piecemeal nibbling at the edges, so to speak, where landless peasants might build a house and enclose a few acres and hope that nobody would dispossess them. Then, a community might petition its landlord to allow the enclosure of upland common so as to improve the quality of the pasture in the new enclosures. Lastly, the lord of the manor might bring about enclosure either by agreement with his tenants or enforce an Enclosure Act upon them, with the sanction of Parliament behind it. Equally, strong landlords might allow enclosure by tenants only in a controlled fashion and at their own pace, so that a regular advance up an escarpment might be the result, as on the East Moor of north Derbyshire in the sixteenth–seventeenth centuries.[96] Concerns about commoners were rarely of moment and even the news that there were only fifty deer left on Dartmoor in 1621 did not deter the makers of 'newtakes'. One environmental consequence of enclosure was the persistence of trees in the now controlled lands on the lower slopes: a farm might have up to 100 non-woodland trees within the holding.

* One hand = 4 inches = 10.61 cm

PLATE 3.8 *A general view of Exmoor. The convertibility of the heather moor (fore-ground) is shown by the many improved pastures on the far hillside, whose environment is not significantly different. The regular patterns suggest the influence of a determined owner during the period of enclosure and thereafter.*

In the eighteenth century, manorial lords rather than yeomen were major agents of transformation. Progressive landlords, fired with the desire for '**improvement**' (the 'sustainability' of the age) might try to deal with large areas. Within the former Forest of Knaresborough (Yorks NR), some 25,000 acres were enclosed by a few landlords in 1770–8 and on the North York Moors from 1773, one landowner tried to drain and grow crops on Kempswithen up to 293 m. The land was drained, limed, walled and put either to grass or corn. In the same region, more piecemeal intaking in the period 1550–1750 is indicated by walls whose lower levels are **orthostats** rather than layered courses (Plate 3.9). It seems somehow symbolic of an upland where one-third of the land was enclosed by Parliamentary Acts promoted by large landowners. The drive to improve in North Yorkshire was so great that the Tabular Hills on the south of the main massif of the North York Moors were watered by channels engineered from the siliceous rocks to the north; these areas being limestone were difficult to settle and cultivate unless water was brought to them. Unlike many reclamations of the eighteenth century, these hills remain arable and improved grassland to this day. In Derbyshire, we can see how moorland became part of a great estate, for the upland (Brampton East Moor) to the east of Chatsworth House is an intercalation of moor, woodland and lakes which serve the great house built on its lower slopes. Even in relatively remote parts of England like Westmorland, the cattle business put pressure on landlords to enclose so as to profit from this trade and this meant enclosure, not least because the regulation of grazing via manorial courts had broken down by the 1770s.[97] But simple models always have problems: permission to enclose did not always result in it happening on the ground.

The changes in what we might see as a political ecology of the uplands varied with region and period. In Wales, the population of the upland parishes grew four times as fast as that of the lowland equivalents in 1530–1620 as grants were made to improving landlords to take in what was often designated as Crown waste. On the other hand, in eighteenth century times the uplands shed people to the towns in spite of some extra employment being created by walling, hedging and other rural construction. This period was marked in the Yorkshire Dales by the building of the barns out in the fields. They stored winter feed but also acted as manure factories by concentrating beasts, just as in the lowlands sheep were folded at night upon fields of turnips.

The ecological effects of these social and political changes revolve around the greater intensity of grazing made possible by the division of the land. In the landscape, great straight stretches of walling divided one holding from another and it became profitable to invest in upgrading the pasture. A major agent in that process was the building of limekilns and where possible opening

PLATE 3.9 *Periods of high prices often leave islands of enclosed and improved land in a setting of moor. These may become 'enduring reclamation' or may eventually revert to rough pasture. Here the moorland vegetation shows the pattern of former enclosures, only two of which remain as improved land. This part of the North York Moors was probably enclosed in the seventeenth century.*

up small seams of coal to fire them. Where coal was absent, then any remaining woodland came under pressure. Drainage also became popular since it allowed the replacement of rushes by more palatable grasses such as *Agrostis* and *Festuca*. These flourished particularly if the land has been ploughed a few times, limed and some of the stones taken out. (They could be piled up at the field's edge or incorporated into the walling.)

Hitherto, the purple moor grass (*Molinia caerulea*) had been a favourite with cattle, though only in the summer, since it is deciduous. It is a chosen species with most animals in milk. Sheep will eat most species, including heather, but get little out of the mat-grass *Nardus stricta*. Ponies, on the other hand, can eat out the sweeter centre of the tuft. There is no evidence that the overall pressure on grazing resources diminished and so the tendency for the less palatable species to increase at the expense of the others was always present. More ecological manipulation came in the form of burning, which encouraged shorter and less woody heather, removed the dead leaves and stems of *Molinia*, and also promoted the growth on wet peaty areas of cotton-grass (*Eriophorum* spp) which gives sheep an early bite when food is scarce in early spring. Some progressive landlords were interested in better stock and while these may have been for example hardier in winter, their capabilities probably included stronger teeth and hence the ability to mow the vegetation much closer. Where stands of gorse (*Ulex europaea*) existed then it was regarded as a good foodstuff once cut and bruised. It also has the advantage of standing above the snow in winter.

Other ecological changes in these centuries included the continued rundown of woodlands. This happened where there was no interest in maintaining coppice for iron-smelting and so management might virtually cease. In Ryedale (North York Moors) ironworking persisted after the closure of Rievaulx Abbey and good management seems to have failed here, for the woods seem to have been repeatedly cut over to the point where there were many low-value pollarded oaks and ashes, fittingly called 'dodderells'.[98] The continuation of rights of turbary, meant that peat-cutting of upland blanket bogs ate into otherwise pastoral areas. The keeping of rabbits on a commercial scale (largely for their fur in pre-refrigeration days) was characteristic of parts of southern Dartmoor and that part of the North York Moors which now lies under Dalby Forest. Their effect on plant species diversity can easily be imagined as being largely negative but it is not impossible that at some densities they provided an open habitat for species that otherwise might have been shaded out.

Dartmoor provided a renewable resource in a way that presaged later, industrial, times. The Corporation of Plymouth acknowledged in the sixteenth century that local sources of water were inadequate and so a channel about 1.8 m wide and 0.6 m deep was constructed into the city during 1590–1 to bring water some 27 km from Burrator. The actual site of the intake is, somewhat ironically, submerged under a modern reservoir.

Minerals and mining

The period between 1500 and 1750 was mostly that of the expansion and consolidation of mineral industries rather than the exploratory setting-up of
medieval and earlier times, but one which pre-dates the immense changes
brought about by the large-scale mining of coal in lowland Great Britain.
Elizabeth I encouraged mineral development in the Lake District by import-
ing German miners but elsewhere the native population were the agents of
rapid growth. As the industries grew, larger companies became involved and
often carved out large territories where they had the monopoly: the London
Lead Company in Teesdale for instance. Thus, demand for lead took over the
environment of whole regions in the nineteenth century and shaped whole
landscapes but the effects before 1750 were not nearly so great and are much
harder to trace than the later overlay. Early in the eighteenth century, Daniel
Defoe's tour of England and Wales[99] went via the Derbyshire Moors and near
Brassington, where on the side of a 'rising hill . . . there were several grooves
. . . which turn out to be shafts into lead veins that are just about wide enough
for a man and a basket with three-quarters of a hundred weight of oar (sic)',
about 38 kg. One of the most lasting remains of this intermediate period may
be the irregular belt of small fields and enclosures, with a few scattered cottages between the open moor and the enclosed and improved land below,
which is still to be seen. The fields now are usually being invaded by moor
plants and their walls are in disrepair unless they have been incorporated into
another holding or form part of a second home. Their reversion to the wild
often started in the nineteenth century when mining became much more
mechanised and shift-working replaced the piecework which enabled a miner
to take time to cultivate a holding or to tend animals.

Moorland industry

Though any farm or settlement might house workers in leather or wool,
changes in the land were almost always the result of mineral extraction, processing and transport. On Dartmoor, the medieval tin industry had exhausted
the most accessible stream deposits by 1400 but some were still being worked
since there were fifteenth-century Acts of Parliament to try and deal with the
silting-up of the estuaries of the Dart and the Plym. Thereafter, the tin was
extracted from veins found by following the outcrop of the ore up from the
stream and then digging it out downwards from the head of the vein, a process
known as 'openwork'. At all events, 252 tons were produced in 1524. In the
eighteenth century the price of tin fell and by 1750 it was not worth recording
the annual output from the Dartmoor **stannaries**: in 1797 there were only
three mines left. A brief revival using adit and shaft mining at sites of former
workings occurred in the Napoleonic Wars, when some mills were powered
by water brought long distances: thirteen kilometres in one case. The end of
the nineteenth century saw the end of the industry on Dartmoor.[100]

On the North York Moors, coal was an important mineral, even though the seams were thin (of the order of 15–20 cm) and it was extracted from thousands of small pits, with the earliest written evidence from 1643. In the Esk valley near Danby, a system of proper shaft mines was run into the hillside. The greatest development, as with iron, was in the nineteenth century. This region was also a centre of alum working (sometimes called 'the first chemical industry') with activity noted in 1604 and sixteen quarries by 1700. Since it takes 30–100 tons of shale to extract one ton of alum, there is a great deal of waste heap production. Fuel is needed since the shale is heaped up and then roasted for up to nine months (sic). Alum was an environmental transformer along the cliff coast near Boulby, at one of the places where moorland had within memory reached practically to the sea[101] and in the seventeenth and eighteenth centuries alum works produced a major transformation of the slopes at Carlton Bank overlooking the Tees estuary, one which remained until late in the twentieth century even though the industry had vanished from North Yorkshire by 1871.[102]

The main trends

'More people' is an underlying influence in the history of the uplands, as everywhere else in Britain. In the hills, it meant more animals to be grazed though in a smaller space since the practice of transhumance vanished. The rabbit was added to the close nibblers' repertory. More mining meant more fragmentation of land use and land holdings and there was an overall expansion of individual rights and practices at the expense of the common and shared but, it must be emphasised, not on the scale of the nineteenth century.

EVERY VALLEY

This seems a good point to highlight evidence mentioned at almost every period so far. The combination of climate, slope and human activities has changed the landforms of the valleys of the uplands in major ways since the beginning of the Later Mesolithic. Sometimes it is clear that increases in rainfall are causative, at others it is obvious that human effects are primary and, more often, there seems to be an interaction of both. The continued removal of upland forest allowed the spread of blanket peats laterally on low slopes through the period of agricultural occupation, in extension of that formed during the hunter-gatherers' occupation. As late as 1000 BC, the peat expanded onto the grassland and open woodlands of Extwhistle Moor at 360 m in the Lancashire Pennines, for example.[103] Lower down, however, the valley slopes of most uplands seem to have been affected by the movement of mineral material. The typical sequence seems to involve the loss of erodible matter from the interfluves and upper slopes and its deposition as fans or as valley alluvium. These deposits may be reworked and relocated several times, working their way down the valley, so to speak, and indeed ending up in the sea. In the Bowland Fells of Lancashire, to take one instance, fans and cones of mineral debris seem to have formed after phases of severe erosion in the periods 5,400–1,900 BP and at 900 BP. Thus the major periods of prehistoric agricultural occupancy are covered as well as the climatic deterioration in the Iron Age; the later phase might coincide with the intrusion of Viking settlement and the introduction of sheep farming on a large scale. To the north, the Howgill Fells experienced a wave of soil erosion, gully formation and the deposition of debris cones in the tenth century AD, again coincident with Scandinavian occupation of the land. The same area experienced Iron Age and Romano-British land use which included vegetation change on the hillslopes but invoked no major disturbance of valley floors or gully development on the hillslopes.[104] In a small upland catchment in the North Pennines (the Thinhope Burn), there was channel and flood-plain metamorphosis in late Roman times and from the eighteenth century (Fig. 3.8). This development (which includes the incision of the channel by 8 m in some places) is linked to shifts to wetter climate, woodland clearance and drainage of the catchment, with distinct emphasis on the role of climatic swings. Between c. AD 530 and the present day, the valley underwent alteration from a stable meandering channel with a floodplain that accumulated fine-grained sediment to a vertically and laterally active boulder-bed stream with a high load of coarse

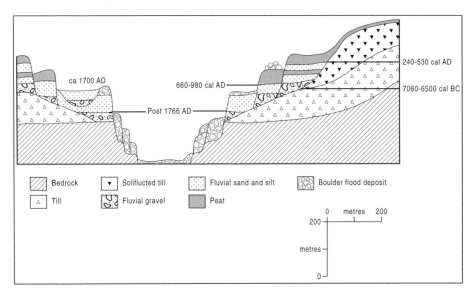

FIGURE 3.8 *A section through the alluvial deposits of the Thinhope Burn in the Northern Pennines. The investigation concluded that there had been slow sedimentation until early historic times but beginning at some time in the second to fifth centuries AD valley floor entrenchment had accelerated markedly.*
Source: M. G. Macklin, D. G. Passmore and B. T. Rumsby, 'Climatic and cultural signals in Holocene alluvial sequences: the Tyne Basin, northern England', in S. Needham and M. G. Macklin (eds), Alluvial Archaeology in Britain, *Oxford: Oxbow Books, 1992, 123–39, Fig. 12.6 at p. 132.*
Reproduced with the permission of M. G. Macklin.

sediment and low sinuosity. Investigations on a moorland on the Fell Sandstone of Northumberland allowed the inference that human-caused disturbance during prehistoric times was an important precursor of accelerated soil erosion but that climatic change in the form of storms and floods c.800 BC was a likely immediate cause of alluviation in the valleys. What is also clear from upland Northumberland is that only major phases of climatic change are likely to bring about basin-wide adjustments in river valleys. Otherwise, the geomorphology is more likely to be the result of conditions in individual reaches.[105] This reminds us of the diversity of geomorphology found in these uplands: sensitivities to external factors are variable.

Reviews of the dating of alluvial materials can however point to some periods of high river activity which are usually tied to other evidence of climatic transitions to, and the incidence of, cooler and wetter conditions. The three centuries 2100–2000 BC, AD 200–300 and AD 1000–1100 had larger and more frequent floods in upland areas, for example. There are other peaks at c.3000 BC, in the first century AD and at 600–900 AD. Detailed work in the Tyne Basin adds times such as AD 1350 and AD 1750 to the list. To anticipate the chronological order of the story, the over-riding conditions of the coldest and wettest phase of the Little Ice Age were between 1750 and 1800 and in the Yorkshire Dales evidence of sediment entrainment and deposition from the autumn and winter floods of 1763, 1771 and 1795 can be traced. There was a second major peak between 1870 and 1910. Once the relatively intensive land use patterns of the Iron Age and after had been established then human actions become distinctly significant in alluviation. What seems to have happened is that humans have increased the rates of runoff and sediment supply to rivers and thereby made them more sensitive to relatively modest changes in climate. As Macklin puts it, '. . . as human activity has progressively transformed the Tyne catchment from a natural to a managed landscape, at the same time it has inadvertently made river valley communities increasingly vulnerable to the vicissitudes of secular hydroclimatic change.'[106]

Nearer the present day, closer dating can be applied to floods and phases of sedimentation or erosion in valleys and numerous examples can be quoted of the contribution of erosion to valley floor sediments, their reworking and incision by streams and their removal downstream (Table 3.1). The conclusion from all of these studies is that any image we have of such processes being only gradual may be mistaken. Gradual transmutation of river valley profiles and cross-sections does indeed occur but the catastrophic event is also very important. A land surface may be 'primed' with loose sediment, but which requires intensive rainfall to move it downslope. Certain sizes of rock fragments may only be moved by a great flood. So a kind of punctuated equilibrium may be found in upland valleys. Downstream consequences include the spread of large quantities of alluvium at times of flood, with possible consequences for agriculture if the sediments contain toxic metals like lead and zinc from earlier periods of mining and smelting. Yet, most of the processes are

TABLE 3.1 Key concepts in upland landform behaviour

Concept	Associated processes
Contingency	At one point in time, the immediate history in the locality is critical
Threshold	A small change in one critical variable may cause change to a radicaly different condition
Complex response	One type of disturbance may produce different outcomes in time and space
Landscape sensitivity	Sensitive areas respond quickly to change whereas other parts respond slowly or not at all
Frequency and magnitude	The frequency of extreme events may exert a considerable effect on landscape changes
Integration	Changes in one place may reverberate over a considerable distance

Source: Burt, Warburton and Allison, 2002 (Note 107)
Copyright Joint Nature Conservation Committee.

amenable to current management possibilities showing that once again, environmental change is embedded in the social and the cultural, even to the point of occasional upland valleys being transformed into landscape gardens by Romantically-minded owners as at Studley Royal in North Yorkshire (Plate 3.10).[107]

PLATE 3.10 *The River Skell downstream from Fountains Abbey (Plate 3.7) converted into a landscape garden with canals, classical belvederes and the usual accoutrements of such eighteenth-century developments.*

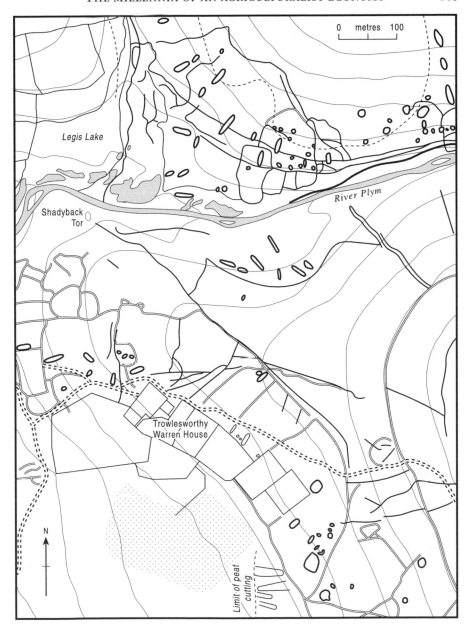

FIGURE 3.9 *A segment of the landscape in the lower Plym valley of Dartmoor, showing its complexity. The irregular enclosures are probably of prehistoric date and the rectangular ones from early modern times, though now largely abandoned. The lenticular shapes are rabbit warrens from the seventeenth to nineteenth centuries. The stippled area north of the river is former tin-working and various water-diversions ('leats') were built to supply it.*
Source: Reproduced by permission from 'The Proceedings of the British Academy', 76 (1990).

Transformations

Through all this time, agriculture persisted. The growing of cereals, usually oats, near villages and hamlets was normal and it symbolised the need for many communities to maintain a local self-sufficiency in a basic food. Animal farming produced a positive balance of saleable products, like the beasts themselves, or usufruct such as butter and cheese. Animals on their way to market might traverse hill lands otherwise only visited as part of transhumance and so induce more intensive grazing in small enclosures in remote places. Mineral extraction took settlement and land use change (as well as other environmental manipulations) close to the sources of the ores: there was not much choice of locality once the decision to develop minerals had been made. These processes led to the transformations of moorland into more intensively managed terrain though none of them was necessarily permanent. A map extract from Dartmoor (Fig. 3.9) shows the way in which layer after layer of change in either direction have remained visible in the landscape. This demonstrates that any environmental component has to adapt to a previous condition which has itself been made by human activity even though there may have been a period of 'desertion' and non-intervention. Any idea that the margins of the moorlands are 'natural' at any one moment is therefore quite untenable.

PLATE 3.11 *The flanks of the Brecon Beacons, with unenclosed moor remaining as common land and a series of enclosures beyond, all of which would once have been moorland but which were subject to improvement at various stages in their history. As is common in Wales outside the Valleys, a strip of deciduous woodland covers the middle slopes.*

In 1750 there was a tesseration of several different uses of land and water resulting from disparate environmental changes: a more differentiated landscape than in the Neolithic but at all times containing land which was reverting from the receipt of greater attention to a more extensive class of uses, a wilder set of ecosystems. We expect some radical alterations in the nineteenth century since that was a period of rapid change in economy and environment (Plate 3.11). From the viewpoint of the twentieth and twenty-first centuries it may look as if the uplands were excluded from the dominant processes of industrialisation and intensification of production but closer examination is needed to see if that was really the case.

Notes and references

1. J. Thirsk, *Alternative Agriculture. A History from the Black Death to the Present Day*, Oxford: Oxford University Press, 1997.
2. I. G. Simmons and J. B. Innes, 'The ecology of an episode of prehistoric cereal cultivation on the North York Moors, England', *Journal of Archaeological Science* 23, 1996, 613–18.
3. I. G. Simmons and J. B. Innes, 'An episode of prehistoric canopy manipulation at North Gill, North Yorkshire, England', *Journal of Archaeological Science* 23, 1996, 337–41.
4. P. Rasmussen, 'Leaf-foddering in the earliest neolithic agriculture. Evidence from Switzerland and Denmark', *Acta Archaeologica* 60, 1990, 71–86; P. Rasmussen, 'Analysis of sheep/goat faeces from Egolzwill 3, Switzerland: evidence for branch and twig foddering in the Neolithic', *Journal of Archaeological Science* 20, 1993, 479–502; J. N. Haas, S. Karg and P. Rasmussen, 'Beech leaves and twigs used as winter fodder: examples from historic and prehistoric times', *Environmental Archaeology* 1, 1998, 81–6.
5. I. G. Simmons and J. B. Innes, 'Mid-Holocene adaptations and later Mesolithic forest disturbance in northern England', *Journal of Archaeological Science* 14, 1987, 385–403.
6. The expressions in *italics* come from the model of introduction of farming in M. Zvelebil and P. Rowley-Conwy, 'Transition to farming in Northern Europe: a hunter-gatherer perspective', *Norwegian Archaeological Review* 17, 1981, 104–28.
7. P. J. Fowler, *The Farming of Prehistoric Britain*, CUP, 1983, gives good treatment of the uplands.
8. J. R. Collis, D. D. Gilbertson, P. P. Hayes and C. S. Samson, 'The prehistoric and medieval field archaeology of Crownhill Down, Dartmoor, England', *Journal of Field Archaeology* 11, 1984, 1–12.
9. R. Young, 'Barrows clearance and land use: some suggestions from the north-east of England', *Landscape History* 9, 1987, 27–33; see also D. Coggins, 'Settlement and farming in Upper Teesdale', in D. A. Spratt and C. B. Burgess (eds), *Upland Settlement in Britain: the Second Millennium B.C.*, Oxford: British Archaeological Reports British Series 143, 1985, 163–75.
10. D. A. Spratt, 'The prehistoric remains', in D. A. Spratt and B. J. D. Harrison (eds), *The North York Moors: Landscape Heritage*, Newton Abbot and London: David & Charles, 1989, 28–44.
11. R. White, *The Yorkshire Dales. A Landscape through Time*, Ilkley: Great Northern Books, 2002.

12. P. C. Buckland, M. Parker Pearson and M. A. Girling, 'Is there anybody out there? A reconsideration of the environmental evidence from the Breiddin hill-fort, Powys, Wales', *Antiquaries Journal* **81**, 2001, 51–76.

13. B. R. Gearey, D. J. Charman and M. Kent, 'Palaeoecological evidence for the pre-historic settlement of Bodmin Moor, Cornwall, southwest England. Part II: land use changes from the Neolithic to the present', *Journal of Archaeological Science* **27**, 2000, 493–508.

14. K. E. Barber, 'Peat-bog stratigraphy as a proxy climate record' in A. F. Harding (ed.), *Climatic Change in Later Prehistory*, Edinburgh: Edinburgh University Press, 1982, 103–13.

15. H. H. Lamb, 'Climate from 1000 BC to 1000 AD', in M. Jones and G. W. Dimbleby (eds), *The Environment of Man: the Iron Age to the Anglo-Saxon Period*, Oxford: British Archaeological Reports British Series **87**, 1981, 53–65.

16. S. Buckley and M. J. C. Walker, 'A Mid-Flandrian tephra horizon, Cambrian Mountains, West Wales', *Quaternary Newsletter* no. **96**, 2002, 5–11.

17. S. Applebaum, 'Roman Britain', in H. P. R. Finberg (ed.), *The Agrarian History of England and Wales*, I.ii A.D. *43–1042*, Cambridge University Press, 1972, 5–277.

18. G. R. J. Jones, 'Post-Roman Wales' in Finberg (ed.), 1972 (Note 17), 281–382.

19. J. Turner, V. P. Hewetson, F. A. Hibbert, K. H. Lowry and C. Chambers, 'The history of the vegetation and flora of Widdybank Fell and the Cow Green reser-voir basin, Upper Teesdale', *Philosophical Transactions of the Royal Society of London* **B 265**, 1973, 327–408 for the later prehistory, with dates BC as uncali-brated radiocarbon years; for the Mesolithic, see R. H. Squires, 'Flandrian history of the Teesdale rarities', *Nature* **229**, 1971, 43–4; idem, 'Conservation in Upper Teesdale: contributions from the palaeoecological record', *Transactions of the Institute of British Geographers* **NS 3**, 1978, 129–50.

20. P. D. Francis and D. S. Slater, 'A record of vegetational and land use change from upland peat deposits on Exmoor. Part 2: Hoar Moor', *Somerset Archaeology and Natural History* **134**, 1990, 1–25; P. D. Francis and D. S. Slater, 'A record of veg-etational and land use change from upland peat deposits on Exmoor. Part 3: Codsend Moors', *Somerset Archaeology and Natural History* **136**, 1992, 9–28.

21. See e.g., D. D. Bartley, 'Pollen analytical evidence for prehistoric forest clearance in the upland area west of Rishworth, W. Yorkshire', *New Phytologist* **74**, 1975, 375–81.

22. R. Tipping, 'Towards an environmental history of the Bowmont Valley and the northern Cheviot Hills' *Landscape History* **20**, 1998, 41–50; R. Tipping, 'The determination of cause in the generation of major prehistoric valley fills in the Cheviot Hills, Anglo-Scottish border', in S. Needham and M. G. Macklin (eds), *Alluvial Archaeology in Britain*, Oxford: Oxbow Books, 1992, 111–21.

23. K. E. Barber, L. Dumayne and R. Stoneman, 'Climate change and human impact during the late Holocene in northern Britain', in F. M. Chambers (ed.), *Climatic Change and Human Impact on the Landscape*, London: Chapman and Hall, 1993, 225–36; L. Dumayne and K. E. Barber, 'The impact of the Romans on the envi-ronment of northern England: pollen data from three sites close to Hadrian's Wall', *The Holocene* **4**, 1994, 165–73; L. Dumayne, 'Iron Age and Roman vegeta-tion clearance in Northern Britain: further evidence', *Botanical Journal of Scotland* **46**, 1994, 385–92; A. Manning, R. Birley and R. Tipping, 'Roman impact on the environment at Hadrian's Wall: precisely dated pollen analysis from Vindolanda, northern England', *The Holocene* **7**, 1997, 175–86.

24. A. Gear, J. Turner, J. P. Huntley, M. McHugh and J. B. Innes, 'Human activity and environment', chapter 3 of B. Vyner (ed.), *Stainmore. The Archaeology of a North*

Pennine Pass, Hartlepool and London: Tees Archaeology and English Heritage, 2001.

25. S. West, D. J. Charman, J. P. Grattan and A. K. Cherburkin, 'Heavy metals in Holocene peats from south-west England: detecting mining impacts and atmospheric pollution', *Water, Air, and Soil Pollution* **100**, 1997, 343–53.
26. H. Riley and R. Wilson-North, *The Field Archaeology of Exmoor*, Swindon: English Heritage, 2001.
27. C. J. Arnold and J. L. Davies, *Roman and Early Medieval Wales*, Stroud: Sutton Publishing, 2000, chapter 9.
28. D. J. Long, F. M. Chambers and J. Barnatt, 'The palaeoenvironment and the vegetation history of a later prehistoric field system at Stoke Flat on the Gritstone uplands of the Peak District', *Journal of Archaeological Science* **25**, 1998, 505–19.
29. See for example, A. Caseldine, *Environmental Archaeology in Wales*, Lampeter: St David's University College Department of Archaeology, 1990.
30. P. J. Reynolds, 'Rural life and farming', in M. J. Green (ed.), *The Celtic World*, London and New York: Routledge, 1995, 176–209. Though mainly about lowland agriculture, this essay is full of interesting material about rural life in the Iron Age of Britain.
31. P. J. Fowler, *The Farming of Prehistoric Britain*, Cambridge: CUP, 1983, 205.
32. W. Linnard, *Welsh Woods and Forests*, Cardiff: National Museum of Wales, 1982.
33. P. J. Fowler, 1983 (Note 31), p. 120.
34. R. Hingley, 'Iron, ironworking and regeneration', in A. Gwilt and C. Haselgrove (eds), *Reconstructing Iron Age Societies*, Oxford: Oxbow Books, 1997, 9–18.
35. They are at about 200 m ASL in a moorland context; ^{14}C dating gives activity between 996–441 cal BC as well as in Roman times: C. J. Arnold and J. L. Davies, 2000 (Note 27).
36. The main sources for this section are: H. P. R. Finberg (ed.), *The Agrarian History of England and Wales, Vol I. ii A.D. 43–1042*, Cambridge University Press, 1972; J. Thirsk (ed.), *The Agrarian History of England and Wales, Vol II 1042–1350*, Cambridge University Press, 1988; J. Davies, *The Making of Wales*, Stroud, Glos: Alan Sutton/Cadw, 1996; R. Millward and A. Robinson, Upland Britain, Newton Abbot: David & Charles, 1990; C. Gill (ed.), *Dartmoor: a New Study*, Newton Abbot: David & Charles, nd [but *c*.1970]; R. Muir, *The Yorkshire Countryside. A Landscape History*, Edinburgh: Keele University Press, 1997; A. J. L. Winchester, *Harvest of the Hills. Rural Life in Northern England and the Scottish Borders 1400–1700*, Edinburgh: Edinburgh University Press, 2000.
37. C. Thomas, 'Livestock numbers in medieval Gwynedd: some additional evidence', *Journal of the Merioneth Historical and Record Society* **7**, 1974, 113–17.
38. Uncultivated and forested.
39. The distance from the 1,000 ft contour on the Pennine flanks near Waskerley to the nearest 'moor' place-name on the western side of Durham City itself.
40. J. Bailey, *General View of the Agriculture of the County of Durham*, London, Richard Phillips, 1810.
41. The place-name 'Ringmoor' [Down] on Dartmoor may be derived from ME 'ridde', a moor cleared of undergrowth. See D. G. Price, 'The moorland Plym – abandoned settlement features of post-prehistoric age', *Transactions of the Devonshire Association* **112**, 1980, 81–93, 90.
42. R. J. Silvester, 'Medieval upland cultivation on the Berwyns in North Wales', *Landscape History* **22**, 2000, 47–60.
43. B. R. Gearey, S. West and J. Charman, 'The landscape context of medieval settlement on the south-western moors of England. Recent palaeoenvironmental

evidence from Bodmin Moor and Dartmoor', *Medieval Archaeology* **41**, 1997, 205.

44. The setting for a BBC sit-com called 'The Last of the Summer Wine' about, basically, the aimless lives of three incorrigible 'characters' in their 60s. It tends to polarise viewers into devotees and deriders. I try never to miss it, though the death of some key actors has lessened the appeal. A colleague once referred to one of my items of dress as my 'Compo sweater'; some readers will know what he meant.

45. B. J. D. Harrison and B. K. Roberts, 'The medieval landscape', in D. A. Spratt and B. J. D. Harrison (eds), *The North York Moors. Landscape Heritage.* Newton Abbot and London: David & Charles, 1989, 72–112.

46. G. Redmonds and D. Hey, 'The opening-up of Scammonden, a Pennine moorland valley', *Landscapes* **1**, 2001, 56–73.

47. R. Trow-Smith, *A History of British Livestock Husbandry to 1700.* London: Routledge and Kegan Paul, 1957.

48. R. Trow-Smith, 1957 (Note 47), 107–108; his source is C. H. Tupling, *Economic History of Rossendale*, Manchester: Cheetham Society Papers lxxxvi, 1927.

49. J. H. Tallis, 'Climate and erosion signals in British blanket peats: the significance of *Racomitrium lanuginosum* remains', *Journal of Ecology* **83**, 1995, 1021–30; J. H. Tallis, 'The pollen record of *Empetrum nigrum* in southern Pennine peats: implications for erosion and climate change', *Journal of Ecology* **85**, 1997, 455–65.

50. A. J. L. Winchester, 'Hill farming landscapes of medieval northern England' in D. Hooke (ed.), *Landscape. The Richest Historical Record*, Amesbury: The Society for Landscape Studies, 2000, 75–84; Supplementary Series **1**.

51. B. R. Gearey, S. West and J. Charman, 'The landscape context of medieval settlement on the south-western moors of England. Recent palaeoenvironmental evidence from Bodmin Moor and Dartmoor', *Medieval Archaeology* **41**, 1997, 195–210.

52. Sir Gawain and the Green Knight, lines 740–6, translation by P. Stone, Harmondsworth: Penguin Classics, 1959. The localities of much of this poem can be traced directly from evidence in the verse, passing through North Wales, the Wirral and Staffordshire.

53. The Green Chapel can be identified as a cleft in the Millstone Grit (Lud's Church Cave) near Gradbach in Staffordshire. See R. W. V. Elliott, *The Gawain Country*, Leeds: University of Leeds School of English: Leeds Texts and Monographs **NS 8**, 1984.

54. J. McDonnell, 'Pressures on Yorkshire woodland in the later Middle Ages', *Northern History* **28**, 1992, 110–25.

55. O. Rackham, *The History of the Countryside*, London: Dent, 1986.

56. A. Fleming, 'Towards a history of wood pasture in Swaledale (North Yorkshire)', *Landscape History* **19**, 1997, 57–73.

57. R. Tipping, 1998 (Note 22).

58. O. Rackham, 1986 (Note 55) p. 130. But J. Somers Cocks hints that a local [Dartmoor] forester called Aethelred who held manors in 1066 and 1086 was part of a Saxon organisation for hunting ('Saxon and early medieval times', in C. Gill (ed.), *Dartmoor. A New Study*, Newton Abbot: David & Charles, nd. [1970], 93–4). In 1300, a woman was Warden of Exmoor Forest.

59. R. Grant, *The Royal Forests of England*, Stroud, Glos: Alan Sutton, 1991. J. C. Cocks (*The Royal Forests of England*, London: Methuen, 1905, 336) says that the Domesday Book entry for Withypool on Exmoor declares that in the days of Edward the Confessor it was held by three foresters.

60. In chapter 8 of C. R. Young, *The Royal Forests of Medieval England*, Leicester University Press, 1979, though referring to a lowland and forested area. But we might expect the practice *a fortiori* in upland regions. There is an amplification of the idea in the development of grouse shooting in the nineteenth century.

61. Part 3.II, trans P. Stone, in the Penguin Classics edition of 1959 (Note 52). The land hunted over was not of course a Royal Forest: it would have been a Chase since the deer belonged to a lesser noble than the king. The castle at which Sir Gawain stays has a wooded park all round it (two miles in circumference) but this is clearly not the locality where the deer were hunted, which was much wilder.

62. D. Austin, R. H. Daggett and M. J. C. Walker, 'Farms and fields in Okehampton Park, Devon: the problems of studying medieval landscape', *Landscape History* 2, 1980, 39–57.

63. A. Raistrick, *The Pennine Dales*, London: Eyre & Spottiswoode, 1968. The quotation is from the Bishop's equivalent of Domesday, the Boldon Book of 1183.

64. L. Drury, 'Durham Palatinate forest law and administration specifically in Weardale up to 1440', *Archaeologia Aeliana* 6, 1978, 87–105.

65. R. B. Manning, *Hunters and Poachers. A Cultural and Social History of Unlawful Hunting in England 1485–1640*, Oxford: Clarendon Press, 1993.

66. C. Platt, *The Abbeys and Priories of Medieval England*, London: Secker & Warburg, 1984. For more examples of rhetoric surrounding the Cistercians in Yorkshire, see N. J. Menuge, 'The foundation myth: some Yorkshire monasteries and the landscape agenda', *Landscapes* 1, 2000, 22–37.

67. A. Dent, 'The last wolves in Yorkshire: and in England?' *Cleveland and Teesside Local History Bulletin* 43, 1982, 17–26.

68. C. Aybes and D. W. Yalden, 'Place-name evidence for the former distribution and status of wolves and beavers in Britain', *Mammal Review* 25, 1995, 201–22; M. Gelling, 'Anglo-Saxon eagles' *Leeds Studies in English* 18, 1987, 173–81.

69. R. A. Donkin, *The Cistercians: Studies in the Geography of Medieval England and Wales*, Toronto: Pontifical Institute of Mediaeval Studies, 1978. By 'opening up', Donkin means the completion of the colonisation process begun in Anglian times; the prehistoric period is not counted.

70. B. K. Roberts, 'Man and land in Upper Teesdale', in A. R. Clapham (ed.), *Upper Teesdale*, London: Collins, 1978, 141–59.

71. M. Williams, 'Marshland and waste', in L. Cantor (ed.), *The English Medieval Landscape*, London: Croom Helm, 1982, 86–125.

72. R. E. Hughes, J. Lutman, J. Dale and A. G. Thomson, 'Quantitative analyses of the 13th/14th century land holding systems of north-west Wales' *Agro-Ecosystems* 5, 1979, 191–211; R. E. Hughes, J. Lutman, A. G. Thomson and J. Dale, 'A review of the density and ratio of sheep and cattle in medieval Gwynedd, with particular reference to the uplands', *Journal of the Merioneth Historical and Record Society* 7, 1976, 373–83. [Both papers contain mountain material as well as moorland locations]

73. J. McDonnell, 'Upland Pennine hamlets', *Northern History* 26, 1990, 21–39.

74. T. M. Mighall and F. M. Chambers, The environmental impact of prehistoric mining at Copa Hill, Cwmystwyth, Wales, *The Holocene* 3, 1993, 260–4; J. G. Evans, *Land and Archaeology. Histories of Human Environment in the British Isles*, Stroud: Tempus Publishing, 1999, ch. 5.

75. A. G. Rigg, 'Lawrence of Durham: dialogues and Easter poem: a verse translation', *Journal of Medieval Latin* 7, 1997, 42–126. The quoted section is from *Dialogues* 2, written at a time when Cumin was Bishop and there was much conflict, so Lawrence was decrying this form of wealth accruing to one of the bad guys.

76. R. White, 2002 (Note 11), ch. 7.

77. S. Gerrard, 'The Dartmoor tin industry: an archaeological perspective', in D. M. Griffiths (ed.), The Archaeology of Dartmoor. Perspectives from the 1990s, Exeter: *Devonshire Archaeological Society Proceedings* **52**, 1994, 173–98; R. A. Fairbairn, *The Mines of Alston Moor*, Keighley: Northern Mines Research Society, Monograph no. 47, 1993; L. O. Tyson, *A History of the Manor and Lead Mines of Marrick, Swaledale*, Sheffield: Northern Mines Research Society Monograph no. 38, 1989.

78. Quoted in A. Raistrick, *The Pennine Dales*, London: Eyre and Spottiswoode, 1968. The date was actually in the seventeenth century but that is not very important here.

79. J. Walton, 'The medieval mines of Alston', *Transactions of the Cumberland and Westmorland Antiquarian and Archaeological Society* **NS XLV**, 1945, 22–33. The 1290 dispute seems to have been over £40-worth of trees cut down outside the normal rights of the miners. Both records together suggest that there was a good area of local woodland.

80. J. L. Drury, 'Medieval smelting in Co Durham: an archivist's point of view', in L. Willies and D. Cranstone (eds), *Boles and Smeltmills*, Matlock: The Historical Metallurgy Society, 1992, 22–4.

81. R. J. Clough, *The Lead Smelting Mills of the Yorkshire Dales*, Leeds: The Author, 1962, 39.

82. T. Gledhill, 'Smelting and woodland in Swaledale', in L. Willies and D. Cranstone (eds) 1992 (Note 80), 62–4. 'Chopwood' is wood cut into short lengths and split, suitable for the smelters' furnaces.

83. J. Bailey op. cit. 1810 (Note 40). Interestingly, the landowner planted the deciduous trees partly because they 'will . . . afford more beauty to the eye.'

84. A. Raistrick and B. Jennings, *A History of Lead Mining in the Pennines*, London: Longmans, 1965.

85. I. Blanchard, *International Lead production and Trade in the 'Age of the Saigerprozess' 1460–1560*, Stuttgart: Franz Steiner Verlag Zeitschrift für Unternehmensgeschichte Beiheft **85**, 1995, ed. H. Pohl.

86. K. A. Hudson-Edwards, M. G. Macklin and M. P. Taylor, '2000 years of sediment-borne heavy metal storage in the Yorkshire Ouse basin, NE England, UK', *Hydrological Processes* **13**, 1999, 1087–102.

87. R. Muir, 1997 (Note 36) p. 121 neatly summarises some of the necessary cautions over the interpretation of *wastea* and *wasta*.

88. R. Tipping, 1998 (Note 22); R. Tipping, 'Climatic variability and 'marginal' settlement in upland British landscapes: a re-evaluation', *Landscapes* **2**, 2002, 10–29.

89. Much of the material on Early Modern times comes from A. J. L. Winchester, *The Harvest of the Hills. Rural Life in Northern England and the Scottish Borders 1400–1700*, Edinburgh: Edinburgh University Press, 2000.

90. E. G. Fogwill, 'Pastoralism on Dartmoor', *Transactions of the Devonshire Association* **53**, 1954, 89–114.

91. H. Riley and R. Wilson-North, *The Field Archaeology of Exmoor*, Swindon: English Heritage, 2001, 90–1.

92. *Report of the Royal Commission on Common Land 1955–58*, London: HMSO, Cmnd 462, 1958; the Commons Registration Act was 1965; for a regionally descriptive but neither historical- nor management-oriented treatment, see W. G. Hoskins and L. D. Stamp, *The Common Lands of England and Wales*, London Collins, 1963 (New Naturalist series **45**). They estimated that England had 1,055,000 acres of common land and Wales 450,000 acres. Two-thirds of the

English common land was upland in the seven northern counties (reckoning Yorkshire as the three Ridings) and 30 per cent of Breconshire (as it then was) was common land, again most of it upland.

93. D. Coleman and J. Salt, *The British Population. Patterns, Trends and Processes*, Oxford: Oxford University Press, 1992.

94. The main sources for this section are those cited in note 19. See also two essays by Joan Thirsk in her collection *The Rural Economy of England*, London: The Hambledon Press, 1984, 'Horn and thorn in Staffordshire: the economy of a pastoral county', 163–82; and 'Horses in early modern England: for service, for pleasure, for power', 375–402.

95. C. W. J. Withers, 'Conceptions of cultural landscape change in upland North Wales: a case study of Lanbedr-y-Cennin and Caerhun parishes *c.* 1560–*c.* 1891', *Landscape History* 17, 1995, 35–47.

96. S. R. Eyre, 'The upward limit of enclosure on the East Moor of North Derbyshire', *Transactions of the Institute of British Geographers* 23, 1957, 61–74.

97. I. D. Whyte, 'Patterns of parliamentary enclosure of waste in Cumbria: a case study from north Westmorland', *Landscape History* 22, 2000, 77–89.

98. R. Gulliver, 'What were woods like in the Seventeenth Century? Examples from the Helmsley Estate, Northeast Yorkshire, UK', in K. J. Kirby and C. Watkins (eds), *The Ecological History of European Forests*, London and New York: CAB International, 1998, 135–53.

99. Daniel Defoe, *A Tour Through England and Wales*, London: Dent Everyman Edition, 1927, 2 vols.

100. F. Booker, 'Industry', in C. Gill (ed.), *Dartmoor. A New Study*, Newton Abbot: David & Charles, nd [1970], 100–38.

101. J. K. Harrison, 'Landscapes of industry', in D. A. Spratt and B. J. D. Harrison (eds), *The North York Moors: Landscape Heritage*, Newton Abbot and London: David & Charles, 1989, 159–83.

102. A. R. Millard, 'Geochemistry and the early alum industry', in A. M. Pollard (ed.), *Geoarchaeology: exploration, environments, resources*, London: Geological Society of London Special Publications 165, 139–46.

103. D. D. Bartley and C. Chambers, 'A pollen diagram, radiocarbon ages and evidence of agriculture on Extwhistle Moor, Lancashire', *New Phytologist* 121, 1992, 311–20.

104. A. M. Harvey and W. H. Renwick, 'Holocene alluvial fan and terrace formation in the Bowland Fells, Northwest England', *Earth Surface Processes and Landforms* 12, 1987, 249–58; A. M. Harvey, F. Oldfield, A. F. Baron and G. W. Pearson, 'Dating of Post-glacial landforms in the Central Howgills', *Earth Surface Processes and Landforms* 6, 1981, 401–12.

105. M. G. Macklin, D. G. Passmore and B. T. Rumsby, 'Climatic and cultural signals in Holocene alluvial sequences: the Tyne Basin, northern England', in S. Needham and M. G. Macklin (eds), *Alluvial Archaeology in Britain*, Oxford: Oxbow Books, 1992, 123–39; M. G. Macklin, B. T. Rumsby and T. Heap, 'Flood alluviation and entrenchment – Holocene valley-floor development and transformation in the British uplands', *Geological Society of America Bulletin* 104, 1992, 631–43; D. G. Passmore, A. C. Stevenson, D. C. Cowley, D. N. Edwards and C. F. O'Brien, 'Holocene alluviation and land use change on Callaly Moor, Northumberland, England', *Journal of Quaternary Science* 6, 1991, 225–32; A. J. Moores, D. G. Passmore and A. C. Stevenson, 'High resolution palaeochannel records of Holocene valley floor environments in the North Tyne basin, northern England', in A. G. Brown and T. A. Quine (eds), *Fluvial Processes and Environmental Change*, Chichester: Wiley, 1999, 283–310.

106. M. G. Macklin, 'Holocene river environments in prehistoric Britain: human interaction and impact', *Quaternary Proceedings* 7, 1999, 521–30.
107. T. P. Burt, J. Warburton and R. J. Allison, 'Eroding upland landscapes? Past, present and future perspectives', in T. P. Burt *et al.* (eds), *The British Uplands: Dynamics of Change*, Peterborough: JNCC Report no 319, 2002, 60–6. The example of the River Skell at Studley Royal is stretching 'upland' a bit, of course.

CHAPTER FOUR

Into an industrial economy

THE MINERAL RAILWAY: 1750–1914

The transition to a mineral-based economy represents a fundamental change in the economy and ecology of Great Britain. The nineteenth century witnessed the completion of the passage from a largely organic, solar-based, economy to one which derived its strength from minerals. In the uplands, the organic economy persisted in the form of plant and animal production but the addition of fossil fuels to the local and national economies wrought considerable changes.

The fundamental element in that shift was coal and the uplands' geology contained only a little of the black gold except in South Wales, where moorland stood above the rapidly industrialising valleys that ran southwards to the coast. The effects of lowland economic growth were however felt in the demands for upland products such as mutton, wool, butter, cheese, water and, eventually, open space. In an era of new knowledge and new technology, the importance of nineteenth century innovations in the uplands lagged behind the lower terrain, with the electrification of many hill villages and farms, for example, coming only much later, in the twentieth century. Local mineral economies based on metals that could be imported cheaply from abroad underwent decline whereas the demand for low-value, high-bulk roadstones and smelter fluxes (from quarries) escalated rapidly. Linkage into a wider world was demonstrated by the effects of war, when isolation produced higher prices with the incentives to reclaim or intensify agricultural production in the hill lands. Such processes fed into the national *Zeitgeist* of 'improvement': a growing nationalism encouraged the feeling that the uplands should contribute more to the wealth of the Kingdom. This socio-political context must have been intensified by the great population increases of the century. The English uplands kept pace and even exceeded the all-England rate of growth in the century after the first census of 1801, but the Welsh upland population grew very slowly after 1851.

The new watchword

Jonathan Swift's assertion earlier in the eighteenth century[1] that the essential servant of his country and of mankind would be the man who '. . . could make

113

two ears of corn or two blades of grass to grow upon a spot of ground where only one grew before . . .', was at the heart of the later desire to make the moorland areas more productive. Gregory King had calculated in 1696 that over a quarter of the country (10 million acres or 4.04 million ha) was waste lands and this was confirmed by the county reports of the Board of Agriculture from 1794 onwards (Fig. 4.1). They calculated that 21.3 per cent

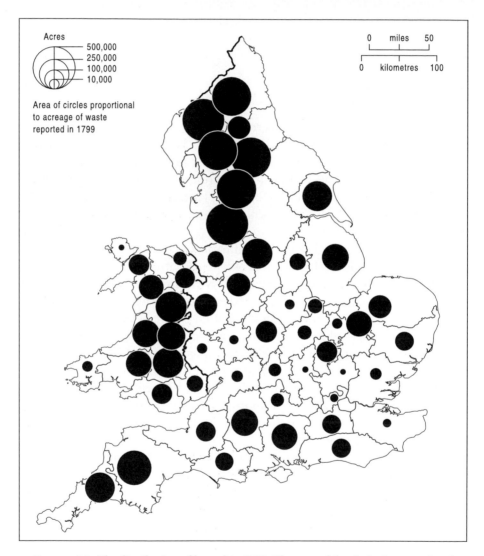

FIGURE 4.1 *The distribution of 'waste' in 1799. The area of the circles is proportional to the acreage of land thus reported (1 acre = 0.4 ha). The predominant contribution of the upland counties is obvious though there is a lot of heath at lower altitudes which would have helped form opinions at the time, especially in the southern counties nearer the metropolis.*
Source: adapted from numerous secondary sources, all ultimately from Arthur Young's General Report on Enclosure *of 1808.*

nd and Wales was 'waste' in 1770, a figure reduced to a fraction of that proportion by 1900. Not all 'waste' was moorland, but the ranking of counties with the biggest reductions in waste during the nineteenth century[2] suggests that it was the dominant constituent of that category, for the list reads, 'Radnor, Flint, Westmorland, Denbigh, Montgomery . . .'. The Napoleonic Wars were a considerable spur to reclamation in Cumberland and Westmorland and most hill areas converted some more wild land to tillage. Even without such stimuli, what gained the approval of authorities like Arthur Young was land enclosed by a stone wall and, as in Craven in the 1770s, a farm called Greenfield, '. . . from the appearance of *green* fields in the midst of black desarts [sic]'.

The necessary pre-condition for many improvement schemes was the availability of capital and so landlords with access to money were in the forefront. They also required large-scale sanctions and the mechanism of Parliamentary Enclosure was deployed to surround large tracts of upland with walls and fences so as to allow better grassland and even cultivation to be pursued. In many areas, however, very large swathes of land were enclosed without there being any actual improvement.[3] For some estates, afforestation was part of the set of measures: in 1780, Thomas Johnes, a friend of Sir Walter Scott, bought an estate (with the name of 'Hafod') in the upper Ystwyth valley which had been despoiled by earlier lead mining and created a house and gardens[4] surrounded not only by improved farms but also by four million trees planted between 1796 and 1913. Larch and hardwoods were taken out of his own nursery and grown at altitudes up to 400 m (1,200 feet).

More common, however, was the regional identification (the county was seen as an important unit) of the waste as needing improvement and various measures being applied to it, making use of advice from the Board of Agriculture and its inspectors like Charles Vancouver and John Holt, from books, and from the journals of agricultural societies that flourished at the time. Thus in 1795 Holt thought that upland Lancashire was full of mountainous tracts, craggy, steep and barren but nevertheless would benefit from a general enclosure act as a precursor to improvement. Vancouver's Devon of 1808 contained a 'red Irish bog' on the high parts of Dartmoor, the improvement of which would make it, 'an important object on the great field of national territory'. Entirely predictable, then, are some of the remarks in Arthur Young's tour of the north of England, published in 1771, where in Teesdale (to take just one example) the Earl of Darlington is praised for improving moors near Newbiggin, 'which used not to yield a farthing an acre rent; but upon inclosing, and then paring, burning, and liming, sowing with turnips, oats and hard corn and laying down with grass seeds, have immediately been advanced to 7s. 6d an acre, at which rent they now remain.' The Earl is regarded as a 'spirited cultivator'. Young thought that Teesdale, as he went upstream from Barnard Castle, was 'most exquisitely picturesque', but has little time for the upper Tees unless enclosure had taken place and crops of turnips and oats taken; to allow the land to remain 'waste' when it could be 'so easily, quickly, and cheaply improved' was a shame. The upper Tees beyond Newbiggin was different, for it was a very different country, partaking much more of the 'terrible sublime', where 'you ride through rapid streams, struggle alongside the sides of rocks, cross bleak mountains, and ride up the channel of torrents as the only sure road over bogs'.[5]

Not all the advocacy in the world could necessarily make all these changes permanent and few inducements were as strong as the increased prices for grain and livestock paid during the years of the wars with France. Some major schemes were clearly the obsessions of particular entrepreneurs: on Dartmoor, the Duchy of Cornwall lost interest in the moor in the 1738–70 period and it was left to individuals like Thomas Tyrwhit to launch schemes for afforestation and better agriculture at Postbridge and Princetown in the 1780s, ending with building the prison at the latter site in 1806 as a more secure investment. In Weardale, the Bishop of Durham led the way with an enclosure of 60,000 acres (24,282 ha) of pasture by 1800, which caused the abandonment of several small settlements.

The case of County Durham illustrates that in some regions the moorland extended well into areas now apparently remote from today's upland core. Between 1750 and 1870, 17 per cent of the county was enclosed amounting to 110,749 acres (44,820 ha), all of which except 5,573 acres (2,255 ha) were common, fell, moor and waste. Not all the land allocated by Commissioners was actually enclosed though in west Durham the lands of the Bishop suffered many instances of illegal encroachment. At St John's Chapel, some 400 m ASL, the 1801 Crop Returns showed that barley and oats were being grown. In this period, the spurs running east from the Pennines were largely converted to tillage, where before waste had stretched down to the outskirts of Durham City.[6] In North Yorkshire, moorland extended west from the Pennines down towards Knaresborough, to altitudes as low as 100–200 m. Within the common land of the Forest of Knaresborough there were two small settlements called Low and High Harrogate, with eighty-eight springs within a radius of 2 miles (3.2 km). After the mid-eighteenth century, better road connections made the area fashionable and sixteen of the springs were used medicinally. After 1770 the Forest was enclosed, though leaving 200 acres (81 ha) of open land for visitors to have an airing and free access to the springs. This amenity gave character to the resort and as The Stray is still there. The Enclosure Commissioners were allowed to carry out landscaping works and to plant trees. But as late as 1800 there was 'marshy heath' at the upper edge of town and only in the nineteenth century did the expected conveniences of pump room, assembly rooms and fashionable shops materialise.[7] Thus we have one of the few examples of moorland being converted into a town, though the section on South Wales provides others, in rather different circumstances. On Exmoor lies one of the few moorland 'follies' (The Mound in Cloutsham), with elaborate bastions and earthworks round a former tree clump.

Daniel Defoe visited Exmoor and dealt with it in short measure: 'The country is called *Exmore*, Cambden calls it a filthy, barren ground, and, indeed, so it is'. A century later, however, an Act for inclosure was passed and in 1815 some 10,000 acres (4,047 ha) of the former Royal Forest passed into the hands of John Knight of Worcestershire, for the sum of £50,000. The farm of Simonsbath was reserved for the Crown, including 12 acres (4.8 ha) for the church and its appurtenances should the area become permanently inhabited. The Commissioners of Enclosure had reported that there were only thirty-seven trees on Exmoor Forest and that the area was largely used only for summer pasture for sheep. Some of this was wet enough to carry cotton-sedge (*Eriophorum* spp) which along with purple moor-grass (*Molinia caerulea*) was burnt by the shepherds.[8] Knight continued to make the pastures available while he formulated his plans for improvement, which included buying out further allotments, to the total of 16,000 acres (6,475 ha) by about 1820 and he circled his estate with a wall which was 29 miles long (and which sometimes interfered with the staghounds) and built 22 miles of roads to replace the rutted tracks of beforetimes, a deer park and a 7 acre reservoir (Pinkery Pond) whose intended function remains obscure. Knight also acquired the mineral rights but these were never exploited in his time.

The great plan was to convert the *Molinia* grassland to plough land. The top of the turf was pared off, burned and then the soil was limed (3 tons to the acre), ploughed and left until the next planting season. Where necessary a deeper ploughing broke up the iron pan of the podsolic soils. Oats, turnips, barley and spring wheat were all successfully grown on the 2,500 acres (1,011 ha) reclaimed by 1845. In later years, the estate was divided into smaller holdings which were let to tenants and so in about 1880 the high water mark of farming, and thus the disappearance of moorland, was reached (Plate 4.1). From an income of £350 in 1818, the former Forest, with three-quarters of its wild land now converted to improved pasture and tillage, yielded some £4,500 per annum. The depressed years at the end of the nineteenth century brought a Royal Commission on Agriculture of 1879–82, and their county reports include a visit to Exmoor, which admits that four-course farming has been difficult to pursue, as was the original intention, but that improved grassland (replacing the *Molinia* and bilberries, there having been very little heather) supported about forty tenanted farms running Cheviot sheep plus a little corn for their own use. The estate continued to run sheep and ponies on the remaining open moor, and this pattern of management persisted up to WWI.[9] Exmoor was no longer in its pre-1800 condition of '. . . a state of nature, wild and desolate as an American prairie'.[10] Exmoor has retained enough decidu-ous woodland (about 10 per cent of the land cover) to support three main herds of red deer and some smaller ones and its beech hedges established after enclosure within and outside the Royal Forest added to wildlife habitats. It also escaped the worst excesses of state upland forestry.

Exmoor may however have been an exceptionally reclaimable set of

PLATE 4.1 *A farm on Exmoor, with some of the early enclosures, probably for cereal growing, around the buildings and then surrounded by large intakes of former moor, which end at a distinct boundary with still-open moor.*

environments in both natural and politico-economic terms, as an example, Parry's data for the North York Moors show that of the 49 per cent of the area of today's National Park (70,246 ha) was rough pasture in 1853; the total change in that category from 1853–95 was −3.6 per cent, and from 1895–1904 was −0.2 per cent. On Dartmoor, the change between 1885 and 1904 was actually +6.3 per cent, comprising 3,267 ha. Clearly the depression of the late nineteenth century was affecting reclamation rates but even before that there had been a retardation from the faster rates of the earlier part of the century.[11] The last smallholding to be taken in from Dartmoor by a pioneering house-holder was in 1870 and was of 30 acres (12 ha) at Nun's Cross Farm where heather was converted to grassland cut for hay.[12] Put in a wider context, the late nineteenth-early twentieth-century story is of increases in upland land cover of an unimproved character: the actual areas are given in Table 4.1 and the rates of change in Table 4.2: the reversion to wilder status in Wales is remarkably high. In neither country however does this mean that even very high areas were necessarily bereft of enclosures: walls especially remained from periods of enclosure. To some extent, the greater area of open land is reflected directly in the sheep numbers for the period: an increase in the absolute numbers but a decrease of 0.36 per acre (0.145/ha) in England and 0.63 per acre (0.25/ha) in Wales (Table 4.3).

In Wales, Parliamentary enclosure affected the uplands less than most other

TABLE 4.1 Land use 1889–1919

Mountain and heath (acres)	1889	1919
Upland England	1,382,519	1,740,368
Upland Wales	818,122	1,215,425

Permanent grass (acres)	1889	1919
Upland England	2,973,288	2,859,259
Upland Wales	1,444,434	1,347,681

(1 acre = 0.4047 ha)
Source: E. J. T. Collins, *The Economy of Upland Britain 1750–1950: an Illustrated Review*. Reading: Centre for Agricultural Strategy, 1978. Reproduced with permission of the Centre for Agricultural Strategy at the University of Reading.

TABLE 4.2 Rates of change 1889–1919

Per cent	Upland England	Upland Wales
Permanent grass	−3.8	−6.7
Mountain and heath	25.9	48.6

Source: E. J. T. Collins, *The Economy of Upland Britain 1750–1950: an Illustrated Review*. Reading: Centre for Agricultural Strategy, 1978. Reproduced with permission of the Centre for Agricultural Strategy at the University of Reading.

TABLE 4.3 Sheep in uplands 1889 and 1919

	1889	1919
Upland England	3,438,349	3,558,819
Upland Wales	2,399,928	2,768,679

Source: E. J. T. Collins, *The Economy of Upland Britain 1750–1950: an Illustrated Review*. Reading: Centre for Agricultural Strategy, 1978. Reproduced with permission of the Centre for Agricultural Strategy at the University of Reading.

parts of Britain. The fringes of the moorland blocks changed most but even here peasants built cottages on the margins of the wastes, like necklaces around the upland pastures (*ffriddoedd,* singular *ffrid*), with a maximum in 1780–1820. Legally enforceable enclosure led as always to dispossession of some commoners and also to disputes about ownership: there were riots caused by enclosure proposals in Ceredigion and Carnarvon during the Napoleonic Wars. These proposals, also peaking in the 1770–1820 period, were mostly about improving landlords' rights rather than improving commons: one Act to enclose 40,000 acres (16,188 ha) of the Great Forest of Brecon deprived the commoners of half of their grazing land.[13] Elsewhere, conservative attitudes to land management prevailed, for in 1808 farmers on Dartmoor were growing potatoes in lazy-beds and it was only in that century that horses replaced oxen on the moorland farms. Very little modernity attended the use (followed by rapid abandonment) of some intakes in the Pennines in 1800–25 while oats and potato prices were high.

The improvement of the wastes was therefore not always permanent since land might be let fall from its improved conditions, whether of grassland or tilled land. Equally, the passing of an Act of Enclosure did not always mean that landowners took action. In some places the marginality of the region was recognised by the exemption of tenants from obligations to fence or wall, for example. Nevertheless, many effects from this period are seen in the moorland environment today, often as reverting pastures being re-occupied by bracken, *Juncus* rushes or coarse grasses such as mat grass and even by heather. Without other stimuli to organic production in the uplands it seem probable that even less attrition of moorland would have taken place. However, in the nineteenth century, the development of mineral-based economies was always likely.

Animals of the margin

Though sheep and cattle are the dominant animals of the pastoralist economy in upland Britain, others have been tried with greater or lesser success: populations of feral goats in the Cheviots, for example, are descendants of attempts at supplementary milk production; ponies on Dartmoor sustain a modest industry with national markets in the leisure industry and export for, it is surmised, meat.

A successfully farmed animal of the moorland edge has been the rabbit, yielding both fur and flesh. In medieval times, it was characteristic of lowland marginal areas such as Breckland, and later its production dominated areas like the limestone upland of Lincolnshire. Large areas of southern Dartmoor bear the traces of warrens but the well-preserved remains on the southern edge of the North York Moors allow a good assessment of their environmental effects. Here, farms belonging to large estates ran rabbit warrens, especially in the eighteenth and nineteenth centuries, with the valuable skins destined for London (in the eighteenth century by sea via Scarborough) and the less costly meat for urban centres such as Hull. The rabbits were kept at densities of about 5–10/ha and the annual yield was of the order of 22/ha. Thus a large warren would have produced perhaps 16,000 rabbits in a normal year in, for example, the early nineteenth century. Maps of the warren areas suggest that they were enclosed areas of rough pasture next to open moorland and different types of resulting features can be seen. These include pillow mounds (usually about $10 \times 5 \times 1$ m but sometimes 100 m long) which were dug up so that the rabbits might burrow in them and then be caught with ferrets and nets. It is possible that the North York Moors examples pre-date the eighteenth century since this was the classic medieval technique. There were also 'extensive' warrens of perhaps 700 ha bounded by walls and streams, with pit traps ('rabbit types') dispersed across them. Lastly, there were farm warrens

where a fenced area had pit traps built into perimeter walls and where additional rabbits from outside might be enticed into the farm by providing
fodder. Examples of this practice survived until the myxomatosis epidemic of
the 1950s.

The environmental impacts of warrening on this scale can be deduced. At
the very local scale, there is the soil disturbance caused by pillow mounds or
by any burrowing activity due to rabbits kept at a high density (Plate 4.2).
Grazing by rabbits is likely to render the enclosures useless for other stock,
although a holding might usually keep sheep and cattle as well: hence the value
of walling off the rabbits as completely as possible. Moving out from the actual
warren, it was essential to have access to woodland for 'green bark' to be fed
to the animals in winter; an area of hazel coppice was the usual source. Roots,
cereals and hay were also in demand for rabbit food. On a still larger scale, at
times of high skin prices, rough grazing and common might be taken in to be
part of a warren, so that rabbit farming became part of the enclosures eating
away at the moorland edge. The Napoleonic Wars were a good time for this,
though fur prices fell sharply after the 1820s and abandonment of warrens
took place. At the end of the nineteenth century, further reclamation for all
purposes became uneconomic, warrening included. As a final consequence,
the abandoned warrens must have been good nuclei for the feral rabbit populations now so obvious on these moors.[14]

PLATE 4.2 *The remains of pillow mounds on a hillside on the Brecon Beacons. These
were dug in order to provide loose earth for rabbits, in this case during the nineteenth
century, though many earlier examples are found in Britain, especially in lowland areas
like the Breckland.*

Mineral wealth

The uplands possess a geological patrimony which invites exploitation. This is coupled with a topography in which valley sides often provide relatively easy entry into deposits and generally low land values which are uncompetitive with mineral wealth. Even on common land, the lord of the manor retained mineral ownership rights and the way was thus made easier for development of mines and quarries of all kinds. Any bell-pit, shallow adit or simple quarry has some local environmental impact and the remains can frequently be seen in almost any upland valley as well as on the plateaux. Here, only the major developments are discussed but smaller features include the Clargill Head mine in the upper South Tyne (near Alston) in the North Pennines, whose rich deposits of silver supplied the Royal Mint in the eighteenth century and part of the nineteenth century as well: 35 oz of silver per ton of lead was extracted in the 1720s.

During the eighteenth and nineteenth centuries, Britain was the world's largest producer of lead. It was perhaps the most important mineral extracted from the moorlands and often processed there as well. Its peak price in London was in 1809, at £31–6–0½ per ton, having risen from about £13 in 1771 and falling to that same level in 1909, when imports became much cheaper.[15] The Peak District, the Yorkshire Dales and the northern Pennines were the main regions of lead mining and in all three places an earlier industry was greatly expanded. The lead mines of Powys also had a boom period in the nineteenth century, once steam engines had been introduced in the 1850s.[16] In the case of Derbyshire, the greatest period of environmental impact of the industry was perhaps just over by 1750, when it became dominated by the larger companies who worked larger mines but fewer of them and after 1850 there was a rapid decline except at Mill Close Mine in Darley Dale which persisted until the 1930s along with some fluorspar mines. Environmental relations in the period 1650–1800 included very long drainage tunnels ('soughs') which emitted mine-water into a main water-course perhaps 5–7 km away, the heaps of black slag from stamping mills and the waste lead from refining plant emitted from chimneys. Extraction of the base ore from the major veins and their side branches ('rakes' and 'scrins') was often carried on from a series of surface pits, giving rise to linear strakes of overburden in the landscape which are still visible today.[17] In the northern Yorkshire Dales, largely under local management, the maximum production period was from the 1820s to the early 1870s; as an extractive industry its environmental impact outran the quarrying and burning of lime for agricultural purposes and the use of limestone blocks for enclosure walls and buildings (Plate 4.3). It was eclipsed after about 1870 by the decline of demand for lead and the increased demand for road aggregate, railway ballast and iron-making fluxes, all of which could be transported by railway from the limestone quarries on the dales' hillsides.[18]

In the Yorkshire Dales and North Pennines, long flues were used from 1747 onwards, with Grinton in Swaledale as the pioneer location. They conveyed

PLATE 4.3 *Hushes on veins of lead ore on the Gunnerside Gill in the Yorkshire Dales. At their foot is the extensive area of waste from the crushing plant (which was water-powered), with the ore storage 'bouse teams' clearly visible. The industry was at its peak in the mid-nineteenth century in this region.*
Source: Yorkshire Dales National Park, YDP100/12. Reproduced by permission of R. White, Yorkshire Dales National Park Authority.

the poisonous fumes to chimneys from which dispersal over a wide area could be combined with the recovery of lead which condensed on the inside of the structure. They might well be 1–3 km long and usually ran uphill, not always in a straight line: the flue from the Keld Heads mill at 200 m ASL in Wensleydale ran through several bends before discharging from a stack at Cobscar Mill at about 380 m.[19] The Rookhope chimney north of Weardale, though, had only one significant bend in its 3.5 km length and 220 m of ascent. The smoke from smelting plants was noxious over a wide area: in the Tyne valley, it created pollution (there was sulphur as well as lead oxides) over an area some 10–11 km around the mill and farmers had to be compensated for poisoned stock. Still visible today are many 'hushes', as described for earlier times. Later versions (as at Spout Gill in the Yorkshire Dales) inserted a dam with a sluice so that multiple hushing could be carried out and so instead of a single use, a hush might have a lifetime of a century.[20] An attempt to convey the immediate and side-valley scale environmental impacts of lead mining and smelting in the North Pennines in the nineteenth century is shown in Figure 4.2.[21] One further result was the contribution of alluvial material to river terrace formation and it is thought that the deposits of the floodplain and lower terrace of the Wharfe derive more from lead mining wash-off than from agriculturally-produced input.[22]

Fuel demands were always likely to bring environmental impact. For instance, a Yorkshire miners' charter of 1737 affirms their rights to take wood from a lord's estate for mining use and for smelting. Yet by the time of the major expansion of lead mining, with production peaks in the 1820s around Alston, in 1850–80 in Teesdale and 1840–1950 in the Yorkshire Dales, wood was no longer a significant fuel. Where possible, large quantities of peat were cut and led to the mills and this was often the main source of fuels for cupola furnaces in the upper parts of the valleys of the Pennines. Lower down the dales, however, it was economic to bring in Durham coal to fuel the use of reverbatory furnaces. Coal had been in use for powering pumping engines in Derbyshire from about 1750 onwards. Thus, the railways came into lead mining areas and extended from the valleys into the moorlands themselves, as with the Stanhope and Shields line of 1834 and the Cromford and High Peak Railway which is almost all over 1,000 ft (300 m) and as high as 1,264ft (379 m) near Buxton. Many other lines reached into the Pennine valleys to bring in coal and take out lead: Leyburn, Alston, Middleton-in-Teesdale, and Allendale were all connected for those reasons.[23] In all these uplands, we can talk of lead-affected environments and landscapes and that of the Yorkshire Dales is shown in Figure 4.3.

The environmental relations of lead mining extend beyond the extraction of the ore and its refining. The extension of lead mining into formerly pastoral uplands brought about nineteenth century changes in the settlement and land use patterns: villages were created like Allenheads at 1,327 ft (398 m) ASL and Nenthead at 1,411 ft (423 m) and there was a miner's house in Teesdale at

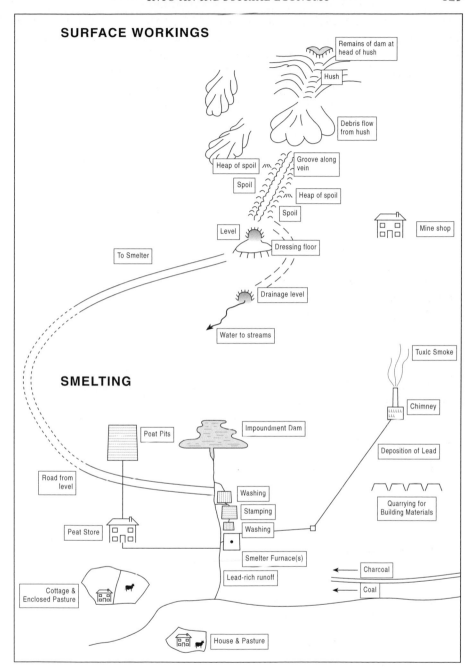

SURFACE WORKINGS

Remains of dam at head of hush

Hush

Debris flow from hush

Heap of spoil

Groove along vein

Spoil

Heap of spoil

Spoil

Level

Mine shop

To Smelter

Dressing floor

Drainage level

Water to streams

SMELTING

Toxic Smoke

Chimney

Peat Pits

Impoundment Dam

Deposition of Lead

Road from level

Quarrying for Building Materials

Washing

Stamping

Peat Store

Washing

Smelter Furnace(s)

Charcoal

Lead-rich runoff

Coal

Cottage & Enclosed Pasture

House & Pasture

FIGURE 4.2 *An attempt to show the diversity of environmental impacts of lead mining in the nineteenth century (see also Plate 1.3). At the top of the diagram are the heaps and gouges at the places where the ore was extracted and below the environmental relations of a smelter. The extraction sites were naturally much commoner than smelters. Peat was a popular fuel in areas where it was common since it was essentially free; if it was lacking then charcoal or coal had to be brought in, though shallow seams of poor coal were often exploited in lead-mining districts.*

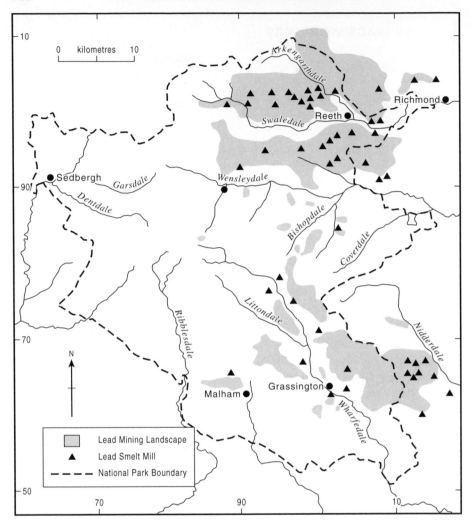

FIGURE 4.3 *The useful concept of a lead-mining landscape applied to the Yorkshire Dales. Within the stippled area all the phenomena of Figure 4.2 would be found and their impact intensified by the density of smelters, remembering that they contributed to contamination of soil, water and air.*
Source: R. F. White, 'The lead industry in the Yorkshire Dales', in A. J. Howard and M. G. Macklin (eds), The Quaternary of the Eastern Yorkshire Dales, *London: Quaternary Research Association, 1998, 54–66. Fig. 26 at p. 54. Reproduced with the permission of the Quaternary Research Association.*

2,000 ft (600 m) ASL. More important however were the scattered cottages and rows of houses (Fig. 4.4) where miners were granted a smallholding to improve their nutrition.[24] It was possible in the North Pennines for a miner to have 3 acres (1.2 ha) of meadow and 3–4 acres (1.2 to 1.6 ha) of upland pasture on which to keep a goat and two cows or a cow and a packhorse. Cottages in the 'planted' village of Garrigill high in the South Tyne carried 6 acres (2.4 ha) of

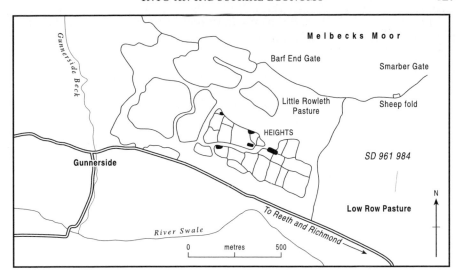

FIGURE 4.4 *A conversion from moorland to enclosure due to lead mining in Swaledale (Yorkshire Dales). On the slope above the river, a set of miners' cottages (in black) is accompanied by some small 'island' enclosures which were used largely to keep domesticated animals; tracks on the moorland connect other sheep-herding enclosures and lead to the miners' work-places.*
Source: T. B. Bagenal, Miners and Farmers. The Agricultural Holdings of the Lead Miners at Heights, Gunnerside, in North Yorkshire, *Keighley: Northern Mine Research Society, British Mining No. 62, 1999, p. 7.*

land plus moorland stints. Smaller holdings than these were common but all shared in being only grass and moor: ploughing was not permitted. Geese and bees thus added to the stock that were kept; and on the larger holdings sheep might be kept. All this was facilitated by enclosure, which made more land for meadow and winter pasture. Improvement became profitable to the point where it could carry three times the number of stock even in locations like Allendale. Considerable work was required to achieve such advances: burning, paring, liming, composting and manuring were all needed to bring the grasses up to a high quality pasture. In Teesdale, there was a tradition of a balanced dual economy of farming and mining but no deliberately created small holdings for miners.[25] This contrasts with Swaledale in North Yorkshire where in the nineteenth century land was recorded as let to miners under a copyhold tenure, with enclosed fields for hay and controlled grazing together with a 'gait' for cattle on the open common. Thus islands of improved land were found at the edge of the moorland.[26] Enclosure also led to better routes for roads, along with fencing and walling, though small quarries were left in common, as was access to peat beds. It allowed the creation of gardens for miners as well, with most employees having them by 1860 in contrast to the eighteenth century. In the Durham and Northumberland dales, the better roads meant that carts could be used to bring up coal from the eastern coalfield, making that fuel ubiquitous by about 1840. Supplementary fuel came from poor coal derived from thin seams in the limestones: it was mixed with clay and formed into turnip-shaped 'cats'. A last addition to the environmental relations of the miners is their propensity to poach game birds, notably grouse and snipe. A Weardale ballad of *c*.1818 entitled 'The Bonny Moor Hen' chronicles attempts by the Bishop to clamp down on such usufructs.[27] By contrast, a privileged visitor might be lyrical about the scene on Alston Moor, where 'the blue shades of evening impress a character of sublimity on the surrounding hills', and where the Weel of the Tees exhibited waters, 'spread in the hollow of a vast and dreary basin . . . a broad river flowing through the midst of a desert'.[28]

In its domination of the industrial landscape of a single region, the extraction of tin from Dartmoor is probably the outstanding example, though there were also lead and copper mines on the metamorphic aureole of the granite during the nineteenth century. After 1750, as in so many mineral industries except coal, the number of establishments declines but they become bigger. On Dartmoor, there was a revival in the 1780–1814 period but steady decline in the rest of the nineteenth century, ending for ever in 1939. The head of the East Dart was permanently lowered by stream works and the 'blowing houses' after 1750 were mostly fired by coal rather than peat charcoal, diminishing their impact on the blanket bogs of the moor. By the time of its demise, this industry had provided visible evidence of its presence in 130 of the 250 square miles of the Moor.[29] On the metamorphic aureole, there were some thirty tin mines working after the eighteenth century but at least half of them were copper mines as well. In the early nineteenth century, 40 per cent of the world's copper came from Devon and Cornwall, though much of the Devon copper came from the Tamar valley rather than the moor. All the copper came from the aureole rather than the granite and there were some forty mines at the peak period between 1820 and 1860, after which there was a sharp decline. The most important was Wheal Friendship mine near Mary Tavy, which started up in 1714 but was re-opened in 1796 and worked copper until 1870, with arsenic being extracted for a further decade. At its apogee, the mine and mills covered 30 acres (12 ha), with the tributary leats powering seventeen waterwheels.[30]

Dartmoor was the only major moorland with a china clay industry, in its south-west corner. Started there in 1833, the clay merited a special mention at the Great Exhibition of 1851. Its main development was on common land at Lee Moor and had coalesced into one company by 1919 whose major environmental impacts have been the unreclaimable holes and heaps in which eight tons of waste are produced for every ton of clay, whose uses are mostly in cosmetics and as a filler in paper. The remains of earlier china clay works in the fastnesses of the moor, as at Red Lake can still be seen and the environmental impacts gauged (Plate 4.4). Some iron, wolfram and even uranium have also been sought within the confines of Dartmoor but only iron came to be a significant industry, especially in the area of Haytor until about 1921.

PLATE 4.4 *Most china clay extraction is from the edges of granitic uplands in south-west England; this isolated pit and spoil heap is at Red Lake well within the southern block of Dartmoor's moorland area.*

Iron working is most strongly associated with the North York Moors. There are seven seams of ironstone within the Jurassic rocks and all were exploited in medieval times. There seems to have been a 150-year interval after about 1650, after which the half-century after 1800 saw a rapid growth of iron working which was exported to Tyneside for smelting: much of it was taken from coastal drifts directly into boats. Of greater environmental significance was the exploitation of the main seam on the hills overlooking Teesside and in Rosedale and in the Esk valley where for example Grosmont had blast furnaces in the 1860s. In all three places, the customary environmental changes to land, water and air took place, including the development of railways and improved roads which accelerated most forms of change. From 1845, Stanhope in Weardale also had a blast furnace, using iron ore from further up the valley, but the richer ores to the east meant that it soon moved to Tow Law, which was then largely in wild country just undergoing enclosure and reclamation.

An example of the secondary effects of lead mining can be seen in the Stiperstones area of Shropshire where lead production peaked between 1865–85 but where only two mines remained by 1916. It produced a very dense settlement pattern and a farmland configuration where land already enclosed was further subdivided and enclosed, to provide holdings for mineworkers. Thus there is today a high proportion of abandoned dwellings near the former mining lands and often these are at high altitudes, up to 450 m ASL.[31]

A ubiquitous feature through all these uplands is the quarry. On Dartmoor, the granite was extracted for many buildings both local and national, the most notorious of which is the prison at Princetown, built in 1804–6 and which sparked off a number of other quarry developments such as those at Haytor in the 1820s where a granite-built tramway carried granite boulders eventually bound for London. But limestone quarries are more common. Where they might often have started as local developments supplying lime for agricultural use, the development of iron smelting and of extensive roadmaking increased the demand enormously. So the ancestors of today's large-scale extraction plants in the Peak District, Weardale (actually a cement plant and closed in 2002), and the southern North York Moors, were founded in the nineteenth century and usually grew as transport systems improved, especially the provision of railways. A massive kiln was constructed at Langcliffe (Yorkshire Dales) to be near the Settle and Carlisle Railway.[32] The early nineteenth century construction of limekilns in the North York Moors stimulated the mining of coal from small collieries in the Danby and Rudland areas, which reached a peak in 1812–13 and persisted until the 1890s, leaving spoil heaps and abandoned shafts which are still visible. The Esk valley and Scugdale carried jet mining after 1800: the latter valley had thirty-nine mines in 1877 but the vogue had passed by the 1890s, when Queen Victoria's mourning attire had rather ceased to lead fashion. As usual, waste production far exceeded the useful material. The same is true of slate. Though slate is associated mainly with the mountain areas of the Lakes and Snowdonia, there were large quarries in the Berwyn Mountains of Denbighshire by 1790, with two of them each producing over one million slates every year. In Dentdale and Weardale, polished limestones called 'marbles' were produced for decorative purposes at varying times from the medieval to the early twentieth century.

Isolation of moorland areas combined with quarrying and some forms of mining to bring about the right circumstances for gunpowder making and on Dartmoor there was a powder mill near Postbridge in the second half of the nineteenth century. Charcoal was one of its elements, which may have been produced from willow and alder trees in the low-lying areas nearby. In the nineteenth century it was thought that naptha (for candles, fuel and mothballs) might be extracted from peat and one of the nascent industries of the Princetown area from 1844 was devoted to this industry: at one point the prison was lit by naptha gas. The largest development of this kind was on the western edge of the moor at Rattlebrook Head in the 1850s where a series of enterprises to extract the peat itself and a form of crude oil (which was to have been piped down to Devonport) were installed from 1850 onwards attracting enough capital to connect the works at 1,750 ft (525 m) ASL by a standard gauge railway to the LSWR line near Bridestowe in 1879. None were successful for long but as late as 1955 attempts to extract the peat industrially were in progress.[33]

Urban water from rural catchments

The uplands had a renewable supply of water, which made them choice sites for the supply of what was becoming a necessity for the cities of the nineteenth century, not just for industrial use but to provide people with clean potable water and a reasonable sewage system.[34] In 1846, though, only 10 out of 190 local authorities possessed their own waterworks and one story of the nineteenth century is of the extension of control by municipalities. The more far-sighted of these saw public benefits not simply in the prevention of disease but in better fire-fighting and fewer floods. Some moors indeed had been exploited earlier for city supplies and a few had been the sites of reservoir construction in the service of canals. The peat cover of some catchments was seen as a bonus since it was thought it would act as a water store at times of high precipitation, something not borne out by recent research.[35] The Leeds and Liverpool canal needed a reserve storage capacity of 1.3 million m^3 and this was mostly met from the moorlands traversed by the canal.[36] Though the story is dominated by the construction of large reservoirs, many smaller ones were made, so that many moorlands around industrial centres (Glossop in the High Peak for example had over fifty cotton mills in the one valley in the nineteenth century) house a mixture of large and small impoundments.

The extension of the burgeoning cities' interests into the uplands to supply both their industries and their citizens brought them into upland valleys and indeed into conflict with each other. Effectively, the law on water rights meant that large-scale development could only be undertaken if the land were bought, along with exclusive rights to its water. So cities often acquired common rights: Leeds for example let 7,000 acres (2,800 ha) of grazing in the Pennines and Bradford became a manorial lord. In the 1860s, Manchester's chain of artificial lakes in the moorland valley of Longdendale were the most extensive in Europe if not the world, with sixteen impoundments covering 854 acres (346 ha)[37] and Lake Vyrnwy in Montgomeryshire (Fig. 4.5) (Plate 4.5) was the largest artificial reservoir in Europe at the time it first started sending water to Liverpool in 1891; its gathering grounds amount to 23,290 acres and the water surface may rise to 1,121 acres (454 ha).[38] (By the end of the twentieth century the Peak District National Park had fifty-five reservoirs within its boundaries, supplying water with an annual value of perhaps £100 million.) Some towns claimed rights but failed either to use them or share with them smaller boroughs; some bigger players bought resources simply to pre-empt others and regional co-operation was rare: Sheffield gained thirteen large impoundments which drew on 30,403 acres (12,304 ha) in the century after 1830 (Fig. 4.6). The Derwent Valley board which in time organised water for Sheffield, Derby and Nottingham was in a class of two, along with the Metropolitan Water Board of London.[39] As early as 1869, Halifax and Bradford were in conflict about the exact location of the watershed between their catchments. The construction of reservoirs drowned and fragmented holdings and the restrictions laid down by public health authorities often

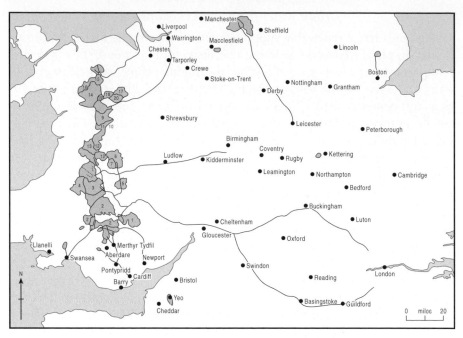

FIGURE 4.5 *The upland water gathering-grounds of Wales and their connections into England as well as the industrial south of the Principality as seen in 1936. (The original Vyrwry catchment is not numbered.) The diagram contains several suggestions not taken up, including the use of canal systems to supply London with Welsh water.*
Source: R. C. S. Walters, The Nation's Water Supply, *London: Ivor Nicholson and Watson, 1936, Fig. 35, p. 112.*

restricted grazing, so many small farms were abandoned. When acquiring the Elan Valley in Wales in 1892, Birmingham boasted that its 45,562 acres (18,439 ha) supported only thirty-one sheep farms, all very strictly controlled since the city had acquired the freehold and the manorial rights.[40] Supposedly to minimise human or animal contamination, afforestation became popular and was first introduced at Oldham in 1885.[41] The growing amenity use of the uplands, discussed later, also came into conflict with the sequestration and sealing-off of land, which amounted to 33,000 ha of gathering grounds in England by 1904 and 2,500 ha in Wales, though some of that is in mountains such as the Lake District rather than moorlands *sensu stricto*. The environmental impact also extended to the downstream course of the river, for all undertakings were required to supply 'compensation water', to keep enough water in the course below any impoundment to supply users lower down. This gains greater importance as time progresses since the maintenance of flow sufficient to allow extraction in the lower courses of rivers is important. Together with the reservoirs, intakes and long-distance transfers constitute a remarkable network of water translocation even within one river basin, such as that of the Humber (Fig. 4.7).

PLATE 4.5 *The great environmental change instituted by upland water catchment is given an imperial and quasi-military expression here at Lake Vyrnwy in mid-Wales. The impoundment was at one time the largest in Europe.*

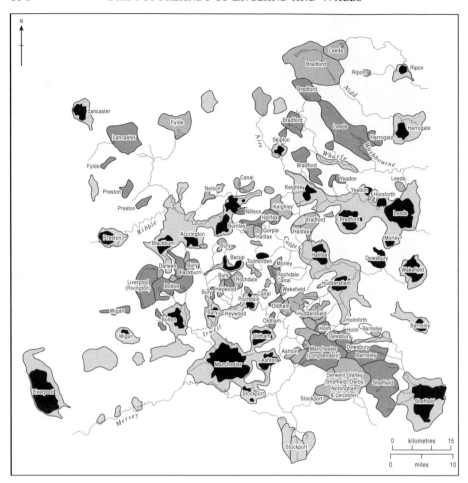

FIGURE 4.6 *The parcelling-up of the Pennine gathering-grounds as depicted in 1936. Urban areas are shown in black, those in place by 1869 in dark stipple and those from the sixty-five years thereafter in a lighter stipple. The general lack of co-operation and hence fragmentation of moorland management is apparent.*
Source: *R. C. S. Walters,* The Nation's Water Supply, *London: Ivor Nicholson and Watson, 1936, Fig. 30 opp. p. 91.*

Some dams were built on peat and were filled with peat, which was an adequate fill provided it was allowed to settle at a slow pace. Happily, few dams were overtopped, though the Dale Dike wall near Bradfield failed in March 1864, just after the impoundment filled. There was a loss of 244 lives downstream and £373,000 was paid in compensation.[42] Where the terrain or the lack of capital precluded the construction of impoundments then open channels of some length might supply water: Drake's Leat on Dartmoor was for example supplemented by the Devonport leat from the neighbourhood of Princetown in the late eighteenth century: a distance of at least 20 miles around the hillsides. The fall of water from such leats may well have led to

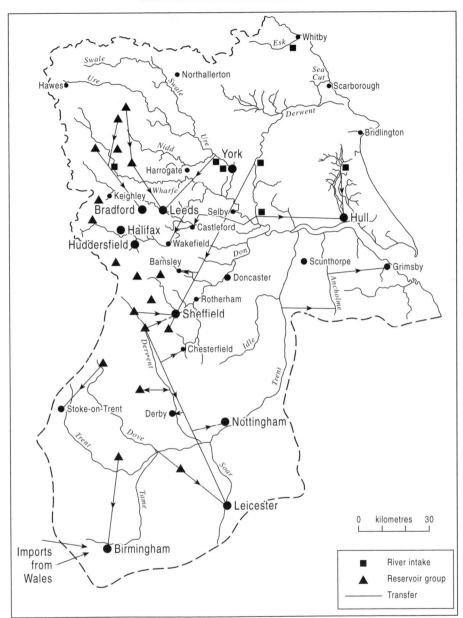

FIGURE 4.7 *In this map of water linkages in parts of the Midlands and Yorkshire, the concentration of 'reservoir groups' is in the uplands but the water is then transferred considerable distances in different directions using pipes and rivers.*
Source: Environment Agency, nd.

the idea that turbines could be installed on Dartmoor rivers to generate electricity. By 1890, Okehampton had harnessed the East Okement to light up 220 bulbs of eight candlepower each, placed in the streets, and a Baptist Chapel. The Dart's power lit up the monastery at Buckfastleigh in 1901 and on Exmoor in 1890 the East Lyn's rapid fall encouraged the first publicly-owned pumped storage plant in the world, powering a turbine which allowed the sale of domestic electricity with a charge of £1 per annum for a 60 watt bulb.[43]

As a footnote to these environmental manipulations, there is the attempt in 1874–86 to establish an iceworks on Sourton Tor on Dartmoor.[44] Ponds totalling about 5–6 acres (2 to 2.4 ha) in area were supplied with water via leats and the entrepreneur waited for nature to act before gathering the ice, storing it until the warmer weather and then selling it to the Plymouth fish trade. The winter of 1879 was a good year but in most others, the Duchy of Cornwall did not get either its annual rent for the land (£10) nor its royalty on sales: an early casualty of global warming perhaps.

Secondary effects

The revolutions of the nineteenth century were so far-reaching that they exerted effects well beyond their immediate impacts. In the Pennines, for example, most upland villages had cottage-based textile industries by the eighteenth and early nineteenth centuries. The first effect of true industrialisation was the management of water since the overshot water-wheel was very efficient if well supplied; this was from about 1795 onwards. Thereafter, the steam engine in towns provided employment which attracted whole families out of the countryside, aided by for example the depression of 1828–31. The rate of loss might well have been higher had not better communications come along in the form of the railways which made possible the cheap extraction and sale of minerals. In environmental terms, railways mostly cut gouges into moorland hillsides (Rosedale, Peak District, Dartmoor) and occasionally ran on embankments (North Pennines) and there is a high viaduct at Ribblehead in the Yorkshire Dales (Plate 4.6); at Fen Bogs on the North York Moors the railway is floated on a bog surface. At all locations the railways increased the risk of fire on heather moor and in valley woodlands with a high coniferous presence.

The wealth created by industrialisation led to recreational use by the rich, as detailed below. In a few cases, the lure of the open terrain (and presumably the cheap land values) led wealthy industrialists to create an estate for themselves out of moorland. The outstanding example is Cragside in Northumberland, near Rothbury. Until 1863, the upper reaches of the Coquet were mostly moor, with a few enclosures on the valley sides. The armaments tycoon William Armstrong built a modest hunting lodge above the Debdon Burn in 1864–6, with a 20 acre (8 ha) garden. Fired with enthusiasm, like many of his guns, he expanded this during 1863–83 to about 1,700 acres (680 ha) and planted about seven million coniferous trees. Lakes were also

PLATE 4.6 *The Ribblehead viaduct on the Settle–Carlisle line. Promulgated as a main line to Scotland, such lines nevertheless had local functions in moorland areas in exporting livestock and minerals and bringing in coal. Construction was often an environmental influence in its own right, with navvies' camps as well as materials for building the line and the debris from tunnels and cuttings.*

created and they powered a domestic electricity plant which made the house the first to be lighted by the newly tamed energy source.

New attitudes and new uses

Uplands were usually a place of recreation for the rich, and hunting had a long pedigree as the chief form of this activity. But improvement, industry and loss of woodland habitat militated against the survival of red deer. On Dartmoor, for example, the deer had gone by 1780 in spite of the likely ability of a unitary landowner such as the Duchy of Cornwall to protect them. Fox hunting on foot was still maintained, partly on the grounds of predator control, but it never gained the social cachet of its lowland equivalent. Dartmoor never supplied good shooting and though some red grouse were introduced in the nineteenth century, the moor was too wet to manage for their perpetuation in large numbers in the way that became possible on the drier moors of eastern England. Walking on the moor for pleasure however is recorded as early as 1788.

Industry into sport

Without doubt the new phenomenon in upland pleasure was grouse shooting. As Turner's picture of 1816 from Beamsley Beacon (near Wakefield in

West Yorkshire) shows, the red grouse (*Lagopus lagopus*[45]) was shot by moorland owners (and probably by many others as well) on a more or less casual basis by walking over heather moor with dogs and putting up the birds (Plate 4.7). Flying low, there was a chance that a shot from behind would bring down a bird, to be retrieved by the dogs. In the 1840s three or four guns with two or three dogs each might bag six brace per gun per day.[46] Since the density of birds was governed by the patchiness of the heather, there was likely to be ample time for a muzzle-loader to be replenished before another bird took flight. A number of intersecting social and technological changes after about 1850 changed this practice and in turn introduced an almost entirely new ecology to the drier moorlands of the eastern side of England, and also of Scotland. The western side of the Pennines variably adopted the new sporting methods and the western marches and eastern Wales contained a few suitable moors. Wet uplands like Dartmoor never became centres of grouse shooting and moor management, since the staple food of the grouse is heather, plus insects for the chicks in the first two weeks of their life. Hence, the best places in England to see grouse moor management at work today are the North York Moors and the eastern flanks of the North Pennines in North Yorkshire and Co Durham.[47]

PLATE 4.7 *J. M. W. Turner's watercolour: 'Grouse Shooting on Beamsley Beacon'. The picture (1816) seems to pre-date intensive grouse moor management by fire: walking up the birds with dogs is more likely and there are no signs of heather having been recently burned. The prominent nose of the central figure suggests that this is the painter himself. As often with Turner, the background topography is vertically exaggerated. Source: By kind permission of the Trustees of the Wallace Collection.*

Important changes in grouse shooting came about in the wake of developments in technology. The indirectly structural was the rapid expansion of the
railway net and, especially, the generation of express trains from the capital to
the moorland areas of the north of England and to Scotland. The City gentleman, or would-be gentleman, could travel overnight to a shoot, spend the day
in the field and be back on the next night's sleeper. Weekends were even more
feasible, though of course there was no sport on Sundays and Christmas Day.
Men who were making serious money in the flush of industrial growth but
who were not from landed families could take the initial steps to establish their
credibility by appearing at the right country houses and displaying appropriate skills, even perhaps getting a tweed-clad daughter's picture into *The Tatler*
and improving her chances of an upward marriage. In this they were aided in
a more direct technological fashion by the advent of the breech-loading
shotgun. During the 1840s and 1850s, French gun-makers had produced
sporting guns which broke in two to allow the breech end loading of an integral cartridge which was fired by a hammer striking a central pin. One such
gun was shown at the Great Exhibition in 1851 and soon after, British gun-
smiths rapidly ironed out its problems and sold it in large numbers. Whereas
before 1858 there were fewer than 200 of these British-made guns in Britain,
there were several thousand by the early 1860s.[48]

The new killing efficiency was wasted on a sparse bird population. It had,
though, been noticed that where sheep had grazed or there had been small
burns,* then the shorter heather supported a higher density of grouse.
Purposeful management to that end was started about 1860, with one legend
having it discovered by accident on the Duke of Portland's estate. Once this
practice appeared to be 'sustainable' then it spread rapidly in the 1860s,
though for some reason there was no burning in 1872 and 1873, with reduced
bags as a result. As well as burning, the land was manipulated by digging drains
into the wettest parts of the moors, so that heather was encouraged at the
expense of cotton grass, bog-moss and purple moor grass. Tracks for ponies
and for carts were built and small 'houses', usually a single-story, single-
roomed stone construction, appeared in remote moorlands, ready for lunch-
time shelter. The actual shooting was done from small waist-high structures
now called butts, made variously of stone and turf or of wood; in each stood
the gun with a loader or even two, so as to keep up the fire rate.[49] Along with
intensive moor management by burning and draining, large bags could be
obtained in good years: in England, the Forest of Bowland (Lancashire) moors
at Littledale and Abbeystead (1,700 acres, 680 ha) yielded 2,929 birds to eight
guns on the opening day of the 1915 season (August 12th was the first day,
December 10th the last) and the first three consecutive days of that year saw
the killing of 5,951 birds; the whole season's bag was 17,078. Broomhead Moor
in Yorkshire (4,000 acres, 1,619 ha) had given up 2,843 birds in one day in

* Small fired areas, not northern streams

1913.[50] In the Yorkshire Dales, Lord de Grey shot 47,468 grouse between 1867 and 1895, along with even higher numbers of pheasant and partridge.[51]

Since the early days of moorland management, immense fluctuations have been a feature of grouse populations. Heather beetle has infested the birds' main food and sheep ticks have killed young birds but the main problem has been the endemic strongyle worm's propensity to erupt into epidemics. Strongylosis is caused by a parasitic threadworm (*Trichostrongylus tenuis*) which prevents the bird absorbing nutrients through its caecum, and mortality is high. Many other factors were considered from early days and the level of predation was always foremost, resulting in some very high mortalities of raptors at the hands of gamekeepers.[52] Such was the power of the landowners that a government enquiry was set up into grouse. This reported in 1911, and its main conclusion was that better burning was required.[53]

The popularity of grouse shooting has been such that it persists as a feature of the economy of some upland regions, with about half a million being shot each year in England and Scotland, generating a gross income of the order of £10 million in the 1980s.[54] Unlike lead mining, therefore, it will be reconsidered below in accounts of the twentieth century.

The red grouse was not the only sporting quarry on the uplands, though no other species provoked such management efforts. Blackcock (*Tetrao tetrix*) was also a popular shoot. This is a bird of the moorland edge, especially where there was rough and sparsely wooded land with bushes, gorse, and rushy ground at, for example, the edges of plantations. It breeds best where tall ground vegetation is higher than 40 cm.[55] The birds might breed in moorlands but they liked to feed in places like stubble fields in early winter just as the grouse season was ending. In about 1869, a shoot on Cannock Chase took 252 birds in one day. If a moorland contained a number of wet areas with tall grasses and sedges, then these might attract snipe (*Gallinago gallinago*) to feed; landowners who drained moors thoroughly were unlikely to retain any upland snipe populations. Some extra diversity was sought by moor managers in the nineteenth century when they noticed that grouse moor management in Scotland encouraged high densities of mountain hare (*Lepidus timidus*) and so brought it southwards into the Peak District, the Borders and North Wales. Numbers have never been high and density is related to the amount of heather, so that populations are dependent upon sheep numbers; they are also badly affected by hard winters.

New appraisals

These fragments of social history direct us to the notion that the role of the uplands in the national consciousness was changing. Until the nineteenth century, those travellers who mentioned the uplands largely did so in a disparaging fashion. Like the Alps that stood between the rich and the promised pleasures of Rome and Venice, they had to be crossed but as quickly as possible, and a warm coat of invective was often a help. During that century, however, a rapid start was made to the changes in evaluation of the uplands that would eventually be crowned with the designation of many of them in England and Wales as National Parks, following the legislation of 1949. This story is carried on in two places in this book: in the material on Conservation (Chapter 5) and in the discussion of Representation (Chapter 7).

THE TWENTIETH CENTURY: 1914–PRESENT

Here we examine the dominant processes of the twentieth century, bringing the story up to date, where 'date' is largely the end of the year 2001, though some additions are found in the Conclusion. But the major themes of the reclamation of moorland, the amount of deciduous woodland, afforestation with conifers, and the extraction of water and minerals are continued here and even given some more detailed treatment in Chapter 6 which looks at a few regions in more detail. In addition, some features unique to the twentieth century are discussed, such as the presence of a nuclear power station and the introduction of windfarms.

Reclamation and reversion

The 'improvement' of moorland can take a number of forms: liming and re-seeding of pasture, conversion to tree cover and maintenance of patches of different-age heather are all examples. From time to time, however, the radical step of enclosing and converting moorland to ploughed arable becomes feasible economically. (It has always been attainable technologically, as accounts of prehistory show.) Here an account is given of trends in the reclamation of moorland for crop production (as evidenced by plough marks in the soil detectable on aerial photographs[56]) in the nineteenth century and twentieth century. Though the story encompasses much of the nineteenth century, it is presented here as a single narrative, not least because the major piece of research on the topic covers the period from the 1830s until the mid-1970s or as recently as 1980.[57]

The background is the general state of the upland economy, starting in the early nineteenth century, when it was flourishing. Land reclamation was by the prevalent political atmosphere in favour of enclosure, by cheap labour for fencing, and the cheapness and ubiquity of lime now that coal was inexpensive, leading to the building of many small lime-kilns along the valley sides of many uplands. Improved breeds of hill sheep with a ready market among lowland fatteners and steady profits from many mining enterprises brought people into the upland areas to work and live. To these was added the great surge in agricultural prices caused by the wars with France, with the island nation needing to produce more of its own cereals. Thus to take in land from the wild and grow corn on it was not only profitable but a patriotic act. The high tide mark of improvement of pasture and conversion to arable was reached about the middle of the nineteenth century. This nexus of favourable conditions declined through the century, with a particular jolt to farming after Waterloo (1815) but with declining profits as free trade became prevalent, even though the demand for food was increasing. Further, many mining enterprises suffered from the cheapness of overseas imports: lead and tin are examples. Hence, the end of the nineteenth century was a time of depression in the hills though perhaps less than in the lowlands, but between 1880 and 1901, for example, many upland parishes lost one-fifth of their population to mining in the Empire or simply to the cities of the plain. In the North Pennines, the degree of depression depended more on the state of mining than on agriculture since animal products found a ready sale in the industrialising lowlands to the east. This depression and desertion was not universal, for many extractive enterprises still flourished; the granite quarries of Bodmin and Dartmoor sent stone to London and many other cities, and Dartmoor uniquely had a large prison in its centre to provide employment and to create demand for food products.

Yet population continued to decline through the period to 1945. Economies were underpinned by state enterprises such as forestry and military training, some tourism, and unreliable subsidies from central government; 'unreliable' in the sense that they were normally regarded as responses to a particular 'crisis' rather than a standing benefaction to a way of life otherwise likely to vanish. Only after WWII was it plain that subsidies to agriculture were to be permanent so that farmers and estates could plan their output to maximise income from such sources, provided they were willing to fill in the forms. But the presence of large numbers of different subsidies to hill farms after about 1950 encouraged some land managers to consider adding to their arable acreage, especially since new strains of barley were available which grew well in upland areas and for which there was a good market as animal feed. The visible effects of successful reclamation schemes brought about a rapid reaction from the conservation and scenic value interests and so the recent story is that of wanting to keep moorland as open land but to try and alleviate the falling prices for its animal products. The introduction of different kinds of subsidies (loosely termed 'agri-environmental') from the EU has worked against more reclamation but little in favour of reversion, either. Population decline has halted but the new inhabitants are not hill farmers nor estate workers in forestry and farming.[58]

One element of diversity may be added to these general trends: that of land ownership pattern. The big estates may have different priorities from a tenant or an individual owner-occupier; the former may have access to more specialised information and services and have different outcomes for land use, such as a higher priority for sports such as grouse shooting. The latter, on the other hand, has always been quick to respond to opportunities such as price-related incentives to reclaim during the early nineteenth century. They may interact, as when sporting profits are used to keep down rents charged to agricultural tenants. The great mining estates such as those of the North Pennines in the ninetcenth century provided a totally different example of maintaining reclaimed land well above any 'normal' economic limit.

Over 200 years, the results can be codified as Figure 4.8. This envisages a central core of unimproved moorland which might have been cultivated in prehistoric times or, marginally, in the medieval period but which in 1800 had not been improved for at least 500 years. During the years 1800–2000, some of this core may have been converted to farmland and stayed that way; some of that land may subsequently have reverted to moorland, yet a small proportion may have been taken in again. This may have happened (exceptionally, it must be stressed) up to four times. Some of the reverted land may stay reverted. A parallel process can be seen for afforestation, though more than one cycle is unusual, given the slow growth of trees. The crucial piece of landscape is the wall or fence which separates the unimproved moorland rough grazing from the farmland of improved and managed grassland or ploughed arable, which is the main focus in this section.

Results on the ground can be extracted from the detailed research by Parry and his team. In the Yorkshire Dales, for example, some 60.1 per cent of the area now within the National Park boundary was rough pasture in 1848 (106,497 ha) and 69.2 per cent in 1980 (122,574 ha). These outer dates conceal periods of different trends. The periods of greatest reclamation werc 1896–1904 (+0.8 per cent) and 1971–80 (+3.9 per cent). By contrast, reversion was greatest in 1848–96 (+5.9 per cent) and 1920–50 (+8.5 per cent),

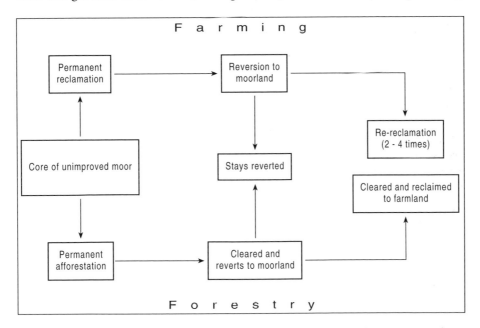

FIGURE 4.8 *A simplified scheme of moorland land use changes in the last 200 years. The key element is the core of unimproved moor, which can then stay unchanged or can be converted in various ways and then may revert. Iteration up to four times has been traced.*

even during WWII it will be noted. In the Brecon Beacons, to take one example, the year 1885 saw some 54.7 per cent of today's National Park area as rough pasture. By 1975, the figure was 48.6 per cent. In between, the moorland area increased most in 1909–48 (+2.4 per cent) and lost most to improvement in 1948–64 (–8.5 per cent). Farmland and woodland gained an impressive 10.3 per cent in 1948–64.[59] The spatiality of the changes in the Yorkshire Dales can be seen in Figure 4.9, which covers the whole period of the data, from 1848–1980. The main contrast is between the large blocks of afforestation, which relate to the acquisition of extensive blocks of land by the Forestry Commission, and the generally smaller piecemeal addition to farmland, tacked on as a fringe to the unreclaimed core. Some of the bigger tracts are the work of large estates rather than individual farmers. The equivalent map for the Brecon Beacons (Fig. 4.10) proclaims the dominant role of afforestation in this upland, with significant blocks of new farmland only in the far west. This contrasts with Dartmoor, where afforestation is less prevalent and where the reclaimed area between 1953 and 1971 is noticeably denser on the drier and lower east side as well as the lower tongue of land extending westwards along the Dart. The exemption of the great central upland blocks is noticeable: the rough pasture zones declined by only 2.8 per cent.

The diagram of land use pathways (Fig. 4.8) allows for shifts in reclamation and reversion to happen more than once and in the Yorkshire Dales data can be assembled to show, for example, that the rough pasture to improved land transfer happened only once to 20,953 ha of land, twice to 6,358 ha, three times to 1,533 ha, four times to 420 ha, five times to 87 ha and, amazingly, six times to 18 ha, a change every 22 years. In the Brecon Beacons, four times is a maximum, happening to only 60 ha.[60] These areas can be identified but of greater interest are those stretches of land which have undergone an enduring reclamation. Over the whole 132 years in the Yorkshire Dales, 6,194 ha of land have been improved and stayed that way; 38.5 per cent of it dated from 1971–80, so that is perhaps a premature categorisation. The periodisation of enduring reclamation in the Brecon Beacons which can be seen in Figure 4.10 seems to show no particular pattern except that any area subject to reclamation may contribute its share of permanency as well as more ephemeral changes. In the Welsh Marches it was the practice always to leave rich meadow and improved pasture alone, while ploughing any newly enclosed moorland fringe in response to high prices or wartime directives: *rhoi croen newydd ar yr hen ffrid* (giving the old *ffrid* a new skin).[61]

One finding of such detailed analyses of change is to distinguish recent reclamation as being either 'primary' – the improvement of land 'never' before improved – or 'secondary', on land which had reverted to rough pasture. In the Yorkshire Dales, only 12 per cent of post-1971 conversion for farmland was primary reclamation and the rest was the secondary re-reclamation of land once improved but allowed to revert. In the Brecon Beacons, about one-third of the farmland created since 1948 has been at the expense of the

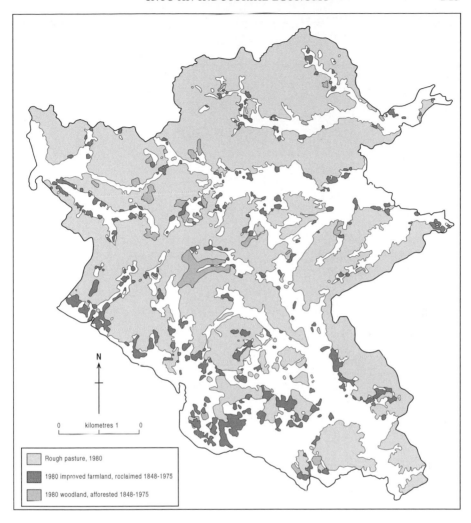

N

0   kilometres 1   0

Rough pasture, 1980

1980 improved farmland, reclaimed 1848-1975

1980 woodland, afforested 1848-1975

FIGURE 4.9 *The conversion of moorland ('rough pasture') in the Yorkshire Dales National Park 1948–75. Woodland (in effect mostly coniferous afforestation) has relatively little significance and improved farmland reclamation is strongest on the southern flanks. As might be expected, it clings to the edge of the higher ground. Land reclaimed for farming under influence of e.g., lead mining has reverted since 1900 and so is not picked up on this representation.*
Source: M. L. Parry, A. Bruce and C. E. Harkness, Changes in the Extent of Moorland and Roughland in the Yorkshire Dales National Park, *Department of Geography, University of Birmingham, Surveys of Moorland and Roughland Change no. 9, 1984, Fig. 4.3 at p. 31.*

moorland core (primary reclamation); about one-quarter of the reclamation for farming that took place in 1964–75 was on land reclaimed in the last decades of the nineteenth century but which had reverted to rough pasture around the 1930s. In 1950–80, however, most of the enduring conversion of rough pasture in the uplands was to forest rather than farmland. In six upland

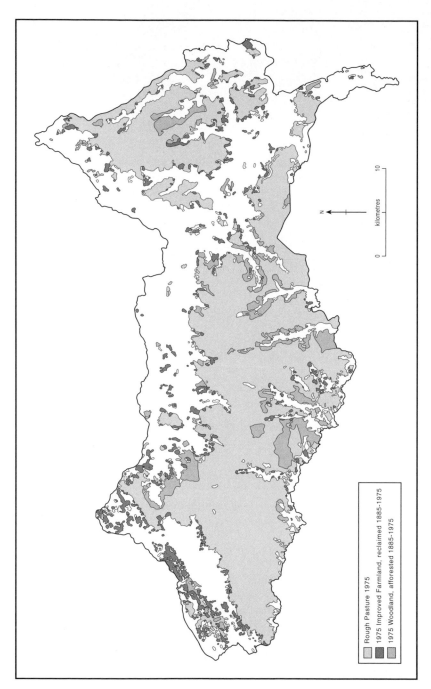

Rough Pasture 1975

1975 Improved Farmland, reclaimed 1885-1975

1975 Woodland, afforested 1885-1975

N

0   kilometres   10

FIGURE 4.10 *The distribution of enduring reclamation in the Brecon Beacons – land reclaimed since 1885 which has stayed that way, with shading indicating the period of enclosure, together with the unimproved core of 'rough pasture'. The dominance of 1965–75 with its favourable subsidy structure seems considerable but the enduring reclamation includes afforestation.*
Source: M. L. Parry, A. Bruce and C. E. Harkness, Changes in the Extent of Moorland and Roughland in the Brecon Beacons National Park, *Department of Geography, University of Birmingham, Surveys of Moorland and Roughland Change no 7, 1982, Fig. 6.2 at p. 37.*

areas, 28 per cent of the conversion was to farmland, whereas 70 per cent was to afforestation schemes. In the same period, regeneration of moorland was 89 per cent from reverted farmland and only 11 per cent from enduring deforestation. The context is of only 15 per cent of the study area going back to moorland in the twenty-year period.[62] The net outcome for these upland areas in the period 1950–80 shows that the area of moorland has diminished by an average of 16 per cent, with the highest proportion in Mid-Wales (−28 per cent) and the lowest on Dartmoor (−3 per cent). The absolute amounts follow the same pattern: a change of cover of 28,400 ha in Mid-Wales and only 1,800 ha on Dartmoor (Table 4.4). The major recipient of change in all cases is forestry,[63] thus adding a temporal stability to the environmental patterns.

TABLE 4.4 Changes in moorland area 1950–80

Area	Actual period	Total area (ha)	'Rough pasture' 1950 (ha)	Net change in 'rough pasture' 1950–1980 (ha)	Percentage change
Mid-Wales uplands	1948–83	130,800	100,900	−28,400	−28
North York Moors NP	1950–79	144,000	68,000	−17,000	−25
Northumberland NP	1952–76	10,600	83,600	−12,000	−14
Brecon Beacons NP	1948–75	135,400	74,100	−8,300	−11
Dartmoor NP	1958–79	95,600	51,000	−1,800	−3

Source: A. Woods, *Upland Landscape Change: A review of statistics*, Cheltenham: Countryside Commission CCP161 1984 Table 10 (Original data is from Parry's work)

The environmental effects of reclamation are well documented but those of reverted land less so. Work on Anglezarke Moor in upland Lancashire was able to use land whose reversion dates were precisely known since Liverpool Corporation acquired the land for a water impoundment in 1847. The findings showed that recently abandoned land had a high variety of plant species but that this declined as the moor took over. Land which reverted about the time of the Liverpool takeover had a typical pH of 4.28 in the 1980s, whereas land that had reverted in the 1927–59 period was at a pH of 5.03. The higher the input of organic matter, the lower the pH, the more acid the soils. In general, the mean pH of the soils corresponds to the length of time since abandonment.[64] The main noticeable changes with reversion are the presence of increased densities of rushes (*Juncus* spp) if the ground is wet; where drainage is free then bracken (*Pteridium aquilinum*) is a colonist and will usually form dense stands.

One factor in the conversion of moorland to other uses is land ownership, where a big estate or landowner is prepared to adopt patterns which smaller owners or tenants would find risky, or the opposite case where an estate is highly conservative in its attitudes to change. This was quite noticeable in the nineteenth century when in the North Pennines there was a dual economy of mining and farming. Much of the mining area would probably have remained unenclosed except for the land allocated to mining families for their own support. The amount generally amounted only to a smallholding but the settlements were scattered and so small patches of land were reclaimed at altitudes otherwise uncultivated: one farmstead in Teesdale was at 610 m ASL. The rents charged were also sky-high. Such policies were pursued by, among others the London Lead Company in upper Weardale and around Nenthead but the Raby Estate in Teesdale stayed static and agricultural in the period to 1850, developing neither the mines nor the reclamation potential to any great extent. In general, though, farming seems to have been under-capitalised, so that reclamation costs were hard to finance. On the Stanhope Estate belonging to the Bishop of Durham, the timber on leasehold lands was reserved to the Bishop, so lessees had no interest in planting on their lands.

In considering these social and economic factors, the role of the physical environment must not be forgotten. Maps of the land use in Lunedale in 1860 show the progressively harder grip of altitude on the potential for all kinds of improvement of any kind of land (Fig. 4.11). The usual consequence of enclosure of moorland was to provide controlled grazing on pastures (enclosed and thus to some extent managed) on the hillsides, and common grazing on the moorland, whose vegetation was not managed but was usually enclosed even if the common rights were not extinguished. Some perspective on recent controversies over reclamation is given by the decision of the then Bishop of Durham to enclose 8,000 ha in Weardale in 1799, largely in order to bring land into cultivation, which would increase its value by (it was estimated) fourteen times and the Bishop's income considerably, via increased rents. In the end, it was agreed that 5,337 ha were unimprovable and the grazing was reallocated to tenants.[65]

Although afforestation was the main beneficiary, the focus on moorland conversion after about 1960 was on agricultural intaking. The debate was thrown into sharp relief by a controversy over reclamation on Exmoor in the 1970s. The national picture was of an intake rate of about 5,000 ha/yr which resulted from an upsurge in the 1950s that had accelerated in the 1960s. During those two decades more land was converted in the Yorkshire Dales than on Exmoor: the controversy arose largely because the post-1949 area of moorland was so small on Exmoor. Although the designation of Exmoor as a

Figure 4.11 (opposite) *Two maps from the Lunedale area of Upper Teesdale in the North Pennines. These show the increasing grip of the physical environment on land use and values in 1860. The upper map is from the high part of the valley (358–485 m ASL) and shows a patchy reclamation of pasture and meadow of predictably low values; the extreme south-west fragment is probably tied to the existence of the mine. The lower map (240–352 m ASL) is of greater opportunities for the development of meadow and better pasture but it also shows the existence of mixed cropping, which has disappeared from this area but which in the nineteenth century was part of the local economy.*
Source: O. Wilson, Landownership and Rural Development in Theory and Practice: Case Studies from the North Pennines in the 19th and 20th centuries. *PhD thesis, University of Durham, 1990, Figs 4.10 and 4.8 at pp. 169 and 167.*

(a)

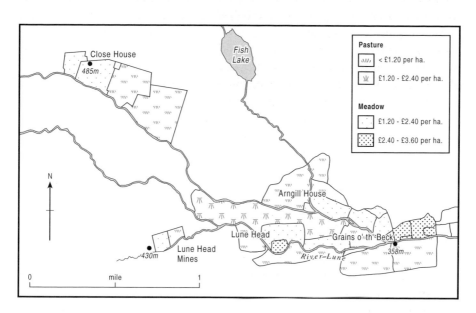

(b)

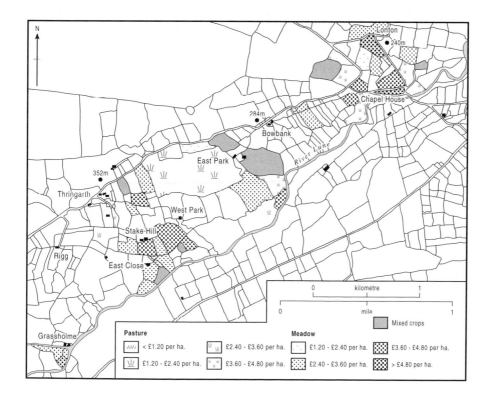

National Park in 1954 was based on its upland-moorland landscape character, any reclamation that took place before 1968 was (a) aided by 50 per cent
grants and **headage** payments from MAFF, and (b) outside the control of the
National Park Authority (NPA), who did not even have to be notified. The
NPA had failed to persuade County Councils to make compulsory purchase
of critical areas of moorland and the one possibility of control using national
legislation was never confirmed by Ministers in this period.[66]

The Countryside Act 1968 empowered ministers to make orders for specific
areas requiring farmers to give six months' notice to the NPA of their intention to convert moor or heath to agricultural land. The Exmoor NPA prepared
a map of areas which it saw as critical and prepared to make blanket orders
under Section 14 of the 1968 Act. But the CLA and the NFU combined to persuade the NPA to enter into a voluntary agreement not to make such orders
provided that its members would notify the NPA of any reclamation proposals within the critical areas. This failed to work in 1976 when two management agreements were proposed which involved reclamation within the
critical areas: one at the Glenthorne Estate (50 ha of moorland, over half the
estate's moorland area) and one at Stowey Allotment, with 120 ha of moor.
There were complications: the Glenthorne Estate belonged to the NPA's Vice-
Chairman, the NPA could not afford the compensation payments, which were
to be indexed against inflation, even if figures could be agreed. Further, the
NPA's dilatoriness in tackling the issue forced the Countryside Commission to
complain formally to the Minister. The Minister for the Environment and his
colleague at MAFF set up an enquiry under Lord Porchester, which reported
in 1977.[67] Porchester recommended a proper notification and approvals procedure and a new map of critical areas. But all was still voluntary, compensation was available from central government only because Exmoor was
regarded as a special case, and MAFF was still paying grants for reclamation.
In parallel, however, the NPA bought 21,938 ha of moor and 308 ha of pasture
between 1963 and 1982, to secure their use.

After 1980, new regulations were introduced on management agreements,
given statutory force by the Wildlife and Countryside Act 1981 (WCA).
Applicants for farm grants were required to notify the NPA, and an elaborate
procedure of response, mediation, appeal and compensation has to be followed.[68] All grant was put at risk if any reclamation was started before the
application was made. After 1989, grant aid for agricultural improvement
was effectively withdrawn, though payments to keep animal numbers high
(especially headage payments for sheep, encouraged by EU Directives)
remain. But the balance has swung decisively in favour of the NPA's ability to
control the outcome of attempts at reclamation. On Exmoor, this has been
estimated at 70 per cent of cases, having been about 15 per cent of cases before
1977. But the withdrawal of grant aid means that farmers willing to embark
on reclamation without grants from MAFF are again subject only to voluntary procedures for notification: their attitude to the NPA is critical.[69]

A retrospective view suggests that there have been two great surges in moorland reclamation, occurring around the period of the Napoleonic Wars of the early nineteenth century and after WWII, especially between 1950 and 1970. The site of most reclamation in the more recent years has been that of land previously reclaimed and reverted and hence at the edges of moorland cores. That pattern can always be skewed, however, by the action of a large estate. Moving nearer the present, and given the end-of-century income levels from hill farming, it might be thought that reclamation for agricultural production is now an issue of the past. Major land use changes might come from woodland which colonised abandoned land, perhaps, or which was planted as a different land use to diversify upland incomes. Yet there are, as so often, unknowns. One is the extent to which global warming, if the models turn out to be right, makes it economic to extend more intensive agricultural production up the hill. The economic context is very complex and a simple warming leads to more cultivation equation is likely to be facile. But warmer uplands with adequate rainfall might be attractive for some crops. A second unknown is the extent to which genetically modified organisms could be tailored to the less favourable conditions in the uplands. That this will be technically feasible is not much in doubt but again whether any company will think it worthwhile is another question. A current assessment of reclamation and reversion would thus emphasise the obvious decrease of wild plants and animals when land is reclaimed, along with soil and nutrient losses; but reclamation has always involved inputs of nutrients as well. In the twentieth century, chemical fertilisers have been added plentifully at times, as have biocides of various kinds. Reversion, on the other hand, involves soil changes of a relatively slow kind, as horizonation begins once more. Colonisation by wild plants may be quite fast, and the disturbed soils may be favourable sites for species such as gorse and bracken, while poorly drained grassland soon shows a good crop of rushes of the genus *Juncus*. Above all, the look of the landscape is most noticed, as the wild character is greater or lesser in area. This, together with any changes in ease of access, forms the crystallising-point of public attitudes towards reversion and reclamation in the twentieth century.

There seems a good case for thinking that the Exmoor controversy created a climate of opinion about moorland reclamation (atypical though that upland was) and now there would be added concern if any reclamation were to be of the primary type and thus cutting into long-established root mats and undisturbed soil horizons. Any conservation measures, such as the compensation of farmers for not ploughing moorland or for being directed by Ministers not to do so, could most effectively (from both environmental and fiscal viewpoints) be directed at the moorland core rather than its 'tidal' fringe.[70] If anywhere in the uplands a causal connection between **gripping**, sheep densities and downstream flooding can be established then the case for land use reform is strengthened.

Forestry in the uplands

Forestry in the uplands is very largely the story of the Forestry Commission and its ambitions. Until the 1980s the care of deciduous remnants of ancient woodlands was seen by the Commission either as irrelevant to their mission or else the business of the Nature Conservancy Council and their designatory powers. The Forestry Commission was interested in heath and moor above all else as a reserve of land for planting.[71]

In 1900 some 90 per cent of all timber and forest products were imported.[72] Between 1850 and 1914, six Royal Commissions, Select Committees and Departmental Committees had sat and reported. An Office of Woods, Forests and Land Revenues had undertaken small scale afforestation on estates it had acquired at Hafod Fawr in Merioneth (1899) and Tintern in the Wye Valley (1900).[73] In 1913 Crown Forests and Woodlands numbered twenty, one of which was 'miscellaneous woods' and the interest in the uplands was small. An exception, perhaps, was the Isle of Man, where 800 acres (324 ha) of 'barren mountain land' were planted up in 1883–91. Overall, national control over a key resource was very light indeed: it was still possible for an estate to clear-fell its woodlands to pay off debts or for the planting policies to be determined by the head keeper in the interests of the pheasant shooting. In such a liberal climate, the intrusion of the State has to be impelled by strong forces indeed and it is a paradox that it was the Royal Commission on Coastal Erosion which successfully launched government forestry in the uplands. The Commission was set up in 1906 but in March 1908 the possibilities for afforestation were added to its remit. The purpose of the tree-planting was largely to provide rural employment. It led in due course to an Advisory Committee on Forestry which took up the idea of large-scale afforestation with some vigour. In 1913, it reported that the scope for afforestation in England was limited to the western and northern counties, and also that Wales was a key possibility. It wanted surveys to be done of possible land below the 1,500 ft (450 m) contour which were currently mountain and moor but used for grazing. The aspiration at the time seemed to be the establishment of not less than four woods in England and Wales of an aggregate area of 20,000–40,000 acres (8,000–16,000 ha). These surveys were duly carried out and some experimental plots established but no land was acquired for large-scale planting. A parallel process of some importance was the government's action in 1909 in setting up a Development Fund, an early form of subsidy to agriculture, forestry and fisheries. Although in early days only 11 per cent of its funds went on forestry (and only 600 acres, 240 ha, were afforested), the precedent of State intervention and financial involvement was established.

Into a situation characterised by leisurely progress and protracted prose, the Great War erupted, and resulted in the felling of 0.5 million acres (202,000 ha) of woodland nationally. The pre-war suppliers of timber were no longer either reliable (like Russia) or required long sea journeys which took up shipping space that was needed for higher-value, lower bulk cargoes. The need to supply some of the many uses of wood in wartime from reliable (and hence renewable) home sources was easily apparent. What would now be called 'sustainable development' was called for. The key document is the report (written in 1917 and published in 1918) of the Forestry Sub-Committee of the Reconstruction Committee. This sub-committee (known as the Acland Committee after its Chairman, Lord Acland) considered the war-time experience of the devastation of private woodlands and the shortages experienced and moved towards targets for land to be afforested with the overall aim of getting a standing reserve of timber which would cover an emergency of three years' duration. (Acland also noted that the only part of the Empire with good timber reserves was Canada and Newfoundland and that representations ought to be made to the Dominion government to have these forests properly managed). The Forestry Act of 1919 set objectives of planting up 200,000 acres (80,000 ha) in ten years, and of having an estate of 1.7 million acres (0.69 million ha) in eighty years. Scotland was to play a key role in furnishing the land though in the early years, land was much cheaper to buy or lease in England and Wales since Scottish sheepwalk and deer forest prices held up. Acland thought that ideally the new forests should be near the manufacturing and coal-mining regions of the country, which suggests that England and Wales would always be key areas for new planting.

The ends having been willed, the means were the establishment of the Forestry Commission. It was established by the Forestry Act of 1919 and the first trees were planted on December 8th of that year in Eggesford in north Devon. The Commissioners were unusual in that they formulated policy and reported directly to Ministers: no officials from an established Ministry interfered with or diverted their policies. As for progress, the forest census of 1924, published in 1928, revealed that there were then 2.9 million acres (1.2 million ha), which grew to 3.4 million acres (1.4 million ha) in 1947–9. The initial planting programme, though ambitious, was not in time for WWII in most respects, when the nationalised forests comprised only about 6 per cent of the cleared and 'devastated' areas. Complete control of the felling of private woods was only brought about in 1939 but it was continued after WWII and is still in place, buttressed by grants and loans to woodland owners. As with the Great War, national stocktaking was a socio-political process (which we shall encounter in other fields as well) and there was a 1943 report on Postwar Forestry by the Commission. It suggested that a national stock of 5 million acres (2 million ha) should be established, of which 3 million acres (1.2 million ha) should come from the afforestation of 'bare land' by which it largely meant moorland and heath. The result was a total stock in 1947 of 3,448,000 acres (1.4 million ha) of which 52 per cent was productive forest, plus another 178,000 acres (72,000 ha) of small woods between 1–5 acres (0.5–2 ha) in size.[74] This supplied about 7 per cent of the national demand, with the Commission in about 1950 having about half of its total land holding under planted forest, in some 480 units. The 1990s holdings totalled 1.089

million ha of which 861,000 ha are forest and woodland, with a value of £1,394 million. The yield is about 4.5 million tonnes/yr of timber from a cut which is 76 per cent of the yearly increment. Since 75 per cent is coniferous plantation, and of that 51 per cent is Sitka spruce, the key contribution of the uplands can be gauged. In the year ending 31 March 2001, only 100 ha of new plantation was established, but 3,500 ha was restocked.[75]

The concept of dedicated woodlands, where owners would trade total control for grants and loans was also brought to the fore and became established in 1947; much national policy-making in the reconstruction period of the 1940s was directed at the private sector.[76] This was a reversal of the view in the 1920s that the contribution of the private sector could be written off. It is an important series of policies but with relatively little representation in the uplands.

The various documents all point in one direction: farmland was not to be planted and cheap land was to be bought or, preferably, leased. Farms might be acquired but only to provide smallholdings for forest workers. In order to fulfil the Commission's policies, the estate had to be assembled from marginal land in the lowlands (mostly likely to be heathland or poor grassland, as in Breckland and the Suffolk Sandlings), and the areas of moorland in England and Wales that were not too wet, steep or high to bear trees of the chosen species. The Acland report was very firm about the ways in which afforestation could improve the 'productiveness' and the population density of large areas '. . . [W]hich are now little better than waste.' Some of the areas were Common, with individuals and groups possessing common rights, and the public was accustomed to have access for recreation to some areas. The objections of either group were dismissed as 'ill-founded' by Acland and his successors.

The planting policy of the Commission had been largely dictated by the need to build up timber stocks quickly after the Great War and to keep a reserve of fast-growing trees. It followed right from the start that the native hardwoods were not of primary interest to it. Instead, fast-growing conifers were the mainstay of planting programmes. Using the expertise of private woodsmen gained in the second half of the nineteenth century and often trained in Germany, the North American species were much favoured and this practice was largely maintained. In 1954, 122 million trees were planted, and 109 million of these were conifers. The native Scots Pine accounted for 25 million but the rest were imported species: Sitka spruce (29 million), Norway spruce (12 million), Japanese larch (16 million), Lodgepole pine (10 million) and Corsican pine (8 million). Given, therefore, that only about 12 per cent of the planting was of native broadleaved species, the new forests were easily perceived as alien in their ecology, environmental impact and appearance in the landscape.

Like any plantation, the trees change the ecology of their immediate surroundings. In many nutrient-poor sites, the husbandry of trees is akin to that of lowland agriculture since the soils are ploughed, fertiliser added and weeding takes place. Ploughing up and down slopes increases the rate of runoff and accelerates erosion.[77] The trees however collect nutrients from cloud water even in the absence of rain so that in mid-Wales, one study found that the plantations gathered 41 per cent more sulphur and nitrogen from the atmosphere than nearby moorland.[78] The planting of Sitka spruce for example, in rows makes for a rapid loss of light on the forest floor, with a minimum at age 11–22 years. The acid soils and low ambient light inhibit soil microfauna and so little breakdown of litter occurs, giving a forest floor which is bare except for fallen material; this may constitute a fire hazard. Thinning of the forest may alleviate these problems and allow some ground vegetation to develop but in times of cost-cutting, the trees grow closely together during their phase of maximum annual increment (at 45–60 years) and on into felling at 60 years plus, unless a smaller size product is required or is acceptable.

There have inevitably been effects on the sediment yields and the water chemistry of afforested catchments. Reservoir operators initially favoured wooded land cover since it reduced the erosion of particulates but the later practice of digging open drains to prepare the soil for planting and to encourage tree growth enhanced the sediment yield, as has road construction and some felling practices. The fine matter has adverse effects on invertebrates and thence on fish; coarser fractions are critical for channel form and stability. In general, a forested catchment has bedload yields per unit area which are five times that of an unforested area, and the network of herringbone patterned drainage ditches is usually the critical factor. Whereas grassland might yield 1.2–2.5 cu m/sq km of gravel bedload and a ditched moorland 8.4–57.1 cu m/sq km, a ditched and afforested upland's contribution might be in the range 8.4–307.7 cu m/sq km. These studies in upland Wales have led to amended prescriptions for ditches. During felling the suspended sediment may rise from 24 tonnes/sq km/yr to 57.1 tonnes/sq km/yr.

In the nutrient-poor ecology of the uplands, the pathways of the chemical elements is always of interest. There is a large input of elements such as sodium and chlorine from rainfall but urban sources contribute the largest fraction of, for example, chromium, manganese, lead and antimony. The average pH of the rainwater is 4.6 and it deposits 9 kg/ha/yr of sulphur and 3.5 kg of nitrogen. The trace elements in rainfall vary less than in the stream water. The latter have seasonal variations and hydrological changes which are not due to variations in rainfall chemistry, showing the importance of the biota in mediating the flow of chemical elements. The chemistry of the streamflow is dominated however by the precipitation, especially by the elements sodium, magnesium and chlorine. The pH of streams normally has a mean of 4.5 but this may drop to 3.5 during periods of easterly winds with snow which collect pollutants; most soils have a low buffering capacity and so the streams reflect the increased acidity, which seems to have been ubiquitous in upland soils in recent years.[79] On the other hand, additional nitrogen and phosphorus inputs do not exert the same eutrophication effects as in the lowlands as it is too cool; forested areas have no greater levels of nitrogen and phosphorus than equivalent grasslands. If any blooms occur, then they are more likely to be caused by phosphorus levels than those of nitrogen.[80]

The upland conifer plantation has attracted a great body of enemies in the wider public. In the inter-war years, their growth caused considerable debate on their role in changing the scene for walkers in the hills, in bringing about some diversions of familiar hill walks during the establishment and thicket stages. Large units were seen as intrusive on account of their colour and especially their shape when the upper boundary of the planted area was ruler-straight, without regard for the contours of the land. Any proposal by the Commission to plant large blocks, especially in the English uplands, was therefore met with a determined campaign led by organisations like the CPRE and the Ramblers' Association.[81] On the other hand, as outdoor recreation became important in the post-war period, people in great numbers have visited the forests and used the parking, picnic sites, guided walks and nature trails. It is an interesting speculation as to whether it is really the visual aesthetics of the plantation that offend or its foreignness, centred on the feeling that conifer forests are dark, sinister phenomena from central Europe (especially Germany), which are invaders of hills supposedly more open to William Blake's 'countenance divine'.

The strength of the opposition was demonstrated in the Lake District in the 1930s, for example, when the Forestry Commission proposed to plant 740 acres (300 ha) of upper Eskdale. In the end, the Commission had to be compensated by public subscription (at the rate of £2 per acre) for not planting there and the land passed into the hands of the National Trust.[82] One outcome was a voluntary agreement between the Commission and the CPRE in 1936 that a central zone of 300 sq miles (777 km^2) in the Lake District would not be subject to planting by the Commission. The agreement, though, was in mountain rather than moorland and no such recognition of public interest was afforded to more prosaic upland scenery.

In the year 2000, the area of woodlands and forests in the UK was about twice that in 1900 (Plate 4.8). Even so, a cover of 9 per cent is one of the lowest in Europe, where 20–40 per cent is more common. In 1924, about 65 per cent of the tree cover was broadleaved; now the same proportion is coniferous. In spite of the economic rationale for the conifers, self-sufficiency for wood products is only about 8 per cent. Regional shifts in recent decades have again emphasised the importance of the uplands in afforestation schemes but upland areas are not involved in the grand schemes of afforestation which fall outside the new National Forests of the lowlands.[83]

The work of historical ecologists has improved our knowledge of all British woodlands and the types of woodland now present can be classified on an historical basis[84] as:

PLATE 4.8 *One of the Forestry Commission's largest acquisitions has been in the Kielder area of Northumberland. The afforested area also contains stands of old-growth woodland (see Plate 4.9), mire systems and heather moor. The Commission now aims to manage these with regard to biodiversity as well as recreation and wood production. The rotation methods of clear-felling and replanting will however still be used.*

1. 'Ancient' woodland where there has been continuous woodland cover (either deciduous or of Scots Pine in the far north of England (Plate 4.9)) since 1600. If the woodland pre-dates the initial fragmentation of woodland in a particular place then it is said to be 'primary'. A number of ancient woodlands have been placed under conservation management by English Nature or charities such as County Wildlife Trusts and the Woodland Trust. Since 1984, the Forestry Commission has also shown interest in its responsibility for them.

2. 'Recent' woodland grown up since 1600. If it has grown on previously unwooded land since the fragmentation, then it is also called 'secondary'. Small areas include coniferous shelter belts, deciduous copses, pheasant coverts and other amenity and sporting plantations, and experimental energy plantations of, for example, willow. Large forests include beech plantations managed for furniture wood in the Chilterns, and the Forestry Commission's estates which are mostly, though not entirely, coniferous.

The period after WWII has seen many of the earlier plantings by the Commission in the uplands come to maturity with the visual changes associated with clear-felling and replanting. All this has happened in an

PLATE 4.9 *'Ancient woodland' south of the Scottish border is associated with deciduous stands. In Kielder Forest there are some specimens of* Pinus sylvestris *which are true natives, although most* P. sylvestris *in England is a relatively recent incomer from the continent.*

atmosphere of heightened public awareness of conservation and amenity issues; one result has been both a greater degree of public use and appreciation of the Forestry Commission's lands and a stronger condemnation of any of their plantings seen as unsympathetic to their local landscapes. The largest areas of planting in Britain are in Scotland, with The Cheviots, Breckland and South Wales as the only really large areas elsewhere. Kielder, though, has one afforested block of over 1,000 km^2. Since the 1981 Forestry Act, about 75 percent of all plantings in the uplands have been by private enterprise.

The ecology of coniferous plantings is well known.[85] In broad terms, the previous environment (often moorland) is gradually replaced with a tree monoculture low in species variety but with a deep *mor* humus layer. In the early phases the bird populations are not reduced but once at the pole stage and thereafter, they are lacking in density and variety of species. Nevertheless, the rides may support populations of nightjars and provide good terrain for sparrowhawks. If there are enough prey species overall, then raptors such as the goshawk, merlin, tawny owl and peregrine falcon may increase in numbers, though the golden eagle, the raven and the hen harrier are adversely affected.[86] Mammals tend to benefit: increased numbers of roe, red and sika deer, pine marten, fox, polecat and wild cat have been observed. A summary of the ecologically negative aspects of afforestation would include:

1. Moorland may survive only in reserves in heavily planted areas such as South Wales. The association between open terrain and the visibility of archaeological remains such as field systems and settlements is lost as well as regional landscape character and well-loved views.
2. Some near-natural habitats have been destroyed: examples include undamaged blanket bog, sand-dunes, and raised bogs.
3. At least 66 higher plant species have been substantially reduced in Scotland, and over 50 species which are locally common only have declined nationally; among bird species, 22 have been displaced or reduced.
4. All invertebrates have been affected by the spraying of trees with organophosphate insecticides to combat pine beauty moth and other pests. However the cumulative area of the forests sprayed in the last fifty years is only 2 per cent of the plantation area.
5. Foxes and crows use the forests as a base to act as predators on moorland birds beyond the forest margin. Many such moorland species will not nest near large blocks of afforested terrain.

Much of the vitriol directed at the forests has however arisen from amenity and recreation groups concerned at the intrusion into open terrain and in particular at the diversion of footpaths and interruption to walking routes. Beyond that, the plantations alter the visual scale in the uplands and much of the feel of wildness is diminished by the fences, roads, planting lines and the other obvious evidence of human presence. The Commission has responded with detailed design criteria involving shape, character, visual force, scale, landscape diversity, and especially the nature of the forest edge. They aim to replace the ruler-straight with the diffuse and have often planted deciduous belts around large areas of conifers. Beyond such simple measures, it is possible to develop a co-evolutionary policy for the conifer forests which increases their value to wildlife and small-scale scenic diversity. The re-establishment of areas of native species' woodlands is one course of action. Another would be to establish cores of uncut conifers as 'old-growth' habitats at the same time as shortening other rotations. Thus there would be more open space in the forests together with a different habitat entirely.[87] The effects of afforestation on islands of mire and other wetlands within the conifer spreads is difficult to evaluate since the effects of the cessation of burning and grazing also change their ecology. Sensitive managment seems to perpetuate the existence of mires and one study suggested that only a zone of 1–3 metres' width round the edge was affected by the presence of lodgepole pine.[88]

The first implacable objection to the upland forests appears to come from one or two recreation and amenity groups and the writers of letters to *The Times*, but they have been reinforced by two trends. The first of these is the use of forestry as a tax haven for the very rich, especially when conservation values were threatened as in the Flow area of Caithness and Sutherland.[89] This has been much less of an issue in England and Wales than in Scotland. The second is where privatisation threatens to remove the compensating worth of recreational access to upland forests: the Commission was forced by the government during the 1980s to formulate a sell-off programme and by 1996 had disposed of 125,000 ha. Only 1,610 ha of that land was subject to access agreements that safeguard recreational use under the new owners. New forestry is likely to move downhill into land spatially marginal to ordinary lowland farming and the recreational potential of such areas seems high.[90] Relaxation of rules in 1985–6 meant that it was easier for agricultural land to move into forest use. Now, the case for continued expansion of forestry rests on an expansion of wood processing (with a strong demand by the industry for softwoods) to obviate some imports, since the UK is the world's second largest importer of wood after Japan, support for remote rural communities, outdoor recreation provision, the reclamation of degraded land and the benefits to wildlife. In support of the recreational role, there is the fact that the upland forests are about the only places in the country for car rally competitions. This is an extreme example of their overall popularity for outdoor recreation, part of which stems from the ability of the planted forests to hide a lot of cars.

With the maturation of the Forestry Commission's estate, the public had for the first time right of access to woods for their general enjoyment, not simply as part of a delimited common rights agreement. Many woods remain in private ownership, though subsidised by public funds without any reciprocal rights of access. It is however, the use of woodland for recreation which will

TABLE 4.5 Woodlands in national parks 1989*

Areas ×10³ ha	Brecon Beacons	Dartmoor	Exmoor	Northumberland	North York Moors	Peak District	Yorkshire Dales
Broadleaf	5.2	5.0	3.3	0.2	7.2	2.6	1.8
Conifer	12.0	3.6	3.2	19.8	28.7	5.0	4.6
All woods	17.2	8.6	6.5	20.0	35.9	7.6	6.4
Woods as per cent of of National Park	12.8	9.1	9.5	19.4	25.1	5.4	3.6

* Only the moorland parks are included and Pembrokeshire is also omitted; it has 1.9 per cent total woodland in its entire area.
Source: G. J. Mayhead, 'Some quantitative data for the woodlands in the National Parks of England and Wales', *Arboricultural Journal* **16**, 1992, 123–32, Table 1, pp. 124–5.

swing views away from the climate of opinion which produced a largely deforested country. If woods are useful for soaking up carbon, for example, then a highly public lead needs to be taken by the Forestry Commission, as indeed by all branches of government in such an all-embracing problem. Our global linkages are explicit in this case, though one calculation is that 1 million ha of new forest would only sequester 3 per cent of the UK's total emissions of carbon dioxide and that carbon storage by UK forests will increase for a few more years but thereafter decline unless another 40,000 ha are planted each year[91] In the uplands, blanket bogs are another great store of carbon and at present it seems as though burning is the major way in which carbon is lost; light grazing by sheep affects carbon storage very little.[92]

Broadleaved woods in the uplands

The values of deciduous woods, usually very small in area compared with the upland conifer forests (Table 4.5), lies in their contribution to biodiversity via species and habitat variety, their presence in the landscape as a 'typical' feature, their productive value as sporting estate, and (though less common) their output of underwood products. They can be maintained for these purposes or unmanaged woods can be 'restored' to a more purposeful set of conditions. They are mostly in private ownership so that the institutional and financial conditions have to be exactly right to encourage woods which, for instance, provide baseline conditions for scientific work as well as conservation areas and scenic ingredients.[93] The historical conditions of their evolution need to be remembered when their management is planned, for many of them are the result of the presence of grazing animals for most of their lives and if it is desired to keep them as they are, then some grazing is probably desirable. In the eighteenth and nineteenth centuries most upland woods were managed as coppice for bark and charcoal: this seems to have been likely in the case of the Dartmoor oak copses like Black Tor Copse, Wistman's Wood and Piles Copse.[94] The re-growth was fenced for 5–10 years but at the end of the nineteenth century, the walls and fences were no longer maintained so that sheep could get in. At that time, sheep had largely replaced cattle in these areas so that cattle have not recently been a significant influence on their ecology. Some effects of grazing may come at times of stress rather than gradually: sheep barked numerous oaks and rowans in Wistman's Wood in the winter of 1962–3, for example. Bark stripping in such circumstances may remove 1.5 kg per day per animal.

As well as sheep, many upland woods now have year-round populations of roe and sika deer and a few in North Wales have feral goats. If the total grazing pressure is light then the bryophytes are not normally eaten and the establishment of tree seedlings may be assisted. Similarly, light browsing has no effect on oak and some trampling by sheep may create sites in which tree seedlings can grow. Many bird species prefer sites with less than 30–40 per cent shrub cover: low ground and shrub cover for instance are good for wood warblers and tree pipits (*Phylloscopus sibilatrix* and *Anthus trivialis*). Low-moderate grazing helps them, as well as pied flycatchers and redstarts (*Ficedula hypoleuca* and *Phoenicurus phoenicurus*). Heavy grazing, by contrast, allows only small populations of small mammals and thus there is little food for predators. The absolute absence of grazing is probably beneficial for established trees but a dense sward of grasses or herbs usually prevents the next generation of seedlings from gaining a foothold.[95]

In 1995, White Papers calling for more deciduous woodlands in the uplands of England and Wales evoked studies of the potential for expansion in upland Wales and in the English National Parks. In Wales, a threefold expansion might concentrate on land currently covered with bracken (about 6 per cent of the upland area); in England the figure was lower, at around 4 per cent. The areas with the lowest woodland cover are, perhaps paradoxically, those where reliance on headage payments to maintain agricultural incomes are highest and those with strong designations for nature conservation. Thus terrain already well-wooded, like the English-Welsh border, might be a better place to seek the overall expansion of woodland. Nevertheless, the benefits from a greater presence of broad-leaved woodland in all uplands would be considerable in terms of biodiversity and scenery. The Forestry Commission has moved in this direction in e.g., northern England, starting with proposals to extend or link existing semi-natural woods (see p. 265).[96] One more possibility is the extension of short-rotation coppice as an energy source. Some 900,000 ha of upland in Wales currently used for sheep might support short-rotation coppice of willow to be used in chip form to generate electricity. Calculations suggest that willow farming could be more profitable than sheep if the subsidies on the animals were removed. The introduction of quantities of fertiliser onto upland catchments is one complication and extensive trials would presumably be necessary as would considerable cultural shifts.[97]

Water and mineral extraction

The 100-plus years of drowning upland valleys for upland water supplies has culminated in two northern developments: in Upper Teesdale and in the Kielder Forest area of Northumberland. The first was the scene of considerable controversy, the latter much less so though not entirely uncontested.

The growth of heavy industry on Teesside in the 1950s was underlain by heavy water users such British Steel and Imperial Chemical Industries (ICI). During 1963–4, ICI radically revised upwards its forecasts of water need. The Water Board (as it then was) then predicted a requirement of 455,000 m³ in 1970, much above its previous estimates. Their eventual proposal was for a regulating reservoir of 312 ha at Cow Green in Upper Teesdale. The site (mostly on Cronkley Fell and Widdybank Fell) was upstream from the scenically famous falls of the Tees at High Force and Cauldron Snout and set among blanket bog and other moorland vegetation, with a top level at 488.5 m ASL. The area was used for sheep, for grouse shoots and had had a barytes mine, closed only in 1959 (Fig. 4.12).

Little known nationally, however, this rather remote area of Upper Teesdale was the site of some botanical rarities, along with some uncommon

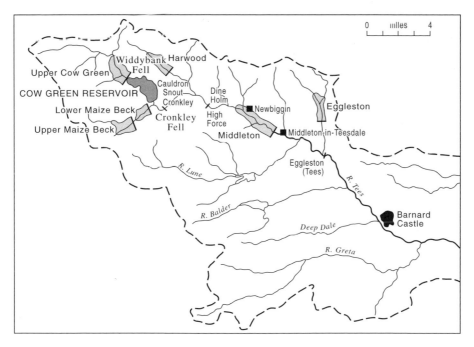

FIGURE 4.12 *The battle for Upper Teesdale in the 1960s. The Cow Green site was eventually selected but it is interesting to note how many others in the same valley system were also 'live' possibilities. We might also reflect on how the battle would have gone if it had been fought in the 1980s. (Thereafter, nobody could have made out the case for the water demand on Teesside.)*
Source: R. Gregory, The Price of Amenity, *London and Basingstoke: Macmillan, 1971, chapter 4, Fig. 3, facing p. 133. Reproduced with permission of Palgrave Macmillan.*

invertebrates.[98] Most of these were Arctic-alpine species at the southern
extremes of their range. The dwarf birch, *Betula nana*, grew sparsely in channels in the blanket bog but many of the species owe their presence to the combination of high nutrient status and lack of competition found on outcrops of
'sugar limestone'. These occur where the Carboniferous Limestone was baked
by contact with intrusive basalts from the Whin Sill. The result is a crumbly
'soil' which most plants find difficult to colonise and so supports the type of
specialised flora of open habitats more often found in periglacial conditions.
Nearby enclosed grassland provided a brilliant show of the spring gentian
(*Gentiana verna*) which attracted a certain number of visitors every year.[99]
The reservoir proposal was to result in the loss of 8 ha of sugar limestone,
which was 12 per cent of the outcrop on Widdybank Fell and 7 per cent of the
total quantity of sugar limestone outcrops in Upper Teesdale. It was thought
to be a possibility that the water body might ameliorate the local climate to the
point where some of the arctic-alpines would be out-competed by plants of
more temperate conditions. The impoundment would, it was argued by the
objectors, directly destroy an association of plant communities unique in
Britain. Indeed, they argued, it was the plant communities and their juxtapo-
sitions that were more important than the presence of rare plants and invertebrates, most of which were found in greater numbers in, for example,
Scotland or Scandinavia. They were, in the eyes of leading botanists, 'surviv-
ing relics of an ancient vegetation type . . . precarious survivors, easily
destroyed and quite irreplaceable.' Such fragments only persisted where,

> a peculiar composition of unstable base-rich soils and special climate
> have prevented the forest or the bog effectively closing over the area, and
> where, more recently, man's actions have spared them.[100]

These considerations were of little importance to the scheme's promoters
but they hit one legal obstacle. Some 300 acres (121 ha) of the site were
common land and for that reason a Private Bill was introduced into
Parliament in November 1965. This gave objectors the chance to mount a vigorous and highly public campaign. This ensured that the Bill was opposed all
through Parliament before finally receiving Royal Assent in March 1967; it
also scared water authorities off Teesdale when further developments were
contemplated.[101]
Construction damage was much more limited than had been feared and
apart from the actual inundation, the main visible environmental changes
have been in the erosion by waves of the southern side of the lake, followed by
the changes in detritus loads in the river, where some 4 million kg/yr of solids
settle out into the bottom. One effect of the latter has been a five-fold increase
in fishing. These recreationists form part of the growth of visitors generally,
from about 2000 per year to about 60–70,000 per year through the 1980s.[102]
Attempts at *ex situ* cultivation of the rarities proved difficult, though expert

supervision at Durham and Manchester was forthcoming. As time progresses further changes may become apparent, as workers have monitored changes in grassland composition on Widdybank Fell over twenty-five years and have concluded that changes in the vascular plant composition (though not that of mosses and lichens) can only be accounted for by shifts in the local climate. In turn, these are almost certainly due to the reservoir causing higher minimum temperatures at the ground surface (especially on very cold nights in autumn and early winter), which also diminishes the duration of snow cover.[103]

As far as is known, the increased number of visitors has not added to plant loss. The main sources of degradation of the plant communities have always been sheep (compared to whose activities, spring picking of gentians is pretty small) and gardeners; the reservoir has little affected these. As with the flower-rich hay meadows lower down the Dale, the main reason for the plants' survival has almost certainly been human activity in terms of grazing densities and regimes, coupled with the fortuitous outcropping of the basalt.[104] The dual economy of farming and mining in the nineteenth century probably provided good conditions for the spread of juniper scrub, which is another unusual feature of the vegetation of Upper Teesdale.[105] While therefore it is true to say that a unique association of communities was disrupted (though no extinctions have taken place), it is also certain that it has been a case of altering the human-made rather more than the 'natural', though that was not a popular thing to say in Durham in the late 1960s. In the end, the decision hinged on the economic assessment that industrial development on Teesside was more important than rare plant communities; the former could also be quantified whereas the latter could not.[106] There is some irony in the fact that the 1980s saw enough contraction of heavy industry on Teesside to make the 'next' reservoir, at Kielder in the North Tyne valley, something of a white elephant. Whether Cow Green need have been built at all is no doubt something which comes to mind occasionally in the minds of those involved who are still alive.

Although not now actually set among moorland, another northern reservoir scheme is worth some comment. The continuing upward projections of demand for water in the north-east (especially for Teesside) brought into play the North Tyne valley. This had been a typical moorland with enclosed farmlands in the valley until its large-scale afforestation, as described above. Sensing that public disapproval of Forestry Commission landscapes was higher than that of open moor, the Northumbrian River Authority began to look at the Kielder area of the North Tyne in the autumn of 1969. The plan was for a reservoir which would hold water that could be released into an inter-basin transfer scheme: from the Tyne via the Wear to the Tees.[107]

As planned, the eventual impoundment was to crest at 185 m ASL and have a surface area of 1,086 acres (440 ha), which made it the largest artificial lake in Europe. All the north-eastern water authorities were in favour. The opposition came from a loose alliance of environmentalist groups, and initially those eleven farmers whose holdings would become unviable, though eventually their compensation deal proved satisfactory. The most powerful 'anti' lobby was a right-wing aristocratic group who campaigned on the grounds of landscape alterations, threats to salmon and trout and to the intrusion caused by the provision of water for industrialised populations elsewhere. The legal process was complicated by the fact that the MP for the North Tyne area was also the Secretary of State for the Environment (Geoffrey Ripon, whose seat centred on Hexham) and normally a natural ally of the right wing. His concern for proper decision-making was such that there were two public enquiries and a certain amount of manoeuvring between the Government and the County Council. But the second enquiry found in favour of the scheme, as had the first and so construction began in 1974. The eventual lake is 7 miles (11 km) long and the transfer aqueduct is 24 miles (38 km) from the Tyne at Riding Mill to Eggleston, about 17 miles (27 km) of which are in the longest rock-drilled tunnel in Europe.[108] The lake and its surrounding area have proved very popular for recreation under the control of a go-ahead set of policies from the Forestry Commission and public opinion would now be much against its loss even though opponents of all such schemes will call it a large white mammoth.[109] Compared with Cow Green, it seems to be more widely beneficial though unnecessary in the current industrial climate. It does mean that in dry summers water restrictions in the north-east are quite rare, and indeed the Tees water sold to Yorkshire in recent years is mostly from Kielder. It is outside the scope of this book to say whether the promise made at the first public enquiry in 1972, that by the year 2001 a Geordie might have two baths per week, has been fulfilled.[110]

If the opportunities available to Geordies are also accessible to Tykes, the difference made by bathing might be less noticeable for water colour has been recognised since the mid-1980s as a problem in the water-gathering grounds of the Yorkshire Pennines. The water is more brown colour than clear in the autumn, more marked after dry summers and even more so after droughts such as 1995. Dry periods accelerate the decomposition of peat but in the immediate post-drought period there is clear runoff from non-peat areas. When runoff is re-established from peat areas then soluble organic compounds from the peat colour the water again. The source of the colour is related to the depth of peat, a low angle of slope, the density of drainage, gripping density and burning frequency, so that hazard maps of catchments can relatively easily and cheaply be compiled. The importance to consumers derives from the production of trihalogenated methanes (THMs) when such water is chlorinated; THMs have been shown to be carcinogenic.[111]

For decades it has been assumed that water was plentiful in upland areas and it was only the costs of its control that guided its use. There is now some shift in that thinking. An increased emphasis on water quality has coincided with apparent changes in climate towards drier summers but wetter winters with more intensive rainfall. At a local level the Environment Agency is compiling

Abstraction Management Strategies for all major catchments, including the uplands (with implementation in the second decade of this century) of which a notable feature will be time-limited licences to abstract water from all sources. Planning documents stress the need for flexible management of supply and demand and the possible effects of climatic change. The maps of potential supply nevertheless identify most uplands as areas of surplus water at any rate in winter, though not necessarily in summer, as in large parts of Wales.[112]

Limestone pavements

The Carboniferous Limestone which underlies so many upland areas in England and Wales sometimes presents its bedding planes parallel to the present land surface and, moreover, is not covered with glacial drift. If the limestone has a poor or non-existent soil cover then it may appear as a scoured 'pavement' of bare rock dissected by a series of narrow fissures. These are parallel to the bedding but may be accompanied by another series at 90 degrees. In northern England the Norse words 'clints' (for the pavement surfaces) and 'grykes' (the fissures) were used for these features and the words have more or less stuck in the English-speaking scientific literature.[113] In the present moorland regions of England and Wales, their distribution seems to be concentrated in north-west Yorkshire from the Malham area northwards towards the neck of high ground connecting the Pennines and the Lake District south of Appleby-in-Westmorland, but north of the Howgill Fells (Plate 4.10). This block takes in the massifs of Ingleborough (723 m) and Whernside (736 m), on the flanks of which there are such pavements and also the full suite of features of **karst** terrain, including solution hollows (**dolines**), limestone caverns and streams which disappear into solution-pipes or 'potholes', leaving dry valleys. Thick outcrops of limestone are often bounded by steep cliffs or 'scars' as around the village of Malham. A band of limestone running 60 km from east to west in South Wales also has some 73 ha of pavements towards its western end, where the outcrop is about 4,500 m in width.[114] The total area in Britain is about 2,600 ha. Where the pavements have a thin soil then a base-rich grassland is developed, and in the grykes a number of plants are sheltered from wind and, possibly more importantly, the teeth of the sheep. In the region as a whole, though, the rainfall ensures the leaching away of much of the mineral capital and there are often pockets of drift, so the grassland is often described as 'mesotrophic' rather than 'calcicolous' and is dominated by plants such as *Festuca ovina*, *Agrostis capillaris* and *Thymus praecox* in a sort of species-rich version of the commoner acid grasslands of the uplands of England and Wales. Calcicoles such as *Helianthemum nummularium* and *Koeleria macrantha* are found on brown-earth soils where the drainage is good. The fissures may contain a largely woodland flora among which the rigid buckler-fern (*Dryopteris villarsii*) is entirely confined to these pavements and a significant proportion of the baneberry population is found in Yorkshire limestone areas. In all, some sixteen species of rare or threatened

PLATE 4.10 *Limestone pavement at The Clouds near Kirkby Stephen in Cumbria. These fissures are unusually deep and thus provide habitat for plants and animals well out of range of the sheep which are pastured on the surrounding moorland. Source: Dr Helen Goldie*

plants in Britain are found in these environments.[115] A range of butterflies includes rarities like the high brown and pearl-bordered fritillaries (*Argynnis adippe* and *Boloria euphrosyne*). The nesting and roosting places afforded to the wren give its Linnean name of *Troglodytes* a distinct meaning. The presence of woodland species is a reminder that some pavements occur in woodlands and others carry scrub.

Stripping of pavements occurred during the nineteenth century in order to build walls and feed lime-kilns but in the absence of mechanisation its scale was quite small. Weathered limestone has been a popular garden feature since the 1870s and so many pavements have been extensively damaged by commercial enterprises, with the problem achieving prominence in the 1960s. A series of protection measures and protected areas was set up after 1981, with the aim of ensuring that existing pavements are not further damaged.[116] Further conservation measures include the avoidance of fertiliser applications around areas of pavement, the control of bracken other than with asulam (which kills all ferns), controlling rabbit populations, avoiding overstocking with sheep or attracting them to the pavement areas with supplementary feeds.

The fallout from Chernobyl

The nuclear accident at Chernobyl in the Ukraine in April 1986 resulted in the deposition of radioactive isotopes in upland Britain. The pathways from the rainfall through to incorporation in marine sediments have all been studied but here the focus is on (a) the fate of the isotopes in water and soils, and (b) its effect upon economically important animal populations, especially sheep. The fallout from this source was additional to that from atmospheric testing of weapons and from releases from Sellafield.

One set of studies investigated the movement of caesium-137 (^{137}Cs) through the forested catchments on Pumlumon (Plynlimon). The initial input in rainfall was measured as 720 Bq m^{-2} of ^{137}Cs and 420 Bq m^{-2} of ^{134}Cs. On the assumption that the rainfall pattern and the deposition of the radionuclides were linearly related, the areas of loss and gain of radioactive particles (and ^{137}Cs binds very closely to clay minerals) could be measured through time. The erosion of fine topsoil, for example, depleted some areas by about 20 per cent in the first two years, some of which was deposited downslope. In the streams, the suspended sediment and the bedload alike carried Chernobyl material, suggesting that both the topsoil and subsoil contributed to the mobile particles, the latter via the drains on this upland. Forest harvesting and mechanical site preparation may have contributed to the bedload component as well.[117]

Cumbria received very high doses, with peak readings in milk of 380 Bq/l of radiocaesium. Sheep appear to transfer much more ingested radioactivity to their milk than cattle: as much as 56 per cent of iodine has been reported. In the uplands, the caesium is almost all 'wet-deposited' and so enhanced activity was noted on an 847 m height in northern England, where clouds encountering the summit dropped more moisture from turbulent air masses. In some uplands, the binding to clay particles mentioned above was lower than expected when peaty soils were dominant and so sheep were much more restricted than had initially been proposed. Transport downslope means that there has been accumulation in the sediments of reservoirs. The best monitoring has been done in West Cumbria (not specifically a moorland area) where high deposition rates (e.g., of over 100,000 Bq m^{-2} of radiocaesium) combined with fallout from the Sellafield plant to give peak concentrations of >300,000 Bq m^{-2}, a native contribution not previously very widely known. In all events, 1,670 holdings in Cumbria were placed under restriction (most of them in upland areas) in 1986, and in the summer of 1988, ten of them were still holding 12,000 sheep that could not be sold off.[118] Regulations here and in Wales (Fig. 4.13) suggest that there will be restrictions until at least 2003, giving rise to the study of options for management for 'clean feeding', all of which preceded the 2001 Foot and Mouth Disease (FMD) outbreak.[119]

Outdoor recreation

The uplands of England and Wales have now become a major focus of the demand expressed by G. M. Trevelyan in 1938 for '. . . regions where young and old can enjoy the sight of unspoiled nature.'[120] Trevelyan did not foresee the great recreation explosion of the post-1950 period in which the uplands have shared. This burst has often been spatially differentiated because of the

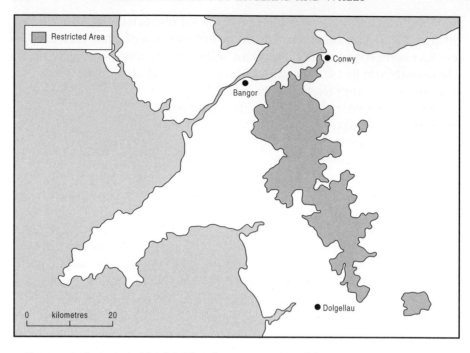

FIGURE 4.13 *Areas in North Wales where movements of sheep are restricted. Though much of the area is mountain, moorlands are also involved in the Restricted Area still in place (in 2000) where sheep and lamb movements are controlled due to the radioactive fallout from Chernobyl in 1986. (The analogous area of south-west Cumbria is all mountainous.)*
Source: A. Nisbet and R. Woodman, 'Options for the management of Chernobyl-restricted areas in England and Wales', *Reprinted from* Journal of Environmental Radioactivity *51, 2000, 239–54 with permission from Elsevier.*

special attraction of the wilder regions (including the National Parks) and has had a distinctive impact due to the nature of the upland environments.

The background to the present, we have to remember, is one in which there has been a large degree of conflict. After WWI there was a surge of interest in having access to open countryside. Films, better transport, more paid holidays, modest accommodation from the YHA, books like the Shell Guides, a fashion for rambling given authority by its association with the growth of the Labour Party and the Fabian Society, all contributed. Since the northern uplands were easily accessible from the industrial cities of those regions, the perceived incompatibilities of, especially, grouse moor management and water undertakings with recreational access formed a locus for confrontations as well as a constant theme in the attempts of the more middle- and upper-class movements to establish National Parks. The pre-WWII attempts to emplace legislation came to nothing: the Access to Mountains bill of 1938 received the Royal Assent in 1939 but was never activated because of wartime conditions. It had a stormy history and was indeed largely repudiated at one stage by its chief advocates since it established trespass as a criminal action, whereas before it has only been a civil infringement.[121] Perhaps the general point is that all these antagonisms prepared the way (as did the national stock-taking during WWII) for different attitudes in the post-1945 years.

Because many uplands in England and Wales are now designated as National Parks, there is a temptation to regard the uplands and the National Parks as synonymous. Large tracts of uplands are nevertheless outside that category: the mid-Cambrian mountains never attained that status in spite of a campaign in the 1970s, and the North Pennines only became an AONB in 1988 after considerable debate. Much of the measurement of recreation demand and impact has been in the National Parks and so caution is always needed in deciding whether some at least of those features have been brought about by the actual words 'National Park'. Another important context reminds us that the greatest outdoor recreation pressure of all is on the coast and that the uplands are thought of as high-quality recreation areas which attract a minority, at least for active pursuits. Hill land is indeed marginal for many popular recreations and the matching of demand and supply is imperfect in terms of, for example, access to hills from London and the south-east. The northern industrial cities are much more fortunate in this respect – as indeed in so many others. The potential for increased use of a region for outdoor recreation was addressed as early as the 1960s in the case of mid-Wales but in the absence of a regional authority, no such co-ordinated development ever took place.[122] Outside the National Parks, therefore, there may be a systematic approach to creating demand (in the shape of a tourism board or other regional development body) but not of managing the recreation opportunities and impacts. Just as it then proves difficult to measure and cost demand, the costs and benefits to other stakeholders are equally difficult to measure and apportion.[123]

If the data for National Parks are not that good, then for other areas they are worse. Further, some designated areas such as AONBs which cover about 13 per cent of England and Wales but only two of which – North Pennines and the Forest of Bowland – are upland areas fell for many years under the general view that their designation makes no difference to environmental management of any kind, an outlook which has been changing as planning authorities have been acting as guides and conduits for the kinds of funding that have come from the EU.[124]

Quantification of outdoor recreation use of the uplands has always been difficult, since most of them are traversed by roads with a high non-recreation use, so that traffic surveys have to be carefully interpreted. Nationwide, a 1995 report suggested that visits to the countryside as a whole had not increased significantly in the previous few years,[125] but a number of estimates suggest a peak in the early 1990s (Fig. 4.14) . There are a few more detailed figures: data from the North York Moors in the 1970–1980 period, for example, suggested that the National Park received 11.3 million visitors per year and that on an

August Sunday there might be 137,500 visitors in the Park area. About 28 per cent of these were simply in transit from Cleveland to Scarborough or Whitby and so were simply road users. About four-fifths of those remaining in the Park sought out twenty-four key locations (the 'honeypots' in the jargon of the time) and a quarter of those were in only fourteen places, three of which were coastal, excluding the main towns. So perhaps one-fifth of visitors who came by car were actual seeking the quiet recreation in wild and open areas that such places are supposed to furnish. The Peak District was host to some 76,000 cars on a summer Sunday in 1970, which was double the number seven years earlier and added up to 11 million in a year. Data for Dartmoor (Fig. 4.14) show a 1984 peak at 8.1 million, followed by a trough that had regained previous values by 1989 and rose steadily to 10.17 million in 1992, the last year of the published figures. Roughly comparable data for some of the English National Parks in 1994 are shown in Table 4.6. In 1971 the number of walkers on the Pennine Way was about 500 per day. A Dartmoor study in 1975 showed that sedentary activities, walking and picnics were the most popular recreations and that most visitors went less than 100 yards from their cars, some up to one mile and a third group from two to five miles. This survey probably underestimated the amount of time that people spend just sitting in their

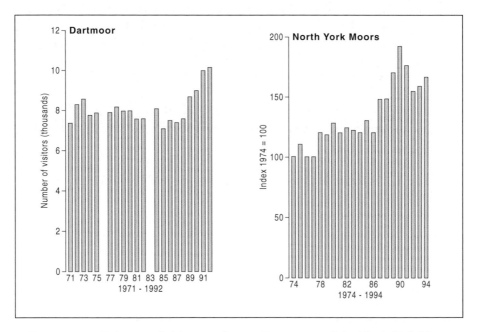

Figure 4.14 *Estimates of visitor numbers to Dartmoor and the North York Moors 1970s–90s. The differences in pattern are not so great that they might not be due to sampling methods or purely local occurrences but there does seem to be a peak in the early 1990s. Note that Dartmoor is depicted in absolute numbers and the Moors as levels above a 1974 index = 100.*
Source: House of Commons Environment Committee, The Environmental Impact of Leisure Activities, *Fourth Report, Volume I, London: HMSO 1995, p. xxii.*

TABLE 4.6 Visitor days and net expenditure English National Parks 1990s*

National Park	Visitor Days 1994	Net expenditure 1997/98 £million
Dartmoor	5.4	2.6
Exmoor	1.4	2.1
Northumberland	1.4	1.6
North York Moors	7.8	2.9
Peak District	12.4	5.4
Yorkshire Dales	8.3	2.8

* Only English National Parks with moorlands are included here. The Peak and Dartmoor figures are considered to be substantial under-estimates.
Source: Countryside Agency *National Parks Fact File*.

PLATE 4.11 *One end of the dam featured in Plate 4.5. The popularity of the place for recreation can be judged from the multiple signs and the CCTV installations. Most of these attractions will have environmental linkages: some on a local scale but the vehicle use is globally significant.*

cars.[126] Detailed surveys are probably not needed to establish a pattern of use which is dominated by the use of the roads and nearby parking areas and in which a minority penetrate the remoter areas on foot, horseback, bike or motor vehicle (Plate 4.11). But that list will alert us to the possibility that environmental impact may be strong, especially once above the limit of enclosed land. Recreation demands, like other forms of consumption, are not constant: the rise of mountain bikes was sudden, for example; by contrast, the popularity of overseas destinations dropped the money from tourism generally in rural areas of south-west England (including Dartmoor and Bodmin Moor) by 20 per cent in the early 1980s. Lack of confidence in possible suppliers meant that grant take-up for involvement in tourism was very low.[127] Growth in wargaming with paintballs and orienteering have not often had an impact on moorlands since they are usually located in woodland and forest, though upland plantation forests are popular for orienteering, as they are for car rallies; both can usually be controlled to minimise most impacts except perhaps noise in the latter case. Another source of use of upland areas is the educational party, from both schools and the post-18 sector. Their use may be for enjoyment or for fieldwork in e.g., biology or geography but either way the addition of parties of people to already heavily used areas (the Malham area of the Yorkshire Dales is the outstanding example) forms another ecological pressure unless the wishes of the young are paramount in which case the cafés bear the brunt of the visit. But the combined totals for organised parties and individuals on their own come to only ten per cent of visitor days to the National Park areas. A grid of incompatibilities of land use in the uplands (Fig. 4.15) suggests that 'specialist recreation' is the least compatible of all uses, whether with other forms of leisure pursuit or with production uses. The exception appears to be the management practice of bracken control, which is not very widespread.

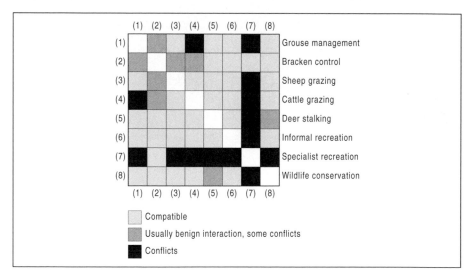

FIGURE 4.15 *A matrix of compatibilities in upland land use for recreation in the early 1990s compiled by a National Park's chief planner. The dominance of 'specialist recreation' is only partly due to the shading chosen: it reflects the demand for 'take-over' on the part of especially those recreations involving machines but to some extent horses as well. (Deer stalking is not a feature of upland land use in England and Wales.)*
Source: Leisure and the Environment: Essays in Honour of J. A. Patmore, *Steel and Glyptis (eds), 1993.* © John Wiley & Sons, Ltd. Reproduced with permission.

It is however essential not to overestimate the use of the uplands. The Countryside Commission reported in 1984 that there was a strong preference among recreationists for 'the wider countryside' and the graphics refer to 'unmanaged countryside' (which includes the coast). Nevertheless, the types of countryside 'stopped at' are headed by the coast at nine per cent, with moorland and woods sharing six per cent and 'hills' three per cent. The category of 'wider countryside' totals 44 per cent. Where 'long walks' are mentioned, then 38 per cent are 'on paths' and only 14 per cent on heathland and moors.[128] So that although visitor pressures in the uplands cannot be denied, any image of the whole population under age seventy flocking to the moorlands has to be resisted. The coasts and the areas near urban agglomerations continue to take on the force of recreation demand.

The measurement of demand may be imprecise but impact is generally visible and can be subjected to some form of repeatable measurement. The biggest impact on environmental quality as a whole is the motor vehicle in its most benign form of providing access to the countryside and needing to be accommodated in car parks and laybys, and producing a need for services like fuel. Its visual impact in the moorlands is of course striking though this is not of itself a factor of environmental manipulation. Only 6 per cent of visitors travel to the National Parks by public transport.[129] More serious in that respect is the appeal of motoring either off-road altogether or in the use of bridle-ways and green roads (such as some of the old cattle drove roads in the Pennines) as playgrounds for four-wheel drive vehicles and their occupants. They, like horses, but much more noisily, can easily churn an unmade surface into a morass which is virtually unuseable by any other recreationist, or indeed by hill farmers using for example quad bikes to manage sheep. The majority of car users however wish to stay close to solid surfaces. As indeed do the strollers (see above) who stay within a short distance of their cars, but there are enough people wanting to walk a few kilometres, as well as the serious hikers, to provide the second most noticed category of impact, which is footpath wear. This class of impact is of course highly noticeable, with popular paths being visible in the landscape from some distance, in clear weather at least. Many such paths have been studied for their degree of wear and classifications produced.[130] In general, paths that cross peat, fine silty soils or clay soils were most susceptible to deterioration compared with well-drained sandy and stony substrates.[131] Highly susceptible to damage are the banks of ancient monuments where these become the path as in parts of Offa's Dyke or the prehistoric Horcum Dyke in the North York Moors.[132] Designated long-distance walks of the 'challenge' variety manifest some of the most worn passages: the Three Peaks path in the Yorkshire Dales in the 1980s had a mean trampled width of 11.4 m over a 20 km length, and a mean 2.7 m of bare area. The mean width of the Pennine Way (based on 20 sample points) was 1.45 m in 1971 and 3.53 m in 1983 and in the Peak District section, some 23/40 km required remedial work, including the airlifting of stone slabs. The bare area in a heavily used part of Edale (Derbyshire) increased by 0.7–0.8 m/yr at one time in the 1970s. In such places, simple drainage measures are often insufficient and so stone pitching, stone flags or even floated stone pitching (on deep or eroded peat) may be needed.[133] The Lyke Wake Walk across the North York Moors underwent 7,680 successful crossings in 1981 and perhaps three times that number walked a part of its length, so that parts of the path are 100 m wide in places.[134]

The grouse moor managers long insisted that walkers disturbed the birds when nesting or just before a shoot; though this has now been abandoned as largely untrue, not every keeper has heard the news. Upland shooting has in some places resulted in bulldozed access tracks which act as foci for soil erosion. Even quiet recreations like sailing are known to disturb waterfowl to the point of them deserting a particular lake. Some reservoirs in upland South Wales have been delimited as non-sailing places or have been strictly zoned in order to maintain wildfowl populations. The popularity of hang-gliding at the Hole of Horcum in the North York Moors led to severe soil erosion at take-off and landing points on steep slopes. A few climbing sites, such as the gritstone Roaches in the western part of the Peak District National Park attract so many climbers as to produce wear of the rocks and when climbs are televised then the extra vehicles and crews create more wear of the environment. Climbing organisations will of course co-operate with English Nature in declaring some cliffs off-limits to climbers when birds are nesting. A summary table from the 1980s of some of the impacts of recreation on upland environments and species is given as Table 4.7.

TABLE 4.7 Disturbance to habitats and species in the uplands

Activity	Sensitive habitats or species	Potential impact
Aircraft	Nesting and non-breeding birds	Disturbance to birds and nesting sites
Caving	Cave formations; bats	Disturbance to geological formations
Mountain biking	Vegetation and soils, especially peats	Damage to vegetation and occasional erosion
4x4 vehicles and trail bikes	Vegetation; birds	Damage to vegetation
Orienteering	Woodlands	Localised damage
Riding	Heath and grass vegetation	Trampling and erosion
Walking and dogs	Birds and sheep	Trampling and erosion on key routes; disturbance of birds; sheep and lamb worrying
Water sports	Lakes; birds	Disturbance to birds
Small powered craft	Lakes; birds	Disturbance to birds and plants

Source: Extracted from House of Commons Environment Committee, *Environmental Impact of Leisure Activities*, London: HMSO, 1995, vol. I, p. xxv.

Beyond the physical and biological impact of outdoor recreation, there are incompatibilities of use. Some of these are permanent: no amount of management will stop boats scaring birds, for example; some are seasonal, as in the case of grouse moors, when walkers are at risk from being shot or may be culpable of starting fires on dry heather moorland. Yet others may be epidemic-related: an outbreak of foot and mouth disease for example must needs close an area to recreation. The incompatibility of motorised vehicles of any kind with walking and horse-riding is well known, as is the way they add to wear.

Environmental management to alleviate some of these effects takes place at different scales. At the most immediate scale, a responsible body (e.g., the National Trust, a National Park authority) will undertake on-site remedial activities such as the repair of paths. Commonly this involves work on the surface and the drainage of a path, with paving inserted on the most-eroded stretches. This last is sometimes controversial since it appears to introduce an urban element into wild environments but it can now and then be pointed out that mine-tracks from earlier use patterns also had some form of surfacing. The practice is now so familiar that it seems likely to stay unless a plastic path surface which never wears out but which imitates the surface in the 1930s can be sprayed along the trackway. In general, there is widespread approval of restoration work except perhaps in the more remote areas. In 1995, though, it was estimated that £7 million needed to be spent on the Pennine Way, with a further £0.75 million per annum for maintenance.[135] These kinds of measures may be combined with a second scale of management, which is an access agreement. Marion Shoard points out that landowners often claim that their ability to exclude others from their land is one of the main benefits of owning it, and this has often been highly evident in the uplands. To minimise trespass and create a co-operative atmosphere, some owners entered into access agreements with National Park authorities in the years after the 1949 Act. The flash-point of Yorkshire grouse moors was defused by an agreement for 14,000 acres (5,600 ha) of Barden Moor and Barden Fell. This receives some 30,000 visitors every year and the agreement provides for the access area to be closed for thirty days each year but that there will be no shooting on Saturdays and (heaven forbid!) Sundays.[136] The conclusion of this type of agreement was helped by the research of scientists like Picozzi, who showed that most people stayed on paths even where the agreement allowed them free rein, that few dogs harassed the birds and that shooting bags did not decline with free public access.[137] In fact the decline of numbers of grouse shot which provoked the study was largely due to poor heather management.

An even wider scale of management is that of a regional plan. Since the reorganisation of local government in 1974, the National Parks have had to submit physical plans to the central government and other planning authorities have had to include upland areas in their structure plans. All have to be updated at intervals of not less than five years. The finance available to implement planning developments has usually been totally inadequate (the 1990 capital outlay in the North York Moors National Park was about 0.6p per visitor[138]) but medium-term aspirations and policies can nevertheless be articulated. These usually place the amelioration of impact at honeypot sites at the head of the priority list, with the second place being restorative work in worn places.[139] The third, climbing rapidly, is the encouragement of public transport in order to reduce the number of cars in the uplands at popular times. In the National Parks in particular, the trend in the last few years has been to acknowledge that not all recreations can be catered for and that there is a distinct move to promote what the 1949 Act called 'quiet enjoyment'.[140]

To conclude: there are some special features of the National Parks since they both attract recreation differentially and have extra powers to control development and use, but they share with all moorland areas the problems of dealing with wear and incompatible uses in relatively fragile environments. The reluctance of the nation to extend planning control to agriculture and forestry has probably never been helpful but it fades into insignificance when compared with the sea-change needed to alter vehicle use and traffic patterns.[141] Yet, the Countryside Agency was in 1999 suggesting that hill farms diversify whenever possible into tourism (since sheep farming provides such a poor living), thus demonstrating once again that the 1949 Act (in the National Park areas at any rate) did not foresee the upcoming conflicts between recreation, the rural economy and landscape preservation.[142] It also would perhaps like to try to slow down any of the trends towards the priority of landscape preservation which have been detected since 1976. In general, the challenge to moorland management generally and the uplands even more widely from recreation is less than that of agriculture (especially in the wake of the big increases in sheep numbers since the mid-1970s and then the 2001 Foot and Mouth Disease outbreak) and probably less than that posed by future climatic change.

An upward hike

After much pressure and argument, the Countryside and Rights of Way Act 2000 (usually abbreviated as CROW) became law. This is largely for the benefit of walkers in 'open' countryside, since cultivated land, improved grassland and golf courses (among other categories) are excluded from it. The statutory right of access is to mountain, moor, heath, down and common land and will come into force once maps of these areas have been drawn up by the Countryside Agency and its Welsh equivalent. Use of vehicles and horses is forbidden, as is bathing. Dogs must be kept on leads from 1 March to July 31 and may be banned altogether from areas used for lambing and where grouse breed. Landowners may exclude or restrict access for up to twenty-nine days per year without permission, though exercise of that right at weekends is discouraged.[143]

It will take three or four years for the maps of access land to be completed and so the Act will come into force only then. The key role of moorland as a focus for the Act is obvious as is the attention paid to the restrictions which were necessary to avoid protracted opposition from farming and shooting lobbies. The effects of bringing more walkers onto moorland and not confining them to paths has yet to be seen; English Nature is concerned about the effects on nesting birds since they are unlikely to be able to enforce exclusion during the spring. It is not usually the landowner and in any case does not have the right staff, in the manner of the gamekeepers of grouse shoots. But given (a) possible climatic change and (b) the general preference of walkers to stay on paths and tracks, then a relaxed view of environmental impact may well be possible.

Military use

The issue of the military use of uplands in land owned or leased by the Ministry of Defence (MoD) has often been focused on the National Parks, though there are significant other areas of moorland which are affected, such as the Warcop area of the North Pennines AONB, near Appleby-in-Westmorland and Mynydd Eppynt in Powys. About 44,000 ha of land in the National Parks are thus used, with concentrations in Northumberland, Dartmoor and the Pembrokeshire Coast, so that this last is outside the frame of consideration here. The Brecon Beacons also has a significant military presence, with the MoD owning just over 1,000 ha, compared with 22,700 ha in the Northumberland National Park and 3,987 ha on Dartmoor (to which must be added 10,000 ha under licence from the Duchy of Cornwall), contributing to a total of about 44,000 ha in the National Parks in 1980. Only the Exmoor National Park is free from MoD ownership, though there was live firing, tank training and 5-in rocket trials during WWII; Bossington Hill above Porlock hosted several installations. The presence of the military confers both benefits and costs on society in general and the local community as well as the environmental systems (Table 4.8).

Use by the military has been evaluated mostly from the point of view of landscape change and effect on visitor use in training areas. Thus the prevention of access when firing is taking place, the digging, fencing, track construction and use, hut and look-out building have been criticised for nearly all military areas in the uplands, with especially strong disapproval of live firing zones, as distinct from 'dry' training areas. The actual use for live firing is usually less than 50 per cent of possible days, with the high 20 percentages being modal on the three live ranges on Dartmoor. Occasionally, MoD installations escape total anathema, as with the original BMEWS installation on Fylingdales Moor in the North York Moors, where the three 'golf balls' of the radar domes eventually inspired some affection; their replacement in the 1990s by a kind of truncated pyramid structure was often deplored and campaigners are on alert for its further development in any 'Son of Star Wars' missile-shield installation by the USA. No such liking has ever been forthcoming for the US constructions at Menwith Hill near Harrogate, however and the

Table 4.8 Costs and benefits of military training areas

Costs	Benefits
Intrusion in 'wilderness'	Training of armed forces
Restriction on access, both real and imagined	Local income, both direct and through multiplier effect
Restrictions on tourist spending	Civilian employment
Direct impact on ecology and on archaeological monuments	Protection of archaeological and ecological sites of value from alternative damaging land uses
Disturbance of grazing regime	Assistance from Services: rescue of people and animals, litter clearance etc
Death/injury from firing and from unexploded ammunition	
Noise	
Impact on landscape by direct damage and by the intrusion of artefacts, roads etc	

Source: S. Owens, 'Defence and the environment: the impacts of live firing in national parks', *Cambridge Journal of Economics* 14, 1990, 497–505, Table 1, p. 498. Reprinted by permission of Oxford University Press.

use of most uplands for low-flying training in warplanes is normally regarded negatively by most residents and visitors alike, though the environmental effects are transitory, so far as is known.[144]

It might be argued, though, that the ecological impact of military use is environmentally more important than the visual changes or the effect on access. For example, a new howitzer on trial on the high ground of northern Dartmoor was alleged to have created craters 3–4 ft (1.5 m) deep, and 20 ft (7 m) across, at 10 ft (3 m) intervals.[145] Owens quotes a Dartmoor correspondent (1989) that shell holes are 15–20 m apart, 1m deep and 2–3 m across, with some having joined together to make large pits with chunks of peat thrown all around.[146] Inspection of some aerial photographs from the 1970s (Plate 4.12) shows a regularly spaced series of circular holes in the deep peat area of north Dartmoor and some of these seem to have coalesced into linear features, one of which has a 'string-of-beads' appearance.[147] Eleven farm animals were killed on Dartmoor in 1984–8, and an area of 13 ha was cleared of ordnance to a depth of 0.5 m in the early 1980s. This took two months and 380 items were recovered.[148] Happily, the use of heavy munitions on the peat areas ceased in 1998. If Dartmoor has been the area of greatest friction over training, then Northumberland is the second, following requests by the MoD to expand training in that region as the Army withdrew from Germany.[149] In 2001, the government approved the construction of 16 km of new roads in the Otterburn area of Northumberland, to allow training with a 45-tonne gun and a Multi-Launch Rocket System. On the whole, the MoD contention that

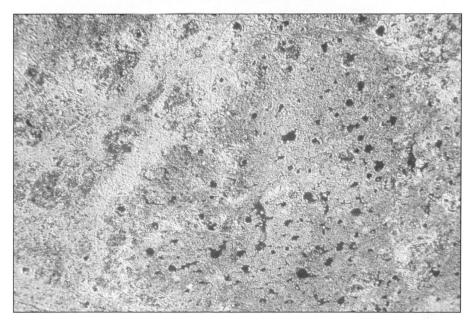

PLATE 4.12 *An aerial photograph of part of the artillery range of northern Dartmoor showing the holes in moorland soils and vegetation created by live firing, some of which have coalesced. The precise scale, date and location are not given in the source.*
Source: Reproduced with the permission of Stephen Johnson, Cyberheritage.

expanding their own use has prevented the 'wrong sort of development' (especially in the National Parks) has not won many hearts and minds. There is a curious environmental link in Europe between ericaceous vegetation and war: ever since the Army dug into the Surrey heaths in the nineteenth century, training on heath and moor has been paramount and the same has been true of the Netherlands and parts of Germany. Perhaps the military planners expected that any invasion by the USSR would best be halted in open country with scrubby Scots Pine.[150]

Quarrying

Reference has already been made to the expansion of quarrying in the uplands during the nineteenth century and this activity is still important. While its role in the local economy has remained roughly constant, its perception in national terms has changed, especially where its presence looms large in National Parks. The situation in the National Parks has been such that most of the comparable data have been collected for those areas and other upland sites are rarely documented environmentally in any detail.

Nobody can deny the antiquity of mineral extraction in the uplands but as in so many instances, the upswing of economic tempo in the nineteenth century changed the scale of activity. The impacts of large quarries are numerous and not all are direct. In this latter category there is collateral damage to nearby habitats from dust, changes to the hydrological regime, alterations in water quality and disturbance to animal species with a low tolerance of this form of human presence. Dust is a big amenity problem for local residents, followed by noise, traffic generation and the appearance of the landscape. A day's working at a big quarry will add 450 lorries and 200 other traffic movements to the usual local pattern. On the B6265 between Grassington and Skipton in Yorkshire, 1,000 HGVs per day are quarry vehicles, out of a total of 6,500 movements. Planning authorities have sought to limit the spatial spread of quarrying and so there are concentrations of 'limestone cliff' environments in various stages of working, abandonment or restoration, for example, to the south-east of Buxton and at Great Rocks Dale, off Wye Dale, all within the Peak District.

The direct effects of quarrying are in one sense obvious: one set of environments (chiefly horizontal) is replaced with another, largely vertical, set. The lifting of the overburden and its destruction is the main ecological impact, though cave systems may also be destroyed. Apart from the value of the mineral recovered, geological exposures are created (some 65 per cent of geological SSSIs are associated with past or present mineral workings) and after-use may produce a new set of habitats for plants and animals. The scale of the impacts is changing. Between 1953 and 1974, there was a 50 per cent reduction in the number of active quarries in National Park areas (Plate 4.13) but their output grew by 500 per cent. In the late 1990s there were quarries in all

PLATE 4.13 *An example of a twentieth-century quarry at Rosebush in the Preseli mountains of western Wales. In attempts to diminish the visual impact of the quarry, large surrounding areas have been planted to coniferous trees.*

the National Parks except Exmoor and most of these were of limestone extracted for aggregate. Other minerals include cement (in the Peak District) and, on Dartmoor only, china clay. The demand for aggregate has resulted in forty-two extraction sites within National Parks, of which half are in the Peak District and Yorkshire Dales, with about 10 per cent of the nation's aggregate coming from the National Parks, of which 42 per cent is in the Peak District, 34 per cent in the Yorkshire Dales and 11 per cent in the Brecon Beacons. There is an estimated forty-four years' supply in the Yorkshire Dales and fifty-two years in the Brecon Beacons, so the activity is long-lasting: this increases its local attraction as a source of employment. There are some thirty dormant and non-operational quarries within the National Parks and a mandatory environmental code of practice adopted by the operators' association may edge that number up a little. Operators, too, have become sensitive to the archaeological significance of any areas they propose to destroy. Exploration for oil and gas, which was growing in the 1960s, seems now in abeyance.

The movement towards fewer but larger quarries has been at the root of much of the controversy that has followed quarry development. Their high profile led, for example, to mineral extraction being identified as the greatest threat to the future of the Peak District between 1951 and 1971. The conflict between demand in the total regional, national and indeed international context, and environmental issues led to a series of statements of varying strength on the degree to which the two were to be accommodated. The Sandford Report of 1974 thought that 'conservation' ought to have priority but only the 1995 Environment Act gave any force to that notion. The subsequent identification by government of what it has called 'sustainability' has reinforced the environmental arguments, notably in the case for extension of a quarry at Spaunton in the North York Moors, where the application was called in by the Secretary of State and then refused. Notably, a few mineral companies have renounced their claims to extraction rights in scenically or environmentally sensitive places, including 0.17 per cent of the Dartmoor National Park at Shaugh Moor[151] and NPAs have served a number of Prohibition Orders on dormant sites. Some dormant sites could still be reactivated: there are examples at Lees Cross and Endcliffe in the Peak District and Blaen Onnu in the Brecon Beacons.[152] There seems little doubt that the main concerns are dust and traffic, with effects on the amenity of residents and visitors, rather than any broader environmental sense.

Windfarms

In a world of worry about global climatic change, the generation of electricity from a renewable source like the wind has a great attraction. Coastal and upland regions are favoured locations for agglomerations of pylon-based turbines; typically each tower is 80 m high and is in a group of perhaps 20–100. Each produces 0.1–0.3 mW and is claimed to have a payback time measured in months, with a lifetime excess of thirty times energy generated to energy consumed. In operation, the carbon dioxide output (tonnes/Gigawatt hour) is 7.4, compared with 1,058 for coal and 823 for natural gas. Since the UK has about 40 per cent of the European wind resource, above the minimum useful velocity of 5–7 m/s, then wind power is set to fill half the target level of 10 per cent of UK generation from renewable sources by 2010. This would save the emission of 14.4 million tonnes of carbon dioxide.

The main objections to clusters of turbines (usually called 'windfarms') are firstly aesthetic and secondly noise. The intrusions caused by construction may last 12–18 months but maintenance is not particularly noisy. The visibility of the turbines is undeniable and usually stretches a long way by definition, since a sheltered valley would not be much use. The noise varies with the wind velocity but is normally a constant soughing sound. It increases with wind velocity but so do some other ambient sounds. A typical windfarm at 350 m generates 35–45 dB (A) whereas the night-time background in rural areas is 20–40 dB(A) and a truck at 30 mph and 100 m distance is 65 dB (A). Both these matters seem subject to the usual varieties of reaction, which range from outright rejection and crusading in favour of the nuclear alternative to an acceptance of both the changes in landscape and an ability to ignore the noise. In terms of impact on wildlife, distinctions are usually drawn between the pre-construction, construction, operation and decommissioning phases, and between places with statutory protection and those without it. In upland areas, the biggest impacts are thought to be habitat damage, toxic pollutants during construction, and vehicle movement during access for maintenance. Thereafter, the creation of new habitat may provide a niche for immigrant species, and any interference with slope and hydrological processes could increase silting and turbidity in the runoff. Noise may also deter mobile species. Any development of windfarms must not contravene the provisions of EU Directives, the WCA 1981 and the CROW Act of 2000, as well as any Biodiversity Action Plans; thus sensitivity to bird migration routes and wader nesting areas is essential.[153]

It is too soon for mature assessment of the environmental effects of windfarms; the first in the UK opened in 1988. Many of the large developments are controversial on aesthetic criteria and at present at least these seem dominant in determining location and size, especially since the fringes of the national parks seem popular sites for developments. Government energy policy is however critical in their economics and so their popularity with large-scale developers may well prove less steady than the velocities encountered on the moorlands. One interesting feature of their environmental linkages is the ability to calculate their energy balance. This sets off the energy generated by them against the energy used in their manufacture, maintenance and scrapping. The most detailed work comes from Denmark and is largely concerned with offshore installations but they are likely to be more energy-consumptive than moorland complexes in England and Wales. Over a twenty-year lifetime, the Danish turbines will supply over eighty times the energy they consume and the energy pay-back time is two to three months but this will vary with the wind velocities of the site.[154] In the UK there seems to be no strategic public policy and decisions are made on a site-by-site basis. Since the applications for development exceeded 200 in England and Wales as early as 1994, the potential for conflict is high.[155] It seems too as if there are concentrations developing in upland Wales and on the margins of the North Pennines. The adoption of zoning, as has often been done for afforestation, seems a way ahead.

A nuclear power station

Given the industrial history of the uplands and their abundance of water and the generally low population densities then it is not too surprising that a nuclear power station was built in such environments. Hence, the only exception to coastal and estuarine locations is Trawsfynydd in North Wales, between Ffestiniog and Dolgellau at about 275 m ASL. It is in the Snowdonia National Park but in a moorland zone rather than in the mountains and its cooling water is derived from a lake formed in the 1920s as part of a hydro project but enlarged in the 1960s.[156] When the plant was planned and built (it went critical in 1964) the environmental relations were submerged beneath the opportunity for local employment, which had been the main local argument in favour of the development when an enquiry was held in 1958. Taken out of service from 1991, the structures will be encased in a concrete shell or 'safestore' with a design life of 135 years by 2004. The radioactive releases were to the atmosphere (mostly ^{14}C and some ^{41}Ar) and to the lake as liquids (tritium, fission products (mostly ^{137}Cs), activation products and actinide). The planned releases and exposures have always been well within statutory limits though the exposure in 1998 of the critical group ('local fishing community') at 0.023 mSv per annum is really quite high by national standards, being exceeded only by Hunterston in the UK, with most values being below 0.1 mSv.[157]

Acidification and eutrophication

Whatever the downsides of windfarms might be, they are point sources of environmental alteration: they can be seen and heard. The same is not true of the major forms of atmospheric pollution that affect the uplands. The effects of the nineteenth century have been known for some time, with the replacement of *Sphagnum* species by *Eriophorum* as a major consequence along with, possibly, more peat erosion. In the 20th century, the concept was developed of critical loads, beyond certain values of which the ecosystems changed in one way or another. The chief focus was the fallout of sulphur compounds such as sulphur dioxide and sulphuric acid, generated by power stations and other industrial plants burning fossil fuels. Switching to natural gas has helped to bring about a drop of 66 per cent in sulphur after 1980, leaving enrichment by nitrogen compounds as a chief source of ecosystem aberration. In the 1990s critical loads were exceeded in 25 per cent of the sites sampled in Great Britain and most of these were in the wetter uplands, especially in the Pennines. In central Wales the low pH in high rainfall has led to a strong lichen decline and the low acidity in stream waters has increased the levels of soluble aluminium with derogatory effects on caddis and mayflies; hence there have been declines in both fish and dipper populations.

One recently detected influence has been that of cloudwater, which contains a ten-fold concentration of most elements compared with 'ordinary' rain. Thus plantation forestry enhances the deposition of solutes to the point where perhaps 50 per cent of them come from clouds, compared with 10 per cent on the open moor. Surface pH on the uplands is well correlated with hydrogen ions and non-marine sulphate from rainwater, with only the cation exchange properties of peat to buffer the ombrotrophic surface waters before they become runoff, unless the rainfall sinks into mineral soils. The peat-mineral soil systems retain ammonium, barium, boron, iodine and lead, for example. (Increased lead in upland Wales was thought to come from car rallies.) There is a net release from the soil-peat system of transitional elements and aluminium, and a net balance of chloride and sodium. The stream water seems to be especially acid during high flow events and little is known about the biology of short-lived acid 'flushes' in notoriously flashy upland streams.[158] Measurements have also been made on the leaching of phosphorus, where a small amount might have a dramatic effect on oligotrophic waters. Heather and *Molinia* loss rates are of the order of 0.03 kg/ha/yr and 0.05 kg/ha/yr respectively and thus much lower than the yield of improved upland grassland at 1.5 kg/ha/yr. As nitrogen leaching increases, so the acid neutralizing capacity (ANC) decreases. So the weather patterns and the location of urban-industrial-transport concentrations are still exerting an influence on the ecology of the uplands; these diffuse sources mean that only national policies will have any effects on, for example, the populations of *Sphagnum* species, the brown trout (*Salmo trutta*) and the dippers (*Cinclus cinclus*) that are being lost.[159]

Still little known is the relative contribution of the moorland and its fringe land-cover types to the semi-permanent store of faecal bacteria that contaminate stream channels, especially in the summer. The levels in the Castleton area of the Peak District National Park are estimated to be high enough to be a health hazard to recreational cavers in the limestone karst system of that region and they certainly exceed the EU Directive on the bacterial quality of recreational waters. The guideline value was exceeded for 45–95 per cent of the time in three input streams and 25–39 per cent in two caverns, over an 84-week sampling period.[160]

Both sulphur and nitrogen deposition would have to be reduced beyond current commitments to avoid the predicted effects on ecosystems. In Wales, for example, the upland soils almost all carry exceedance levels for sulphur and they are deemed to be critical in 26/282 of the 10 km^2 squares of the Principality. Agreed reductions in Europe should result in halving the sulphur deposition across Wales between 1996 and 2010. This is predicted to reduce the proportion of streams with no ANC from 19 per cent in 1995 to 10 per cent in 2030. To produce no streams without a positive ANC by 2030, a reduction in sulphur deposition of 80–5 per cent will be needed.[161] The reduction in nitrogen will take much longer since it has become recognised much later and it is less easy to eliminate from upland ecosystems. Even more recently, the average ozone concentration of the atmosphere has been seen to increase, with exposure being highest in the uplands. The air has become one of the most potent sources of environmental change in these places.

1750–2002: Main themes

One pervading theme of these years is the ever-deeper incorporation of the UK into global systems of economic relations. Though the new national economy might demand more metals and more meat, it might also get them cheaply from abroad. Hence the collapse of the lead mining industry, for example, as other countries competed successfully to supply that particular metal. Meat from overseas also came cheaply once refrigerated steamships could ensure a reliable supply. Notwithstanding this exposure to international

PLATE 4.14 *The moorland environment as preferred by many people at the end of the twentieth century. In this scene in north Wales the dominant vegetation is heather moor without any enclosures nor roadside fences. There are coniferous plantations but they are mostly some distance away from the access points. These forests are popular with recreationists (and they hide cars very well) but less so with advocates of biodiversity and some walkers' organisations.*

TABLE 4.9 Important themes in active environmental management 1750–2000

Theme	*1750*	*1850*	*1950*	*2000*
Grazing: sheep	Important along with cattle	*Increasing*	*Increasing*	*A dominant use of the hill lands; Quadbike common on moors*
Grazing: cattle	*High demand from urban-industrial areas*	Constant: imports are cheap	*Numbers increase with greater ease of dairy transport*	Quotas diminish interest in many breeds of cattle but still important where enclosed land available to farm
Other mammals				Some interest in farming of a specialist kind: e.g., llamas, red deer. Relies on enclosed land not open moor
Sporting birds	Shot by walking up	*Onset of moor management by keepering and fire*	Still a major influence though keepering less intensive	Under pressure from conservation groups to allow predators to flourish; still a major source of income and often a control on sheep numbers
Water catchment	Sporadic	*Penetration of reservoir construction and catchment control into the hills*	*Demand still high*	Major developments for flood control as well as urban-industrial supply
Woodlands	Deciduous woods at low ebb	Sporadic afforestation with conifers	*Effects of FC widely seen*	FC attempts to diversify in uplands; deciduous woods encouraged
Minerals: metals	Important source for home and abroad	*High demand from industry*	**After 1900, cheaper to import and so industries decline rapidly**	A few sporadic enterprises for e.g., fluorspar, re-working of lead spoil
Minerals: quarrying	Local use dominant	*Railways beginning to penetrate and demand increases*	*Roadstone and fluxes for iron/steel industry all important; china clay unique to SW England*	Demand still high but environmental impact strictly controlled

Military	Restricted to a few travellers and eccentrics	*WWII a turning point in the use of moorlands*	MoD still making strong case for expansion though also concessions in some places e.g., Dartmoor
Low-impact recreation: walking etc	*Opening up via railways and then buses: cult of the moorland hiker; mass trespass on grouse moors*	*Diversification of leisure time and the upsurge of private transport*	Pressures on popular places lead to paving paths
High-impact recreation, with machines and horses	(button fill)	(button fill)	*Largely confined to this period: trailbikes, FWD, mountain bikes, hang-gliding tend to conflict with other recreations and values. Horses need to be separated.*
Energy generation	Plans for HEP are rare due to transmission losses	One nuclear station in Wales	*Nuclear plant closes; windfarms becoming popular due to subsidy structure*
Nature conservation	Cataloguing ambitions include the hill flora and fauna	Ecology thinks of uplands as natural then semi-natural; reserves established	*Key theme of biodiversity and its relation to possible climatic change*
The scenery	Needs reclamation and improvement	*Re-valuation of the wild*	Still a major lineament of any planning for the uplands

Key:
Button fill: absent
Italic text: Times of particular rates of increase
Bold text: Times of rapid decrease

markets, some upland 'products' could supply the home market either because they were cheaper or because they had something of a monopoly position, especially in times of war. Stone for road-building and other low-value building purposes and limestone for steel furnaces are examples of high-bulk materials cheaper to transport without a marine phase, once the railways began to penetrate the uplands as in the Peak District and the fringes of Dartmoor. Wood is an interesting intermediate, for it is cheap to ship in ordinary times but a low priority in wartime: hence the foundation of the Forestry Commission. Yet other products cannot be supplied from outside the national polity. Water is the most obvious example: there is no world market in water.[162] Recreation is something of a hinge activity: going overseas has become progressively cheaper but the acquisition of private transport has kept up a strong demand for day and weekend trips to the hills. Likewise, the military can to some extent train overseas in the territories of allies but needs near-to-home areas on cost grounds. At the extreme lies the conservation of nature, with upland England and Wales containing species and habitats deemed to have regional and continental uniqueness worth conservation. Unlike all the other 'products' they have value, but not price. None of these demands is exclusive to the twentieth century but all have been intensified during it, leading to an ever-greater parcelling out and demarcation of land into pieces with a specific function, sometimes with legislation or subsidy structure to match (Plate 4.14).

What emerges environmentally, is firstly the constant importance of vegetation management for grazing and for sport: principally, that is, for the production of sheep, cattle and grouse; and secondly that into that pattern has been inserted a demand for preservation of the scenery (which is at least partly vegetation management translated into visual amenity) and the conservation of wild plants and animals, most of which is summarised in Table 4.9. Both these processes must be examined at greater length.

NOTES AND REFERENCES

1. In chapter 7 of *Voyage to Brobdingnag*, 1726.
2. M. Williams, 'The enclosure and reclamation of waste land in England and Wales in the eighteenth and nineteenth centuries' *Transactions of the Institute of British Geographers* 51, 1970, 55–79.
3. In the post-war years from 1815–50, according to the Cambridge Agricultural History volume (p. 55) commons were enclosed to the extent of 38,000 acres (15,200 ha) in Cumberland, 30,000 (12,000 ha) in Westmorland, 13,000 (5,200 ha) in Northumberland, 20,000 (8,000 ha) in upland Durham, 'possibly over 50,000 (20,000 ha) in the West Riding [of Yorkshire] and about 50,000 (20,000 ha) in the North Riding.
4. The house had walks, rockeries and grottoes and is said to have been the inspiration for Coleridge's *Xanadu* and was painted by J. M. W. Turner in 1798. In 1796, William Blake had engraved a folding map of the estate. The house was demolished in 1958, though some features remain.

5. A. Young, *Six Months Tour through the North of England 1771.* Facsimile edition: Augustus M. Kelley, New York: 1967. [The spelling has been modernised by the replacement of the 'long s' by 's']

6. R. I. Hodgson, 'The progress of enclosure in County Durham 1550–1870', in H. S. A. Fox and R. A. Butlin (eds), *Changes in the Countryside: Essays on Rural England 1500–1900,* London: Institute of British Geographers, 1979, 83–102.

7. P. Hembry, *The English Spa 1560–1815: a social history,* London: Athlone Press, 1990.

8. S. Sidney, 'Exmoor reclamation', *Journal of the Royal Agricultural Society of England* series 2, **14**, 1878, 72–97.

9. C. S. Orwin and R. J. Sellick, *The Reclamation of Exmoor Forest,* Newton Abbot: David & Charles, 1970. This is a revised and updated edition of C. S. Orwin's book of that title published by OUP in 1929.

10. Sidney, 1878 (Note 8). He was referring to the pre-enclosure days. In 1630, Westcote had said of Exmoor that, 'I see a spacious coarse barren and wild object, yielding little comfort by his rough complexion'. Quoted in H. Riley and R. Wilson-North, *The Field Archaeology of Exmoor,* Swindon: English Heritage 2001, 1.

11. M. L. Parry, A. Bruce and C. E. Harkness, *Changes in the Extent of Moorland and Roughland in the North York Moors.* University of Birmingham Department of Geography Moorland Change Project Surveys of Moorland and Roughland Change 5, 1982; M. L. Parry, *Changes in the Extent of Moorland and Roughland on Dartmoor. A Preliminary Survey to 1971.* University of Birmingham, Department of Geography, Moorland Change Project, Surveys of Moorland and Roughland Change **6**, 1982.

12. H. Harris, *The Industrial Archaeology of Dartmoor,* Newton Abbot: Peninsula Press, 1992, fourth edition, 204.

13. D. Thomas, *Agriculture in Wales during the Napoleonic Wars: a Study in the Geographical Interpretation of Historical Sources,* Cardiff: University of Wales Press, 1963; J. Davies, *The Making of Wales,* Stroud/Cardiff: Alan Sutton, Cadw, 1996.

14. The outstanding account is A. Harris and D. A. Spratt, 'The rabbit warrens of the Tabular Hills, North Yorkshire', *Yorkshire Archaeological Journal* 63, 1991, 176–201. Other useful material is in J. Sheail, *Rabbits and their History,* Newton Abbot: David and Charles, 1971; J. Sheail, 'Rabbits and agriculture in post-medieval England', *Journal of Historical Geography* 4, 1978, 343–55; C. D. Linehan, 'Deserted sites and rabbit warrens on Dartmoor, Devon', *Medieval Archaeology* **10**, 1966, 113–44.

15. G. R. Porter, *The Progress of the Nation* (ed. F. W. Hirst), London: Methuen, 1912.

16. Work by the Clwyd-Powys Archaeological Trust at www.cpat.org.uk/projects/longer/mines/mines.htm Accessed on 27 October 2002.

17. T. D. Ford and J. H. Rieuwerts (eds), *Lead Mining in the Peak District,* Bakewell: Peak Park Planning Board, 1968.

18. C. Hallas, *Rural Responses to Industrialization. The North Yorkshire Pennines 1790–1914.* Bern: Peter Lang, 1999, ch. 7. The book is almost entirely social and economic in its treatment.

19. R. T. Clough, *The Lead Smelting Mills of the Yorkshire Dales,* Leeds: The Author, 1962.

20. The major source of this section has been A. Raistrick and B. Jennings, *A History of Lead Mining in the Pennines,* London: Longman Green, 1965.

21. This includes the kind of data provided in Figure 1 of R. A. Fairbairn, 'An account of a small nineteenth-century lead mining company on Alston Moor', *Industrial Archaeology Review* 4, 1980, 245–56; and in Maps 1 and 2 of A. E. Shayler, J. K. Almond and H. L. Beadle, *Lead Mining and Smelting in Swaledale and Teesdale*, Guisborough: Cleveland Industrial Archaeology Society Research Report no. 2, 1979.

22. A. J. Howard, M. G. Macklin, S. Black and K. A. Hudson-Edwards, 'Holocene river development and environmental change in Upper Wharfedale, Yorkshire Dales, England', *Journal of Quaternary Science* 15, 2000, 239–52.

23. Raistrick and Jennings, 1965 (Note 20).

24. C. J. Hunt, *The Lead Miners of the Northern Pennines*, Manchester: Manchester University Press, 1970. Most of this paragraph derives from Hunt's work.

25. B. K. Roberts, 'Man and land in Upper Teesdale', in A. R. Clapham (ed.), *Upper Teesdale*, London: Collins, 1978, 141–59.

26. T. B. Bagenal, *Miners and Farmers. The Agricultural Holdings of the Lead Miners at Heights, Gunnerside, in North Yorkshire*, Keighley: Northern Mine Research Society, Monograph No. 62, 1999. He interprets a document from the early 1860s as dealing with the overgrazing of the commons by sheep from households without common rights and that miners were therefore keeping sheep, though the evidence on the ground and in other documents is scarce.

27. The 'bonny moor hen' was the grouse. The Bishop is called '. . . the fat man of Auckland and Durham the same' and he sent in 'land stewards, bum bailiffs, and game-keepers too'. A number of pubs in the area bear that name.

28. T. Sopwith, *An Account of the Mining Districts of Alston Moor, Weardale and Teesdale, in Cumberland and Durham; comprising descriptive sketches of the scenery, antiquities, geology and mining operations, in the upper dales of the rivers Tyne, Wear and Tees'*, Alnwick: W. Davison, 1833 [in fact this is the third edition].

29. F. Booker, 'Industry', in C. Gill (ed.), *Dartmoor. A New Study*, Newton Abbot: David & Charles, nd, 100–38; S. Gerrard, 'The Dartmoor tin industry: an archaeological perspective', in D. M. Griffith (ed.), *The Archaeology of Dartmoor. Perspectives from the 1990s*, Exeter: Devon Archaeological Society Proceedings 52, 1994, 173–98.

30. H. Harris, 1992 (Note 12).

31. D. J. Panett, D. Thomas and R. G. Ward, 'Farm patterns in the Stiperstones mining district. I. Field patterns and historical analysis', *Field Studies* 3, 1973, 763–82.

32. R. White, *The Yorkshire Dales. A Landscape through Time*, Ilkley: Great Northern Books, 2002, 92.

33. Their cuttings incidentally provided me with the material for the first pollen profile I ever constructed: and a very dull sequence it is, too.

34. There was a 15 to 17-fold increase in the aggregate value of British (and German) water production in the years 1855–1913, quotes J. A. Hassan, 'The growth and impact of the British water industry in the nineteenth century', *Economic History Review* 38, 1985, 531–47.

35. M. G. Evans, T. P. Burt, J. Holden and J. K. Adamson, 'Runoff generation and water table fluctuations in blanket peat: evidence from UK data spanning the dry summer of 1995', *Journal of Hydrology* 221, 1999, 141–60.

36. G. M. Binnie, *Early Dam Builders in Britain*, London: Thomas Telford, 1987.

37. J. A. Hassan and E. R. Wilson, 'The Longdendale water scheme 1848–1884', *Industrial Archaeology* 14, 1979, 102–21. From above, the lakes look almost like an ornamental cascade of impoundments; some (e.g. Torside) have a strong re-creational use.

38. F. Hibbert, 'A description of the Liverpool Corporation water supply undertaking – 1847–1947', *Water and Water Engineering* 51, 1948, 202–11.

39. J. Hassan, *A History of Water in Modern England and Wales*, Manchester and New York: Manchester University Press, 1998.

40. J. Sheail, 'Government and the perception of reservoir development in Britain: an historical perspective', *Planning Perspectives* 1, 1986, 45–60. On the Manchester gathering grounds in The Lake District, even the shepherds became employees of the Corporation.

41. There was an incipient revolution in the 1950s when it was suggested that afforestation actually reduced water yield since the trees transpired water. Subsequent research suggests that it is interception by the tree canopy which is more important and that loss from that process might be 403 mm/yr in an unforested catchment but 854–1,309 mm/yr in a catchment which was 60 per cent covered in trees. However, the forest may not maintain its highest rates of evaporation through the whole growth cycle. It may decline as the trees reach their economic zenith and then for short periods (during and after clear-felling) water yield may be even greater from forested catchments than from grassland areas. See J. A. Hudson, S. B. Crane and J. R. Blackie, 'The Plynlimon water balance 1969–95: the impact of forest and moorland vegetation on evaporation and streamflow in upland catchments', *Hydrology and Earth System Sciences* 1, 1997, 409–27.

42. R. C. S. Walters, *The Nation's Water Supply*, London: Ivor Nicholson and Watson, 1936.

43. J. B. Harris, 'Hydro-electricity in Devon. Past, present and future', *Transactions of the Devon Association* 127, 1995, 259–86.

44. H. Harris, 'The Sourton Tors iceworks, North-west Dartmoor, 1874–86', *Transactions of the Devon Association* 120, 1988, 177–200.

45. The species *Lagopus lagopus* is endemic to the British Isles. Some authors attribute a Scottish and an Irish subspecies (*L. l. scoticus* and *L. l. hibernicus*) but they are indistinguishable in the field.

46. A. Vandervell and C. Coles, *Game and the English Landscape*, New York: Viking Press, 1980. An upland squire started to walk up a moorland at 3:30 am in August 1828 and gave up at 6 pm, having shot forty-four grouse. '. . . a very considerable achievement,' says B. Vesey-Fitzgerald, *The Vanishing Wild Life of Britain*, London: MacGibbon & Kee, 1969, 105.

47. With care. Gamekeepers do not like outsiders walking across managed moors during the nesting season in April nor any time before shooting days, starting on August 12th. Where possible, they will include public rights of way in their exclusion zones. Grouse moors are the hottest points of contention between large landowners and the 'right to roam' movement.

48. Incidental to our story, the coming of this type of gun also allowed pheasant shooting in the lowlands to upstage the partridge shooting which had been dominant for a long time.

49. In the early years of their use they were called batteries. This information, along with a lot of other didactic material and some details of large bags can be found in classics such as Lord Walsingham and Sir Ralph Payne-Gallwey's *Shooting. Moor and Marsh*. London: Longman Green, 1889, 3rd edn. The Badminton Library of Sports and Pastimes, edited by the Duke of Beaufort, assisted by A. E. T. Watson. (The series included a volume on driving contributed by His Grace himself.) In *Shooting. Moor and Marsh* (the other volume of *Shooting* is *Field and Covert*), 'Grouse' forms chapter 1.

50. B. Vesey-Fitzgerald, *British Game*, London: Collins New Naturalist, 1946. The date of 1915 for the Lancashire bag raises questions about the impact of the

Great War on previous culling levels and management. Not about the accuracy of the fire, it seems.

51. A. Done and R. Muir, 'The landscape history of grouse shooting in the Yorkshire Dales', *Rural History* **12**, 2001, 195–211.

52. Though for England and Wales I know of no tally like that of 1837–40 for Glen Garry where 1,795 birds of prey were killed, as well as 109 owls and 475 ravens. The raptors included sea eagles, goshawks, honey-buzzards and gerfalcons. Egg-collecting also affected raptor populations badly in the nineteenth century. See R. Perry, *Wildlife in Britain and Ireland*, London: Croom Helm 1978.

53. (Lord) Lovat, *The Grouse in Health and Disease*, Final Report of the Committee of Inquiry on Grouse Disease. London: Smith Elder, 1911, 2 vols.

54. J. H. Lawton (ed.), *Red Grouse Populations and their Management*, London: British Ecological Society, Ecological Issues no. 2, 1990.

55. D. Baines, 'The implications of grazing and predator management on the habitats and breeding success of black grouse *Tetrao tetrix*', *Journal of Applied Ecology* **33**, 1996, 54–62.

56. A study at Lour in the Tweed valley of southern Scotland suggests that some ridge-and-furrow at about 250 m ASL is due to improvement of rough pasture for sheep grazing in the nineteenth century and not for cereal cultivation. Parallel studies further south are needed. See S. Carter, R. Tipping, D. Davidson, D. Long and A. Tyler, 'A multiproxy approach to the function of postmedieval ridge-and-furrow cultivation in upland northern Britain', *The Holocene* 7, 1997, 447–56.

57. This was M. L. Parry's project at Birmingham University in the 1980s on Moorland and Roughland Change, published as a series of Reports from that University's Department of Geography and referred to individually. They were focused on regions which had received designation as National Parks.

58. Net migration within Britain in 1990–1 brought 36,450 people to the category of 'most remote rural', which was the largest absolute total increase in seven types of district where there was net immigration. This category and 'remote rural' were the highest proportionally as well, with increases of 0.77 per cent and 0.61 per cent respectively. (Inner London lost 1.24 per cent.) See A. G. Champion, 'Studying counterurbanisation and the rural population turnaround', in P. Boyle and K. Halfacree (eds), *Migration into Rural Areas. Theories and Issues.* Chichester: Wiley, 1998, 21–40.

59. These data are from M. L. Parry, A. Bruce and C. E. Harkness, *Changes in the Extent of Moorland and Roughland in the Brecon Beacons National Park*, Department of Geography, University of Birmingham, Surveys of Moorland and Roughland Change no. 7, 1982 ; M. L. Parry *et al.*, *Changes in the Extent of Moorland and Roughland in the Yorkshire Dales National Park*, Department of Geography, University of Birmingham, Surveys of Moorland and Roughland Change no. **9**, 1984. Each Report is sensitive to local conditions and explains carefully the problems of interpreting the evidence. It also explains the methodology used to determine the data replicated here.

60. Parry *et al.*, 1982, 1984 (Note 59).

61. A. D. Rees, *Life in a Welsh Countryside. A Social Study of Llanfihangel yng Ngwynfa*, Cardiff: University of Wales Press, 1950. Most of the data come from the late 1940s.

62. A. Woods, *Upland Landscape Change: a Review of Statistics*, Cheltenham: Countryside Commission, CCP 161, 1984.

63. Woods, 1984 (Note 62).

64. P. Allenby and G. Gerrard, *Soil and Vegetation Characteristics of Reverted Farmland on Anglezarke Moor, Lancashire*, School of Geography, University of Birmingham Working Paper series no. **45**, 1988.

65. O. Wilson, *Landownership and Rural Development in Theory and Practice: Case Studies from the North Pennines in the nineteenth and twentieth centuries*. PhD thesis, University of Durham, 1990; O. Wilson, 'Landownership and rural development in the North Pennines: a case study', *Journal of Rural Studies* **8**, 1992, 145–58.

66. This was an Article 4 Direction Order under the provisions of a General Development Order. An Article 4 allows an authority to require planning permission to be granted even in areas and for developments where this is not normally required and is usually applied in conservation matters.

67. Lord Porchester, *A Study of Exmoor*, London: DoE/MAFF, 1977.

68. In 1977 the standard sum payable on Exmoor for profits foregone on land where ploughing was feasible was £73.87/ha/yr; in 1981, £90.52/ha/yr. If ploughing was not feasible but fertilising and fencing were, then the 1977 figure was £41/ha/yr (L. F. Curtis, ' Reflections on management agreements for conservation of Exmoor moorland', *Journal of Agricultural Economics* **34**, 1983, 397–406).

69. I. Brotherton, 'Factors affecting the conclusion of management agreements in National Parks', *Landscape Research* **14**, 1989, 27–34; I. Brotherton, 'On the rise and demise of conflict in Exmoor', *Journal of Environmental Management* **30**, 1990, 353–70.

70. M. L. Parry, A. Bruce and C. F. Harkness, 'The plight of the moorlands', *New Scientist* **90**, 1981, 550–1.

71. For an excellent cultural-political overview of this century see J. Tsouvalis, *A Critical Geography of Britain's State Forests*, Oxford: Oxford University Press, 2000.

72. A biography of Gladstone by Roy Jenkins (*Gladstone*, London: Macmillan, 1995) shows that felling trees was a favourite leisure activity of the grand old man in the last third of the nineteenth century. Gladstone's diary records such an event on many of the days he spent on his own, and others', estates.

73. N. D. G. James, *A History of English Forestry*, Oxford: Blackwell, 1981. This forms the source for most of the administrative material in this section, with a few details from H. L. Edlin, *Trees, Woods and Man*, London: Collins New Naturalist, 1956.

74. E. J. T. Collins, *The Economy of Upland Britain 1750–1950: an Illustrated Review*, Reading: University of Reading Centre for Agricultural Strategy, 1978.

75. Forest Enterprise, *Annual Report 2000–2001*, London: The Stationery Office, 2002. In 1996–7 the Forestry Commission was reorganised to be a Commission with two executive agencies (Forest Enterprise and Forest Research) and from 1999–2000 each country of the UK has had distinct forestry policies and funding.

76. G. Ryle, *Forest Service. The First Forty-Five Years of the Forestry Commission of Great Britain*, Newton Abbot: David and Charles, 1969. There is a short but effective history of the Commission and its work in J. Sheail, *An Environmental History of Twentieth-Century Britain*, Basingstoke and New York: Palgrave, 2002, ch. 4.

77. W. O. Binns, 'The hydrological effect of afforestation in Great Britain', in G. E. Hollis (ed.), *Man's Impact on the Hydrological Cycle in the United Kingdom*, Norwich: Geo Abstracts, 1979, 55–69. Some work suggests that the overall nutrient cycle of the site is not much affected by conifer growth and harvesting if a

complete cycle is taken into account: G. Page, 'The effect of conifer crops on soil properties', *Commonwealth Forestry Review* **47**, 1968, 52–62.

78. B. Reynolds, D. Fowler and S. Thomas, 'Chemistry of cloud water at an upland site in mid-Wales', *Science of the Total Environment* **188**, 1996, 115–25.

79. J. C. I. Kuyliensterna and M. J. Chadwick, 'Increases in soil acidity in North-West Wales between 1957 and 1990', *Ambio* **20**, 1991, 118–19.

80. The data in this and the previous paragraph are from the Plynlimon study in mid-Wales: C. Kirby, M. D. Newsom and K. Gilman (eds), *Plynlimon Research: the First Two Decades*, Wallingford: Institute of Hydrology Report no. **109**, 1991, ch. 6.

81. CPRE now stands for the Council for the Protection of Rural England but it was 'Preservation' in the 1930s. Some discussion of this and similar organisations is in Chapter 5.

82. J. Sheail, *Nature in Trust. The History of Nature Conservation in Britain*, Glasgow and London: Blackie, 1976.

83. J. Strak and C. Machel, *Forestry in the Rural Economy*, Paper no. 12 of the Forestry Commission, *Forestry Expansion: A Study of Technical, Economic and Ecological Factors*, Farnham: Forestry Commission, 1991.

84. G. F. Peterken, *Natural Woodland. Ecology and Conservation in Northern Temperate Regions*, Cambridge: Cambridge University Press, 1996.

85. G. F. Peterken, 1996 (Note 84).

86. Nature Conservancy Council, *Nature Conservation and Afforestation in Britain*, London: HMSO, 1986; S. J. Petty, *Ecology and Conservation of Raptors in Forests*, London: HMSO, 1998, Forestry Commission Bulletin **118**; the decline of the raven (breeding in every British county in 1900) is caught up with other land use changes as well as afforestation. Like the red kite, it had been an urban scavenger until about 1800. See P. G. Moore, 'Ravens (*Corvus corax corax* L.) in the British landscape: a thousand years of ecological biogeography in place-names', *Journal of Biogeography* **29**, 2002, 1039–54.

87. J. Rodwell and G. Patterson, *Creating New Native Woodlands*, London: HMSO, 1994, Forestry Commission Bulletin **112**; G. Peterken, D. Ausherman, M. Buchenau and R. T. T. Forman, 'Old-growth conservation within British upland conifer plantations', *Forestry* **65**, 1992, 127–44.

88. R. S. Smith and D. J. Charman, 'The vegetation of upland mires within conifer plantations in Northumberland, Northern England', *Journal of Applied Ecology* **25**, 1998, 579–94; see also R. S. Smith, A. G. Lunn and M. D. Newsom, 'The Border Mires in Kielder Forest: a review of their ecology and conservation management', *Forest Ecology and Management* **79**, 1995, 47–61.

89. These arrangements have now been withdrawn. The trees tend to remain.

90. D. Taylor, 'Land availability for future afforestation', in G. R. Hatfield (ed.), *Farming and Forestry*, Farnham: Forestry Commission, 1988, 40–3. The price of land is the major determinant of what is bought for planting up.

91. M. Cannell and J. Cape, *International Environmental Impacts: Acid Rain and the Greenhouse Effect*, Report no. 2 of Forestry Commission, *Forestry Expansion: a Study of the Technical, Economic and Ecological Factors*, Farnham: Forestry Commission, 1991.

92. M. H. Garnett, P. Ineson and A. C. Stevenson, 'Effects of burning and grazing on carbon sequestration in a Pennine blanket bog, UK', *The Holocene* **10**, 2000, 729–36.

93. G. F. Peterken, 'Woodland conservation in Britain', in A. Warren and F. B. Goldsmith (eds), *Conservation in Perspective*, Chichester: Wiley, 1983, 83–100; C. Watkins, *Woodland Management and Conservation*, Newton Abbot and

London: David and Charles, 1990; Rodwell and Patterson, 1994 (Note 87); G. F. Peterken, *Woodland Conservation and Management*, London: Chapman and Hall, 1993, 2nd edn: p. 338 *et seq*; R. C. Steele and G. F. Peterken, 'Management objectives for broadleaved woodland conservation', in D. C. Malcolm, J. Evans and P. N. Edwards (eds), *Broadleaves in Britain. Future Management and Research*, Farnham: Forestry Commission/Institute of Chartered Foresters, 1984, 91–103.

94. A recent account which combines all available types of evidence could reach no fresh conclusions about the origin of Piles Copse. See C. A. Roberts and D. D. Gilbertson, 'The vegetational history of Pile [sic] Copse 'ancient' oak woodland, Dartmoor, and the possible relationships between ancient woodland, clitter and mining', *Proceedings of the Ussher Society* 8, 1994, 298–301. An analogous investigation is J. P. Barkham, 'Pedunculate oak woodland in a severe environment: Black Tor Copse, Dartmoor', *Journal of Ecology* 66, 1978, 707–40. Recent changes in the growth patterns of the trees in Wistman's Wood are set out in E. P. Mountford, C. E. Backmeroff and G. Peterken, 'Long-term patterns of growth, mortality, regeneration and natural disturbance in Wistman's Wood, a high-altitude wood on Dartmoor', *Transactions of the Devon Association* 133, 2001, 227–62.

95. F. Mitchell and K. J. Kirby, 'Impact of large herbivores and conservation of semi-natural woods in the British uplands', *Forestry* 63, 1990, 333–53.

96. The vision document for England (Forestry Commission, *A New Focus for England's Woodlands*, Cambridge: Forestry Commission, 2001) is strong on the spread and use of broad-leaved woodlands but does not separate out the uplands at all. Indeed, there is a great deal of emphasis on the new woodlands of lowland areas.

97. R. J. Heaton, P. F. Randerson and F. M. Slater, 'The economics of growing short rotation coppice in the uplands of mid-Wales and an economic comparison with sheep production', *Biomass and Bioenergy* 17, 1999, 59–71.

98. Two nineteenth-century visitors identified some of the rarities but merely thought them to be 'interesting plants': J. Backhouse and J. Backhouse, 'An account of a visit to Teesdale in the summer of 1843', *Phytologist* 1, 1843, 892–5.

99. A. R. Clapham (ed.), *Upper Teesdale*, London: Collins, 1978, especially chapters 1–3, 5–6.

100. H. Godwin and S. M. Walters, 'The scientific importance of Upper Teesdale', *Proceedings of the Botanical Society of the British Isles* 6, 1967, 348–51. Only four papers are cited, which suggests that scientific work on the flora had not been very plentiful. They omit, however, the overview account by C. D. Pigott, 'The vegetation of Upper Teesdale in the North Pennines', *Journal of Ecology* 44, 1956, 545–86.

101. Initial opposition came from an individual (Dr Margaret Bradshaw), then the Botanical Society of the British Isles and the County Naturalists' Trust and was followed by biologists at the Universities of Sheffield and Cambridge and thereafter by the biological establishment. The Nature Conservancy was hamstrung by the fact that its Director had said early in the discussions that the Conservancy would not object to Cow Green. Further, some of the biologists examined in the Parliamentary process were less than impressive in their knowledge of what they were trying to protect. The story is grippingly told, with extracts from the various proceedings and correspondence in R. Gregory, *The Price of Amenity*, London and Basingstoke: Macmillan, 1971, chapter 4.

102. T. J. Bines, J. P. Doody, I. H. Findlay and M. J. Hudson, 'A retrospective view of the environmental impact on Upper Teesdale of the Cow Green Reservoir', in

R. D. Roberts and T. M. Roberts (eds), *Planning and Ecology*, London and New York: Chapman and Hall, 1984, 395–421.

103. B. Huntley, R. Baxter, K. J. Lewthwaite, S. G. Willis and J. K. Adamson, 'Vegetation responses to local climatic changes induced by a water-storage reservoir', *Global Ecology and Biogeography Letters* 7, 1998, 241–57.

104. The specialist account of the vegetation history is by J. Turner *et al.*, The history of the vegetation and flora of Widdybank Fell and the Cow Green reservoir basin', *Philosophical Transactions of the Royal Society of London* B 265, 1973, 327–408; a more popular account by Turner is in Clapham, 1978 (Note 99), with a special section on the history of the rarities at 97–101. See also R. H. Squires, 'Flandrian history of the Teesdale rarities', *Nature* 229, 1971, 43–4; R. H. Squires, 'Conservation in Upper Teesdale: contributions from the palaeoecological record', *Transactions of the Institute of British Geographers* NS 3, 1978, 129–50. The account of land use history by B. K. Roberts ('Man and land in Teesdale' in Clapham 1978 (Note 99), 141–59) supports this in general terms, though no specific linkages are made for the Cow Green area. Even the hard-rock-soil-flora correlators end by admitting that the open habitats have been maintained by 'continuous grazing of the small areas of limestone grassland from prehistoric times': G. A. L. Johnson, D. Robinson and M. Hornung, 'Unique bedrock and soils associated with the Teesdale flora', *Nature* 232, 1971, 453–6. The reservoir may also have destroyed the habitat of Peg Powler, the green-haired mermaid of the Tees, who had an insatiable appetite for children and whose presence was indicated by frothy substances on the river, known as Peg Powler's suds.

105. O. L. Gilbert, 'Juniper in Upper Teesdale', *Journal of Ecology* 68, 1980, 1013–24.

106. M. C. Whitby and K. G. Willis, *Rural Resource Development. An Economic Approach*, London: Methuen, 1978, ch. 13, 'A cautionary case study'.

107. Anon, 'Kielder Water and associated schemes' *Water Services* 83 (no. 1002, August), 1979, 645–51.

108. D. Stoker, *A Short History of the Kielder Water Scheme*, Newcastle upon Tyne: Northumbrian Water Authority, 1982.

109. In 1985, one geographer suggested that Kielder's importance was to be seen largely in recreational terms, though in future it might attract water-consuming industries to the region. In 2002, these latter had still to appear: see V. Gardiner, 'The Kielder water scheme: changing emphases in multipurpose water management', *Geography* 70, 1985, 254–7. See also V. Gardiner. 'The Kielder water scheme – financial and environmental implications of demand forecasting', in V. Gardiner and P. Herrington (eds), *Water Demand Forecasting; Proceedings of a Workshop, Leicester, 1985*, Norwich: Geo Books, 1985, 119–23. Some of the early calculations about Kielder's value supposed that hydro-power would be generated, which has not in fact happened.

110. 'Geordies of the future will have all mod cons', *Newcastle Journal* 10 February 1972, p. 5. A teletext item in 2002 reported that north-east men now had at least four or five baths per week.

111. G. Mitchell and A. T. McDonald, 'Catchment characterization as a tool for upland water quality management', *Journal of Environmental Management* 44, 1995, 83–95.

112. See for example Environment Agency, *Water Resources for the Future. A Strategy for the North East Region*, Leeds: Environment Agency, 2001 and its equivalent document for Wales, of the same date. There is also an EU context in a Water Framework Directive (2000/60/EC).

113. The classic introduction of its time was M. M. Sweeting, 'The weathering of limestones', in G. H. Dury (ed.), *Essays in Geomorphology*, London: Heinemann,

1966, 177–210. The context of the British Isles is explored in P. Vincent, 'Limestone pavements in the British Isles: a review', *Geographical Journal* 73, 1995, 283–315; some measurements are given in H. S. Goldie and N. J. Cox, 'Comparative morphometry of limestone pavements in Switzerland, Britain and Ireland', *Zeitschrift für Geomorphologie* 112 *(Suppl)*, 2000, 85–112.

114. T. M. Thomas, 'The limestone pavements of the north crop of the South Wales coalfield with special reference to solution rates and processes', *Transactions of the Institute of British Geographers* 50, 1970, 87–105.

115. S. D. Ward and D. F. Evans, 'Conservation assessment of British limestone pavements based on floristic criteria', *Biological Conservation* 9, 1976, 217–33.

116. The 1981 legislation allows the issue of a Limestone Pavement Order. See H. S. Goldie, 'The legal protection of limestone pavements in Great Britain', *Environmental Geology* 21, 1993, 160–6; H. S. Goldie, 'Major protected sites of limestone pavement in Great Britain', in I. Bárány-Kevei (ed.), *Environmental Effects on Karst Terrains*, Szeged: Attila József University, 1995, 61–92. The Limestone Pavement Action Group (LPAG) has a lovely pamphlet *Limestone Pavement. Our Fragile Heritage*. Windermere: LPAG, nd.

117. C. Kirby, M. D. Newsom and K. Gilman (eds), *Plynlimon Research: the First Two Decades*, Wallingford, Institute of Hydrology Report no. 109, 1991. See also D. L. Higgett, J. S. Rowan and D. E. Walling, 'Catchment-scale deposition and redistribution of Chernobyl radiocaesium in upland Britain', *Environment International* 19, 1993, 155–66; P. J. P. Bonnett and P. G. Applebey, 'Deposition and transport of radionuclides within an upland drainage basin in mid-Wales', *Hydrobiologia* 214, 1991, 71–6.

118. Two useful sources in a very scattered literature are D. C. W. Sanderson and E. M. Scott, *Aerial Radiometric Survey in West Cumbria, 1988: Project N611*. London: MAFF, 1989; B. Wynne, 'Sheep farming after Chernobyl', *Environment* 31, 1989, 10–39. MAFF issues yearly press releases with the latest data on restricted areas.

119. A. Nisbet and R. Woodman, 'Options for the management of Chernobyl-restricted areas in England and Wales', *Journal of Environmental Radioactivity* 51, 2000, 239–54.

120. In Trevelyan's foreword to a CPRE/CPRW report on *The Case for National Parks in Great Britain* (1938). The chairman was Sir Norman Birkett. Some interpreters see the rise of the CPRE and the National Trust as protectors of the 'old' rural against the hordes from the city.

121. There is no detailed rehearsal of the controversies that preceded the 1949 Act in this book. For such particulars see e.g., T. Stephenson, *Forbidden Land. The Struggle for Access to Mountain and Moorland*, Manchester: Manchester University Press, 1989; M. Shoard, *A Right to Roam*, Oxford: Oxford University Press, 1999.

122. This is in Annex 2 of the *Reports* of the *The Countryside in 1970* conference (London: Royal Society of Arts/Nature Conservancy, 1965), Study Group no. 6, 'Recreation: Active and Passive', pp. 6.18–6.21.

123. M. C. Whitby, D. L. J. Robins, A. W. Tansey and K. G. Willis, *Rural Resource Development*, London: Methuen, 1974, ch. 6. The Countryside Agency estimated average daily expenditure (excluding accommodation) in the parks in 1994 as £9.70 per person and as higher than day visitors to the countryside in general. See note 129.

124. J. Blunden and N. Curry (eds), *A People's Charter? Forty Years of the* **National Parks and Access to the Countryside Act 1949**, London: HMSO, 1990, chapter 7. Note that this book deals with all the topics of the 1949 Act, including nature conservation.

125. House of Commons Environment Committee, *The Environmental Impact of Leisure Activities*, Fourth Report, Volume I, London: HMSO 1995. Bike trails near Dolgellau in North Wales are estimated to bring in £4 m/yr into the local economy according to a progress report for the Welsh Assembly. (Forestry Commission, *A Woodland Strategy for Wales*, 2001, accessed at www.forestry. gov.uk on 14/11/01)

126. D. Haffey, 'Recreational activity patterns on the uplands of an English National Park', *Environmental Conservation* 6, 1979, 237–42; J. A. Patmore, *Recreation and Resources*, Oxford: Basil Blackwell; J. A. Patmore, 'A case study in National Park planning', in P. Cloke (ed.), *Rural Planning: Policy into Action?* London: Harper and Row, 1987, 88–101. Patterns in the North York Moors may have changed by now because of the popularity of Goathland, following its star role in the TV series, *'Heartbeat'*. Hawes in Wensleydale has also experienced a TV-fuelled surge in popularity.

127. Since pony-trekking accounted for some at least of the grant applications, the uplands may have been as forward as anywhere in trying to raise income. But the data are not very precise. The situation in the south-western uplands was detailed in A. Cowen, G. Lavers and M. Dower, *Integrated Rural Development. A Study of Bodmin Moor, Dartmoor and Exmoor*, Totnes: Dartington Institute, 1984.

128. Countryside Commission, *National Countryside Recreation Survey: 1984*, Cheltenham: Countryside Commission, 1985, CCP 201, see Figs 17 and 19.

129. Countryside Agency, *Visitors to National Parks*, Research Notes CCRN1, 1998, accessed at http://countryside.gov.uk/research/notes/ccrn1.htm

130. Table 2.1 of M. Liddle, *Recreation Ecology. The Ecological Impact of Outdoor Recreation and Tourism*, London: Chapman and Hall, 1997, gives some comparative data of the pressure exerted on the ground surface by different recreation 'users': a barefoot human weighs in at 297 g/cm^2, a man in boots at 206, the shoes of a mounted horse at 4,380, a trail bike at 2,008, a loaded Toyota 4WD at 1,686 and sheep at 690–941 g/cm^2. The shoed horse with rider tops the list, with a Jeep second at 2,240 g/cm^2.

131. N. Bayfield, 'Effects of extended use on footpaths in mountain areas of Britain', in N. Bayfield and G. C. Barrow (eds), *The Ecological Impacts of Outdoor Recreation on Mountain Areas in Europe and North America*, Wye (Kent): Recreation Ecology Research Group, 1983, 100–11.

132. G. Lee, 'Managing the archaeology of the North York Moors National Park', in B. Vyner (ed.), *Moorland Monuments: Studies in the Archaeology of North-East Yorkshire in Honour of Raymond Hayes and Don Spratt*, London: CBA Research Report 101, 1995, 140–52. It is tempting to think that trials of endurance like the Lyke Wake Walk are civilian versions of the way in which uplands are central to the idea of making a (male) soldier sufficiently tough. (Though women do both.) See R. Woodward, '"It's a Man's Life!": soldiers, masculinity and the countryside', *Gender, Place and Culture* 5, 1998, 277–300.

133. The National Trust has been a pioneer in these approaches to moorland and mountain management and restoration.

134. N. G. Bayfield, A. Watson and G. R. Miller, 'Assessing and managing the effects of recreational use on British hills', in M. B. Usher and D. B. A. Thompson (eds), *Ecological Change in the Uplands*, Oxford: Blackwell Scientific Publications, Special Publications of the British Ecological Society no. 7, 1988, 399–414.

135. House of Commons Environment Committee, 1995 (Note 125).

136. M. Shoard, *A Right to Roam*, Oxford: Oxford University Press, 1999. She quotes a number of examples of successful agreements but ignores any which have failed.

137. N. Picozzi, 'Breeding performance and shooting bags of red grouse in relation to public access in the Peak District National Park, England', *Biological Conservation* 3, 1971, 210–15.

138. Patmore, 1987 (Note 126).

139. In 1966, the then Ministry of Land and Natural Resources put forward the idea of Country Parks to take the pressure off National Parks. A few were actually established *inside* National Parks and notably they are usually able to absorb a lot of cars and people in resilient mixtures of e.g., woodland, gardens and meadow. See for instance P. J. Cloke and C. C. Park, 'Country Parks in National Parks: a case study of Craig-y-Nos in the Brecon Beacons, Wales', *Journal of Environmental Management* 12, 1981, 173–85. The notion that not all agriculturally peripheral areas want to turn to tourism for their economic salvation is pointed out by M. Kneafsey, 'Tourism, place identities and social relations in the European rural periphery', *European Urban and Regional Studies* 7, 2000, 35–50.

140. D. J. Pearlman, J. Dickinson, L. Miller and J. Pearlman, 'The Environment Act 1995 and quiet enjoyment: implications for countryside recreation in the National Parks of England and Wales, UK', *Area* 31, 1999, 59–66.

141. We might note that the interaction of the 1999 UK provisions on Environmental Impact in England and Wales (Statutory Instruments 1999 No. 293 and 1999 No. 2228) with the EU Directive 85/337/EEC (as amended by Directive 97/11/EC) may result in a stringent environmental assessment of any proposals to bring uncultivated land and semi-natural areas into intensive agricultural production. The Second Consultation Papers were issued during 2001 by DEFRA and by the Welsh Assembly.

142. No reason why it should have done: legislation is normally conservative and cannot speculate about e.g., technological changes. See also R. Simmonds, 'Conflicts in the countryside', *Journal of the Royal Agricultural Society of England* 158, 1997, 138–43.

143. Details of the Act and the campaign for it could be conveniently found on the Ramblers' Association website (www.ramblers.org.uk) in May 2001; at the same time the CLA had no mention of it.

144. Although the owner of a rental cottage on the side of the Usk valley in South Wales, where jet fighters regularly flew below the level of the house, said she loved them and would very much like to fly them. Penned sheep being dipped nearby seemed less certain.

145. A. MacEwen and M. MacEwen, *National Parks, Conservation or Cosmetics?* London: Allen and Unwin, 1982, 240. It is not clear whether it was the gun or the shells that produced this impact: presumably the latter, unless the gun was moved from time to time.

146. S. Owens, 'Defence and the environment: the impacts of live firing in national parks', *Cambridge Journal of Economics* 14, 1990, 497–505.

147. These are amateur photographs taken by Mr Stephen Johnson of Plymouth and accessed on the internet on 23/07/00 at http://web.ukonline.co.uk/ stephen. johnson/air/hole.jpg (±vertical) and http://web.ukonline.co.uk/stephen. johnson/air/hole2.jpg (oblique). Mr Johnson kindly allows free reproduction of these pictures.

148. S. Owens, *Military Live Firing in National Parks*, London: UK Centre for Economic and Environmental Development, 1990.

149. There were two government inquiries into MoD land holdings and use: (Lord) Nugent, *Report of the Defence Lands Committee*, London: HMSO, 1973, Cmnd 5714; (Lady) Sharp, *Dartmoor: a Report into the Continued Use of Dartmoor by*

the Ministry of Defence for Training Purposes, London: HMSO, 1977. I have seen the last word in the title of Lady Sharp's report mis-spelled to amusing effect.

150. The prosaic explanation is no doubt to do with remoteness and with land prices but there could be something deeper. (Montgomery chose Luneberg Heath for the surrender of the Wehrmacht in 1945.) At all events, the use of moorlands for low-flying training by the RAF and NATO allies seemed not to help in Kosovo in 1999, where aircraft were kept very high in order to avoid the deadly fire of the Serbs.

151. This section relies upon J. Weston, J. Glasson, E. Wilson and A. Chadwick, 'More than local impacts: aggregate quarrying in the National Parks of England and Wales', *Journal of Environmental Policy and Management* **1**, 1999, 245–68. Discussions of the mineral extraction problems in the National Parks can be found in the Sandford Report (1974) and in MacEwan and MacEwan *op. cit.* 1982. For china clay on Dartmoor, see Anon, 'Dartmoor saved from china clay quarries' *Viewpoint* no **32**, 2001, 4–5. [*Viewpoint* is the magazine of the Council for National Parks]

152. Anon, 'Dormant quarries', *Viewpoint* no. **29**, 2000, 4. In 1999 there was an application to re-work limestone on a large scale on Stanton Moor (at Endcliffe and Lees Cross quarries) in the Peak District under a 1952 mineral permission. A decision was due in the spring of 2003.

153. Countryside Council for Wales, *Energy: Policy and Perspectives for the Welsh Countryside*, CCW 1992 Policy Document **030** [bilingual text]; English Nature, RSPB, WWF-UK and BWEA, *Wind Farm Development and Nature Conservation*, Peterborough: English Nature, 2001.

154. Søren Krohn, *The Energy Balance of Modern Wind Turbines*, Copenhagen: Vindmølleneindustrien Background Information Note no. **16**, 1997. Available in October 2001 at www.windpower.dk

155. M. M. Hedger, 'Wind power – challenges to planning policy in the UK', *Land Use Policy* **12**, 1995, 17–28; D. Lindley, 'The future for wind energy development in the UK – prospects and problems', *Renewable Energy* **5**, 1994, 44–57; R. W. Coles and J. Taylor, 'Wind power and planning – the environmental impact of windfarms', *Land Use Policy* **10**, 1993, 205–26; A. Hull, 'Windfarms in the U.K.: the arguments for and against', *Ekistics* **382/394**, 1997, 112–16.

156. The hydro-power was still being generated in 2002.

157. These data are published in the *UK Strategy for Radioactive Discharges 2001–2020* on the website of DEFRA.

158. M. S. Cresser, R. Smart and M. Stutter, 'Water quality in the British uplands', in T. P. Burt *et al.* (eds), *The British Uplands: Dynamics of Change*, Peterborough: JNCC Report no. 319, 2002, 48–59; see also the relevant chapters in S. T. Trudgill (ed.), *Solute Modelling in Catchment Systems*, Chichester: Wiley, 1995.

159. C. J. Curtis, T. E. H. Allott, B. Reynolds and R. Harriman, 'The prediction of nitrate leaching with the first-order acidity balance (FAB) model for upland catchment in Great Britain', *Water, Air and Soil Pollution* **105**, 1998, 205–15; G. W. Campbell and D. S. Lee, 'Atmospheric deposition of sulphur and nitrogen species in the UK', *Freshwater Biology* **36**, 1996, 151–67; C. Neal, J. Wilkinson, M. Harrow, L. Hill and C. Morfitt, 'The hydrochemistry of the headwaters of the River Severn, Plynlimon', *Hydrology and Earth System Sciences* **1**, 1997, 583–617; J. Wilkinson, B. Reynolds, C. Neal, S. Hill, M. Neal and M. Harrow, 'Major, minor and trace element composition of cloudwater and rainwater at Plynlimon', *Hydrology and Earth System Sciences* **1**, 1997, 557–69; C. Curtis, T. Allott, J. Hall, R. Harriman, R. Helliwell, M. Hughes, M. Kernan, B. Reynolds and J. Ullyet, 'Critical loads of sulphur and nitrogen for freshwaters in Great

Britain and assessment of deposition reduction requirements with the First-order Acidity Balance (FAB) model', *Hydrology and Earth System Sciences* **4**, 2000, 125–40; M. C. F. Proctor and E. Maltby, 'Relations between acid atmospheric deposition and the surface pH of some ombrotrophic bogs in Britain', *Journal of Ecology* **86**, 329–40; P. M. Haygarth, P. J. Chapman, S. C. Jarvis and R. V. Smith, 'Phosphorus budgets for two contrasting grassland farming systems in the UK', *Soil Use and Management* **14**, 1998, 160–7; C. D. Evans, A. Jenkins, R. C. Helliwell and R. Ferrier, 'Predicting regional recovery from acidification: the MAGIC model applied to Scotland, England and Wales', *Hydrology and Earth System Sciences* **2**, 1998, 543–54; Countryside Council for Wales, *Energy: Policy and Perspectives for the Welsh Countryside*, CCW 1992 Policy Document **030** [bilingual text]. Much of our knowledge derives from long-term work by the then Institute of Hydrology and this is put in context in J. A. Hudson, K. Gilman and I. R. Calder, 'Land use and water issues in the uplands with reference to the Plynlimon study', *Hydrology and Earth System Sciences* **1**, 1998, 389–97.

160. J. Tranter, J. Gunn, C. Hunter and J. Perkins, 'Bacteria in the Castleton Karst, Derbyshire, England', *Quarterly Journal of Engineering Geology* **30**, 1997, 171–8; C. Hunter, J. Perkins, J. Tranter and J. Gunn, 'Agricultural land-use effects on the indicator bacterial quality of an upland stream in the Derbyshire Peak District in the U.K.', *Water Research* **33**, 1999, 3577–86.

161. C. E. M. Sefton and A. Jenkins, 'A regional application of the MAGIC model in Wales: calibration and assessment of future recovery using a Monte-Carlo approach', *Hydrology and Earth System Sciences* **2**, 1998, 521–31.

162. Not strictly true, of course: Singapore buys water from Indonesia and Malaysia, for example.

CHAPTER FIVE

Vegetation management and nature conservation

Vegetation dynamics and management

The area under consideration is that of open moorland. That is, land which is outside the limits of improvement at present or in the last fifty years, where there is no attempt to seed or fertilise in order to change the species' mix or its growth rates, though 'improvements' in drainage may be attempted. The whole is enclosed by the ring of fences or walls at its lower edge and the moor itself may well be internally divided by walls from a period of Parliamentary enclosure in the eighteenth or nineteenth centuries. Some may also have been recently sub-divided by fences aimed at restricting the movements of sheep, either for own-ership reasons or to locate more precisely the swards which they graze.

'Open moorland' tends to be 'vegetation' to biologists, a form of 'land use' to geographers and planners, and 'upland pasture' to agronomists. For all it forms an intersection of the wild genes of plants and animals (as with heather and grouse) with more domesticated processes dominating the manipulation of ecosystems – the economics of sheep farming and grouse moor manage-ment, for instance. So just as the outer boundary of open moorland is a human artefact which has varied through time, the internal manipulation pressures have undergone evolution and indeed revolution (Plates 5.1 and 5.2). We presume too that the genetics of the wild species have also undergone change but no longitudinal evolutionary studies exist. A functional grouping of wild species, land cover and land use (along with its domesticates) is thus found at any one moment, though any part of this is subject to change from human or natural forces.[1] One major influence has been the designation of most moorlands and their fringes as Less Favoured Areas (LFAs) under EU provisions and the subsequent, though now (2002) superseded, encourage-ment of sheep production. In England, typically 45 per cent of breeding sheep have been in LFAs and 75 per cent in Wales.[2]

Plant and animal communities

The plant and animal communities currently exhibit regional differences based on natural factors, of which altitude is a key variable since it determines the length of the growing season. The annual rainfall is also an important

PLATE 5.1 *One of the commonest moorland vegetation types: acid grassland in the North Pennines dominated by species of* Agrostis *and* Festuca. *Damp soil means also the presence of some rushes (*Juncus *spp) and heather (*Calluna vulgaris*). This area is only lightly grazed; in September it has just about stopped growing for the year.*

PLATE 5.2 *Heather moorland in mid-Wales. The vegetation type is the focus of much conservation concern since it disappears under heavy grazing by sheep. This stand is only lightly grazed and has not been burned for some time; it is even-aged so probably was a fire-patch some years ago. A damper foreground shows some flowering heads of cotton-sedge (*Eriophorum *sp).*

factor; this also varies with altitude although there is an east-west gradient within Wales as well as across Wales and England. Thus the density of heather necessary for a good grouse moor is more easily produced and maintained on the 'dry' moors of the eastern side of England. Slope angle is also a natural feature which affects drainage and hence the mix of wet-tolerant and intolerant species. In a similar fashion, the rock type and its covering of glacial drift may affect the number of small hollows where water can accumulate, and the geology also determines spring lines which are the sources of small areas of wet 'flush' vegetation where the nutrients in the water create more base-rich conditions. It is not perhaps surprising therefore that given a mix of natural conditions, the history of human use, the present uses and external influences such as air pollution, that even in some quite small areas there is a high diversity of vegetation types at present. Yet overall biogeographical trends brought out by a separate consideration of each upland can be discerned: *Calluna-Vaccinium* 'heath' (dry moorland) is found on the steeper slopes at the higher altitudes, often on west-facing slopes. Below it in South and Central Wales, there is a zone of gorse (*Ulex gallii*) with *Calluna*, and with the grass *Agrostis curtisii* on Dartmoor, where it is a regional response to high levels of burning and grazing. In the Peak District, plant communities dominated by *Calluna* accounted for about 32 per cent of the moorland vegetation and a further 25 per cent were dominated by species of *Vaccinium*.[3] Different parts of Exmoor have different representations of the various vegetation communities: the central moors, for example, are mainly *Molinia*-dominated, with *Calluna* moorland much more abundant on the northern stretches and towards the coast. The southern heather moors are however the most invaded by bracken.[4] On all the uplands the blanket bogs are dominated by *Calluna-Eriophorum* or *Scirpus-Eriophorum* communities. These are typically on the low slopes to the east of the western edges of the uplands, on the dip slopes of the main strata. It is also possible that in regions like Wales, the human impact has been to bring out the natural patterns (those of climate, for instance) rather than make them disappear. Table 5.1 shows the way in which the main upland vegetation types are dominant in the different regions of England and Wales, using the National Vegetation Classification categories.[5] The comprehensive account by Thompson and Miles[6] provides a context for the many regional studies of recent vegetation distribution on the uplands. To their main categories we can add minor communities like mine tracks and workings, which often contain an interesting flora whose members may be genetically tolerant of very high levels of, for example, lead in the soils: species such as *Thlaspi alpestre* and *Minuartia verna* occur in such places. The moorland vegetation may also be interrupted by springs and flushes. Here, the presence of mosses, liverworts and herbaceous plants supports a greater variety of e.g., insects than the heather, bilberry and cotton-sedge moors. Such areas may also be important foci for grazing by sheep. The commercial interest of sheep and grouse producers led to a comparative study of moorland species' biological productivity and

TABLE 5.1 The broad regional representation of plant communities making up moorland

NVC type name	International importance	South west England	Mid and south Wales	North Wales	South Pennines	North Pennines	North York Moors
Dry heaths							
Ulex gallii-Agrostis curtisii heath	GB	3	2	0	0	0	0
Calluna vulgaris-Ulex gallii heath	GB	2	2	2	1	0	0
Calluna vulgaris-Deschampsia flexuosa heath	I	0	0	2	3	2	3
Calluna vulgaris-Erica cinerea heath	I	1	2	2	0	1	1
Calluna vulgaris-Vaccinium myrtillus heath	I	2	3	3	2	3	2
Calluna vulgaris-Arctostaphylos uva-ursi heath	W	0	0	0	0	0	0
Vaccinium myrtillus-Deschampsia flexuosa heath	W	1	3	3	2	2	1
Calluna vulgaris-Vaccinium myrtillus-Sphagnum capillifolium heath	GB	0	2	2	0	1	0
Wet heath							
Scirpus cespitosus-Erica tetralix wet heath	I	2	2	2	0	1	2
Erica tetralix-Sphagnum compacium wet heath	GB	1	2	2	1	1	2
Blanket mire/bog							
Sphagnum auriculatum bog-pool community	W	2	2	1	0	0	0
Sphagnum cuspidatum/recurvum bog-pool community	W	1	2	1	1	2	0
Scirpus cuspitosus-Eriophorum vaginatum blanket mire	GB	2	2	2	0	0	0
Erica tetralix-Sphagnum papillosum raised and blanket mire	I	0	3	3	0	2	0
Calluna vulgaris-Eriophorum vaginatum blanket mire	I	0	3	3	1	3	0
Eriophorum vaginatum blanket and raised mire	I	0	3	2	3	3	2
Acid grasslands							
Deschampsia flexuosa grassland	W	2	2	2	2	2	1
Festuca ovina-Agrostis capillaris-Galium saxatile grassland	W	3	3	3	3	3	2
Nardus stricta-Galium saxatile grassland	W	2	3	3	2	3	2
Juncus squarrosus-Festuca ovina grassland	GB	1	3	3	1	2	1

Regional representation: 1 = rare/very local; 2 = locally frequent/locally extensive; 3 = widespread/locally extensive; 0 = not known to occur. International importance: GB = communities with no, or rare, close equivalents outside Britain; I = communities with local equivalents internationally; W = communities with well-developed equivalents internationally.
NVC: National Vegetation Classification

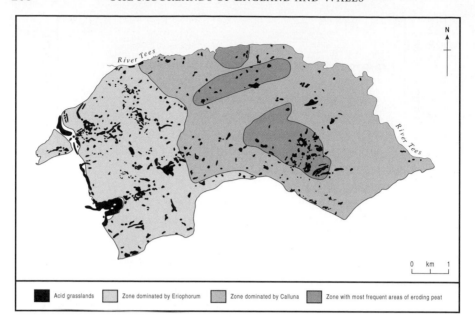

■ Acid grasslands	■ Zone dominated by Eriophorum	■ Zone dominated by Calluna	■ Zone with most frequent areas of eroding peat	

FIGURE 5.1 *A simplified vegetation map of the Moor House NNR in the North Pennines. The western escarpment shows in the continuous stretches of acid grassland, the highest ground has blanket bog with* Eriophorum *(interrupted by grassland on drier areas) and then large stretches of heather on the slopes away down to the Tees. It is on this* Calluna *moor that most areas of peat erosion are found.*
Source: *A. Eddy, D. Welch and M. Rawes, 'The vegetation of the Moor House National Nature Reserve in the Northern Pennines, England',* Vegetatio *16, 1968/9, p. 240. With kind permission of Kluwer Academic Publishers.*

biomass within six upland regions and their relation to heather grazing regimes, yielding new information on, for instance, the response of heather to different grazing intensities at pioneer and mature stages of its life cycle.[7]

A much simplified result of more local mapping can be seen for the Moor House NNR in the Cumbrian Pennines, which lies between 300 and 850 m (Fig. 5.1). Detailed vegetation mapping found eight major types of vegetation, some of which could be subdivided (Table 5.2). The distribution confirms the controls exerted by altitude and slope, with the western escarpment carrying much of the drier *Festuca* and *Nardus* grasslands, with the area sloping away to the east mostly covered with peat, some of which is eroding. The stratification of limestones and sandstones affects the distribution of the flushes which are important for grazing animals and for the production of insects.[8] A Dartmoor study pointed out a pretty universal feature: that plant communities on the moor were arrayed along two axes. The first was dominated by the values for soil moisture, soil acidity, the depth of soil or peat, and the angle of slope; the second by the intensities and incidence of burning and grazing. The second axis, we can note, is largely under human control (Table 5.3). The first is more 'natural' but some elements of it can in time be affected by the

TABLE 5.2 Vegetation types of the Moor House NNR, Northern Pennines

Major types	Included categories
Blanket bog 1–4 m in depth	
	Calluna-Eriophorum
	Scirpus-Eriophorum
	Eriophorum
	Eroding bog
	Recolonised peat complexes and peat-edge vegetation
Poor fens and flushes – low base status	
	Sphagnum-Juncus
	Sphagnum-Carex nigra
Calcareous springs and flushes	
Base-rich springs and flushes	
Grassland and dry-ground vegetation	
	Juncus squarrosus-Nardus
	Species-poor *Nardus*
	Festuca grasslands
	Dry species-rich *Agrostis-Festuca* grassland
Scree	
Bracken (*Pteridium*)	
Made ground and mine spoil	

Source: A. Eddy, D. Welch and M. Rawed, 'The vegetation of the Moor House National Nature Reserve in the Northern Pennines, England', *Vegetatio* 16, 1968/9, 239–84. With kind permission of Kluwer Academic Publishers.

second. Burning may shift the mix of plant species, for example, and thus affect the soil pH derived from the litter.[9]

Blanket bog

If asked to identify the most characteristic feature of moorlands, most ecologists and many walkers would say 'blanket bog'. A skin of peat from 50 cm deep to 6–7 m or occasionally even deeper is formed on many low slopes with a rainfall of 1,000 mm/yr and above and even on slopes of 20° if the precipitation is above 2,000 mm/yr (Plate 5.3). Under these conditions decay rates of organic matter are slow and so production exceeds decay to the point where accumulation occurs. The initiation of this type of peat accumulation has been extensively investigated: some areas seem to have got away in the 7,500–5,000 BP period and are usually regarded as responses to a climatic change in the direction of greater oceanicity. Thereafter many basal peats are in times of known human occupation and the presence of charcoal in many such sections suggests that the maintenance or creation of clearings in woodland allows the formation of podsolic soils which then become waterlogged. Low-intensity fires seem the most likely to lead to peat growth and these are seen as the most likely to be associated with prehistoric patterns of land management. Lateral spread from initial sites is also probable.[10]

TABLE 5.3 Vegetation types of Dartmoor

Vegetation type	Characteristics with strong 'natural' element	Characteristics with strong human element
Blanket bog Calluna-Molinia moorland	Prevalence of *Molinia* characteristic of Dartmoor blanket bogs, forming pure stands. 13,760 ha on the moor. *Eriophorum* also an important peat-former in the recent past on the wettest areas	A fire-produced type seen by Vancouver in 1808, so produced by swaling. Grazed early summer only.
Grassland	Heather commonly present among the grazings though short and sparse; *Nardus* not so common on Dartmoor as *Agrostis, Festuca, Galium saxatile, Luzula campestris* and *Potentilla erecta.*	Vancouver described similar communities; some burning occurs.
Bracken	Extensive invasion of suitable soils	Favoured by heavy grazing, burning and the removal of cattle from the stock mixture
Vaccinium moorland	Localised and especially common on boulder-strewn areas	Stable communities though susceptible to treading.
Valley bog	Great variety of herbs and mosses depending on mineral content	Grazing by sheep at unrecorded levels. Occasional large dogs.
Heath	Peripheral and mostly burned-over; *Agrostis setacea* often dominant after fire but not grazed; heather in areas not recently fired	Swaling on Dartmoor not much mentioned in 18th century; average now is about 1,000 ha/yr: very few old stands of heather over the whole moor.
Grassland with gorse (*Ulex*)	Heavily grazed and burned. Gorse an important fodder for ponies especially in winter	Thought to be more extensive in the past 200 years

Source: D. S. Ward, A. D. Jones and M. Marton, 'The vegetation of Dartmoor', *Field Studies* 3, 1972, 503–33.

Blanket peat comprises about 85 per cent of the peatlands of England, Wales and Scotland, with 75 per cent of it in Scotland. Its somewhat assertive presence has led to many evaluations of its status and value: 1 ha of blanket peat might contain about 19,000 m^3 of water, for example, and it was for many years a piece of received wisdom that peat cover acted as a sponge and retained water which might otherwise run off rapidly and cause lowland flooding. More recent research however suggests that only a small proportion of this water is involved in seasonal exchange with the extra-paludal environments and so there is a limited capacity for temporary storage. If, of course, the peat erodes away then it will not store anything and so some concern about peat

PLATE 5.3 *Blanket peat on Holme Moss in the Dark Peak. Erosion of the 1.5–2.0 m skin of peat shows how the plant remains have accumulated to form the 'blanket'. Much of the foreground is re-vegetated peat which has eroded from the main mass of peat in the background. Since blanket peat forms mostly on slopes of low angle, it contributes to the flat horizons of many high moors.*

erosion is always in evidence among upland managers: this is discussed later in the chapter.[11]

Wildlife management

Most studies emphasise the key roles of management past and present in bringing about today's detailed patterns of vegetation, even though these must fit within the overall environmental limits and indeed in Wales have been said to have accentuated climatic patterns rather than run contrary to them.[12] Many ecological surveys now pay attention to the animal communities other than the main economic species such as sheep and grouse, to a lesser extent ponies and, since some of the literature includes Scotland, red deer. Hence the bird populations of the uplands have been the object of a great deal of study, much of it motivated by the effects of sheep grazing pressure, grouse moor management and afforestation.[13] Obviously, grouse as an economic crop have been central to much research, but the wilder birds of the uplands have not been neglected, possibly because some of them are intimately associated with the wild, remote quality of the upland environment which is part of their cultural attraction.[14] Some fifty-eight species breed in association with the uplands of Britain as a whole and Table 5.4 associates these with their habitats as well as giving population data and trends in abundance. The management

TABLE 5.4 Breeding birds confined to moorland habitats[1]

Species	Population[2] (pairs)	Trends	Main habitats	Annex I[3]
Red kite (*Milvus milvus*)	40	Increasing slowly	nests in trees, hunts in open, Wales	Yes
Hen harrier (*Circus cyaneus*)	500–600	Increased 1950–70 then decreased	Heather moor	Yes
Peregrine (*Falco peregrinus*)	850	Recovered from pesticides	crag nesting	Yes
Merlin (*F. columbarius*)	550–650	Decline	Heather moors	Yes
Red grouse (*Lagopus lagopus scotica*)	50,000	Decline	Heather moors	
Curlew (*Numenius arquata*)	50,000	Increasing	Moors and grassy margins	
Black grouse (*Tetrao tetrix*)	20,000	Increases with afforestation	Margins of heather moors	
Golden plover (*Pluvialis apricaria*)	30,000	Decline	Open land with short vegetation	Yes
Dunlin (*Calidris alpina*)	10,000	Decline in southern regions	Blanket bog	
Short-eared owl (*Asio flammeus*)	1,000	Fluctuates	Heather moors and young forests	Yes
Raven (*Corvus corax*)	5000	Some declines	Open land with carrion	
Ring ouzel (*Turdus torquatus*)	10,000	Some declines	Rocky areas and heather	
Wheatear (*Oenanthe oenanthe*)	80,000	declines in lowlands	Grass with rocks and walls	
Whinchat (*Saxicola rubetra*)	30,000	declines in lowlands	Damp areas and bracken	

Notes
1. Some species are found in mountain areas as well
2. These data refer to Great Britain as a whole
3. Annex I of the EEC Directive, which is a measure of rarity and vulnerability

of moorlands for birds centres first on the knowledge that there is a great deal of difference between the flora and fauna (and hence the food sources) of mineral soils and peat soils. The availability of grassland close to blanket bog increases the period that invertebrates are accessible to upland birds and may provide critical food types such as earthworms. As for flushes, their productivity of arthropods is about eighteen times that of heather moorland and some managers produce them artificially by blocking drains or making a hole with some limestone chips in it.[15]

The mobility of birds means that not all their yearly cycle need be spent on the uplands, for their food sources may be such that in winter another habitat may be sought. The vegetation cover of the hills is not usually enough to secure the perpetuation of most upland species except the red grouse. Increasingly, however, the uplands are the sole habitats for species of wider range but whose habitats elsewhere have been converted to a different land cover. Examples are the lapwing (*Vanellus vanellus*), whose lowland food sources from certain types of agricultural land have diminished and the snipe (*Gallinago gallinago*), victim of land drainage, a fate which has affected the redshank (*Tringa totanus*) to some extent. Such moors as are 'retreats' for these species will become important conservation sites under EU Directives on wild birds and on habitats.[16] Apart from red grouse, few birds have ranges which virtually confine them to moors. Heather moorland, however, does seem to have very strong associations with the tree pipit (*Anthus trivialis*) and the merlin (*Falco columbarius*) but on the other hand to be a repellant for the dunlin (*Calidris alpina*) and the skylark (*Alauda arvensis*). The upland grass-lands carry a number of species which seem to have no preferences for either the acid, neutral or base-rich facies of the vegetation: examples include the snipe and skylark, curlew (*Numenius arquata*) and yellow wagtail (*Motacilla flava*). Equally wide-ranging are the raptors of grasslands: peregrine and kestrel, and four species of owls, of which the short-eared owl (*Asio flammeus*) is the most often seen.

The raptor and scavenger populations of moorlands are much affected by food supplies that are independently affected by management for grazing and grouse. To a large extent, too, they have been classed as 'vermin' and extir-pated wherever possible, as the account of contemporary grouse moor man-agement shows. One species which however has increased its representation has been the red kite (*Milvus milvus*). Between 1946 and 1993, the number of breeding pairs increased from seven to 113 in its heartland of mid-Wales.[17] The red kite prefers areas rich in native oaks and other broadleaved trees, in which it nests. These woods are found mostly at the boundary between enclosed pasture and the open ground/afforestation zone and their distribu-tion seems to be the key (once persecution has diminished) to the distribu-tion of kites. The current extent of coniferous afforestation does not seem to have affected the number of the red kites and indeed may have been benefi-cial in the sense that afforestation reduces the amount of sheep land on which is found the poisoned baits for foxes which kill about 60 per cent of the red kites found dead. The spread of the red kite has also been made easier by its wide ranging for food (to 7 km from a nest site) and its wide food tolerances, since the species has been found scavenging on rubbish dumps and at slaugh-terhouses. There are now 3,200 km² of Wales which form breeding territory for the red kite, of which 16 per cent is under trees of all kinds. The extension of afforestation has proved no barrier to the success of the merlin (*Falco columbarius*) in Wales, where it has moved from breeding mostly in moorland to plantations, which now host about half the breeding pairs of the Principality's population of 60–70 pairs of that small falcon.[18] Kites, ravens and buzzards all feed largely on carrion and the ravens lay eggs in proportion to the amount of sheep carrion in late winter and spring and some buzzards will visit abattoirs in winter. Sheep numbers have meant that a 10 km^2 patch of upland Wales carries at least 190 kg/yr of ovine carrion, though with some annual variation.[19]

British natural history has always been greatly interested in birds but the reasons why upland avifauna bred there, for example, were of little interest until it was realised that the invertebrates of moorland were their chief source of food. Pearsall's classic synthesis included material on invertebrates and thereafter they were not overlooked in accounts of the ecology.[20] The difference in invertebrate faunas between peat and mineral soils is fundamental (different genera of spiders for example are found in each), and the importance of wet flushes for insect populations was indeed mentioned by Pearsall. These flushes are a source of diversity in heather moorland, where the soil fauna reflects the growth phase of *Calluna* found at a particular site: this applies especially to ants and spiders. The overall role of invertebrates seems to be summed up in the finding that there are about thirty species of invertebrate fauna to one vertebrate and that for every sheep on the peat moors, there is 5–10 times its weight in invertebrate biomass. So the invertebrates are important in transferring **net primary productivity** to the peat and since they have ten times the weight of nitrogen and phosphorus in their bodies than is found in the vegetation, they are crucial in recycling and concentrating nutrients. They add to the **biodiversity** of these habitats since the uplands house some twenty-one species of rare invertebrate taxa and overall are not a species-deficient group in the uplands. On the peatlands, there are no earthworms, and so the invertebrate species become the chief food of the birds. For example, there is a vast synchronised spring peak of emergence of craneflies (Diptera) which is the more pronounced at higher altitudes and the breeding of nesting species is timed to coincide with this event, which also returns nutrients to the soil and plants on a large scale.[21] In the case of the meadow pipit (*Anthus pratensis*), nesting is near the interface of peat and grassland so that the first brood can eat the insects from the moorlands' emergence phase and the second brood from the grassland insects. This accounts for the main presence of merlins below 550 m ASL. As there is not much insect food in the uplands after July, the fledged young of species such as lapwing, curlew, redshank and golden plover all move lower down the hills.[22] As noted earlier, the insect fauna of the wet flushes is critical in the early life of grouse chicks, when insects are their main food. The role of invertebrate fauna in the ecology of heather moorland and its interaction with sheep grazing is much less well understood than the behaviour of higher plants, birds and mammals, though the diversity of some invertebrate groups is high when compared with that of higher plants, for instance.[23]

Animal production: general

Many statements in the section above hint at high intensities of human involvement with moorland ecology. Not only were human communities involved in the creation of moorlands, from hunter-gatherer times onwards, but they have been crucial in manipulating all the plant and animal populations ever since, either by design or as a by-product of other management aims. So today, both dry and wet moors are the sites of the production of a number of animals, the first three of which are important commercially:

1. Sheep of various breeds are the mainstay of the upland economy, found in very large numbers. Within the LFA area of England, for example, ewe numbers increased 40 per cent between 1976 and 1993. There was a move towards cross-bred animals which are physically heavier than their predecessors.
2. Grouse on the drier areas of eastern England and as far west as the Forest of Bowland as well as on the rain-shadow uplands of eastern Wales and the Borders.
3. Cattle, which were once very plentiful grazers of the uplands in summer but now less often found on moorland pastures. Numbers on the lower enclosed lands are still high.
4. Rabbits are plentiful on drier soils (though affected cyclically by myxomatosis) and able to reduce any vegetation except bracken to a close-cropped sward. No longer systematically 'warrened'.
5. Other animals such as ponies, which form a breeding stock for foals to be sold off, and goats, which are feral and little cropped.

It is unlikely that the density of Dartmoor ponies for example affects species composition since there are now fewer than 3,000 of them, compared with 30,000 in 1950. There are no subsidies for ponies which makes them less attractive economically than sheep and cattle. They spend the entire year on the lower, drier, common land and probably only affect the vegetation if they gather at spots where they know they will be (illegally) fed by visitors. Goats as found for example on The Cheviot no doubt help to ensure the lack of regeneration of anything woody. Goats are not common on moorlands in England and Wales. A few were introduced to Rough Tor on Dartmoor early in the twentieth century but they were soon extinct. The Cheviots on the other hand have had local herds since time immemorial though some (like the College Valley herd) have been the beneficiaries of translocation: in this case in 1860. Their numbers fluctuate since they are shot when they damage crops, a fate also applied to any found in Kielder Forest though the Kielder herd at Whitelee near the Border is seen as part of an integrated land use and conservation scheme administered by the Northumberland Wildlife Trust. The antiquity of the Northumberland herds is founded on the legend that they were set loose by the monks of Lindisfarne and they have certainly entered folk-lore as good animals to keep alongside cattle:

> Rot nor poke nor loupin' ill
> 'll no come where there's a goat on th'hill

They were said also to prevent abortion in cattle and as human food might be referred to as 'rock venison'.[24]

Animal production: sheep

Sheep are the dominant crop of the moorlands and are discussed in every account of the uplands' past and future. They are at the heart of many a concern expressed by those with a strong interest in environment (especially wildlife) and also by others who recognise their centrality in animal production from the higher lands and as the foundation of a particular way of life.

One reason for the dominance of sheep on moorlands is the variety of plants they will eat. Heather moorland is obviously included, as are the various grasslands, both acid and base-rich, the peatlands covered with cotton sedge and deer-sedge, small flush areas, and mine wastes. Given the grazing habits of sheep, which are to nibble very close to the ground or peat surface, often to stay within a particular area of even open moor (the condition known as 'hefting') and to graze differentially within that heft, then it is clear that the impact of the sheep populations upon plant growth is likely to be strong. In addition, any supplementary feeding causes the sheep to concentrate in the places where the food is placed, bringing greater impact upon the soil surface and on nearby plants, which receive a disproportionate amount of the cropping pressure. Before the eighteenth century, it seems that shepherding kept the sheep on the move but in later decades reduced levels of day-to-day control allowed the sheep to concentrate on the most palatable swards. In the second half of the twentieth century (and especially after 1970) more and more upland sheep received supplementary feed in winter since it became easier to deliver it to the hill, for example with quad bikes. In some regions, sheep production systems are combined with shooting estates, with the one subsidizing the other. Thus in the North Pennines in 1991–3, low-pressure grazing areas typically held 40 sheep/km^2 grazed in the summer only, whereas year-round heavy grazing moors were stocked at over 100 sheep/km^2. (One estate managed for both grouse and sheep had 142 sheep/km^2.) All these densities can be seen as the extension of the earlier trends documented for several upland regions in the period to 1965, which show the numbers beginning to rise as the subsidies for upland farming favoured headage payments for sheep (Plate 5.4). There were large increases in sheep numbers between the mid 1970s and the late 1980s and the numbers in the 1990s generally remained far higher than in the 1970s though with some declines since the mid-1980s. Wales in particular has seen rises since the mid-1950s to very high densities and it has been accompanied by northern England.[25] Thus it is reasonable to say that there are more sheep on the uplands now than in any time since reliable records were kept (Fig. 5.2).

Sheep have without doubt dominated the research into **production ecology**, where the contribution of heather moor, blanket bog and *Agrostis-Festuca* grasslands are often quite contrasted. The animal production is linked to the nutritional quality of the herbage rather than the quantity of it and there may be as many as fifty times the number of sheep on the grassland as on the bog, even though the former produces only 60 per cent of the dry weight of foliage of the peat vegetation. The Moor House (Northern Pennines) studies showed that a ewe of 50 kg eats some 500 kg/yr of herbage (both measured as dry weights) which means that about 60 per cent of the grassland is grazed off, if the maximum density of sheep is about 5.0 ewes/ha.[26] By contrast, even high densities of grouse consume only about 6 per cent of *Calluna* shoots. (The only other two resident vertebrates, the meadow pipit and the frog, are insect eaters.) But the sheep take off about 5 mg/m^2/yr of nitrogen, whereas burning mobilises 249 mg/m^2/yr, stream water 294 mg/m^2/yr and peat erosion 1,463 mg/m^2/yr. On the blanket bog surfaces, the sheep stimulate the growth of six species of higher plants, most notable the cotton-sedges, *Eriophorum* spp., and reduce the growth of two others, of which *Calluna* generates the most concern. Heavy grazing on peat surfaces often reduces the cover of both *Calluna* and *Eriophorum* so that the area of bare ground is increased. Looked at overall, any density over one ewe/ha can

PLATE 5.4 *Sheep on Stainmore in the North Pennines. These animals are pastured at relatively low densities since their moor is also a grouse moor. Since they are given supplementary feed they cluster around cars and can therefore puddle the soils to the point of faster run-off and perhaps of erosion.*

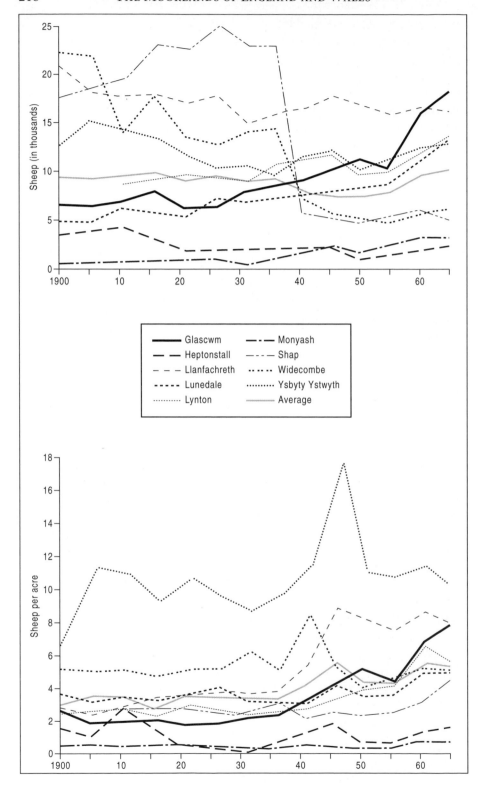

produce ground bare of vascular plants and also bare of bryophytes and hence open to increased erosion.[27] There is a complex relationship between sward height, grazing intensity and species composition: close grazing (3.0–4.5 cm) during the growing season resulted in the spread of *Nardus stricta*, whereas the removal of grazing increased the quantity of *Molinia* and *Deschampsia flexuosa*, together with heather.[28] Research in southern Scotland has demonstrated that sheep and cattle have different diets. They graze at different heights, for instance, sheep can select more readily from fine-grained mixtures of species, and cattle can graze on the taller and more fibrous plants; *Molinia* seemed to be equally attractive to both domesticates.[29]

When time depth is added to such work, the relatively ephemeral nature of some recent plant communities can be seen. On Exmoor, for instance, the past millennium has witnessed an alternation of dominance between heather moor and grass moor with some *Molinia*, while the *Calluna*-dominated 'grey moor' developed originally from grass moor.[30] Such changes are usually associated with grazing pressures; above the Bowmont Valley in The Cheviot, moor-burning ceased in the eighteenth and nineteenth centuries and along with increased numbers of sheep, the two factors reduced a species-rich grass heath to a species-poor community. Equally, heather communities had been established in 1200–1300 under conditions of burning and low sheep numbers, with the beasts taken off the hills in winter; these too did not easily survive the nineteenth century.[31]

Compared with lowland agriculture, the management of the upland pastures has used a small repertoire of techniques. Fertilisers, for example, are often not used at all but if applied are confined to small areas of the most heavily grazed grassland; re-seeding with all the attendant inputs is rarely used at all on the out-bye land, though may be applied below the moorland edge where defined ownership makes the investment accrue clearly to the purchaser of the materials (Plate 5.5). Drainage may be applied to common land where it is judged that the response of the vegetation will be to provide more herbage for the sheep: thus open drains are dug in peat in order to encourage heather and perhaps *Deschampsia* and *Molinia* at the expense of cotton-sedge. Burning on grouse moors is largely controlled in favour of the grouse production aims but elsewhere is used to clear off old vegetation and encourage young growth. The tolerance of moorland plants to burning varies greatly: those plants with subsurface rhizomes generally re-grow the most rapidly.[32]

FIGURE 5.2 (opposite) *Diagrams of sheep numbers in upland parishes in the twentieth century. The top graph is of absolute numbers and the lower frame of densities per acre (1 acre = 0.4 ha). Though there are regional variations, the average in both graphs shows a cumulative increase, one which continued after 1970. The low-figure parishes are probably those where grouse moor management was also important.*
Source: D. F. Ball, J. Dale, J. Sheail and O. W. Heal, Vegetation Change in Upland Landscape, *Bangor: Institute for Terrestrial Ecology, 1982, Fig. 4.6.*
By permission of the NERC Centre for Ecology and Hydrology.

PLATE 5.5 *The profitability of sheep has led to the building of lambing sheds near the pastures and their use for storage of supplementary feed, the supply of which may change grazing patterns. Note the vehicle track onto the hill. Radnor Forest in the Welsh Marches.*

The commonest management tool, of course, has been stocking density and the impact of the heavy densities has been of most concern in the last decade or two.

Plants show different tolerances of levels of grazing, which are summarised in Table 5.5. Such work has shown, for instance, that there are strong differences between the mineral soils and the peats. Blanket bog is not a preferred grazing habitat but when it is a source of forage then there are differences between cattle and sheep. Cattle eat more cotton-sedge leaf than sheep except perhaps in April, when the sheep also eat large amounts of cotton-sedge flowers. Cattle will also graze *Molinia* since they will eat more dead material than the sheep but in general avoid peaty soils since they produce poaching and thus vegetation loss. On drier soils, cattle will eat *Nardus* more readily then sheep, whereas *Calluna* is much more acceptable to sheep than cattle. In general, sheep diets are more variable than those of cattle, both within species and between species. The sheep, for example, eat more dead material than cattle in the autumn and can graze even *Nardus* sward since they can eat in between the *Nardus* tufts. They are less good at getting the fresh material in between the dead leaves and stems of *Molinia*. So while, in general, cattle will graze tall, fibrous swards, sheep are better at selection from fine-scale mixtures; the main difference comes at the variations in height at which different species from within the canopy are grazed by the two herbivores.[33] In the case of heather moor, the patch size seems to be important. In the Peak District, studies showed that sheep spend less time browsing large patches of *Calluna* but use much more of that plant in burnt and discontinuous stands, eating 13.7 per cent of the current year's growth in discontinuous heather but not more than 8.5 per cent in large patches.[34] The pressures of grazing may often become heavier if burning is carried out as well. Thus fire encourages new growth of heather, whose cover it restores two to three years after the burn but if repeated often will encourage dominance by *Molinia* and *Scirpus* instead. Rapid regrowth from rhizomes means that bilberry (*Vaccinium myrtillus*) is often dominant just after fire, at the temporary expense of *Calluna*. Both the common species of *Eriophorum* are tolerant of burning, as is the deer-sedge *Scirpus*. Bracken is not only tolerant of fire but aggressive in its invasion from underground rhizomes of any plant community deprived of aerial cover and root nets by fire. *Nardus* withstands burning, as does wavy hair grass (*Deschampsia flexuosa*) and all the species of *Juncus*.[35]

As an environmentally-aware landowner, the National Trust has worked with fewer sheep per unit area. Erosion on the Kinder Plateau in Derbyshire (Peak District National Park) led to a managed reduction in stock levels and in about ten years this resulted in the recolonisation of bare and eroding ground, especially with the grass *Deschampsia flexuosa* but also with *Vaccinium myr tillus* and heather.[36] Nobody expects this to be other than a successional phase and regrowth of a cover of heather is in general the hoped-for outcome. The lighter stocking regime being advocated to conserve heather moorland may however be detrimental to stands of bilberry and so to protect that plant some zonal variations in grazing pressure are desirable.[37]

Grazing: general observations

The detailed and local effects of burning, draining and grazing of a community may be difficult to predict, but intensive work has identified a number of trends which are summarised in Table 5.6. The detail of this Table yields two important generalities. First, as can be seen from the last column, it does not take long even under moderate grazing densities (which *de facto* are largely of sheep) for vegetation types to become vulnerable to change, and that is generally in the direction of a greater representation of less palatable species. Secondly, given the cessation of burning and grazing, we might have a scrub vegetation (possibly followed by woodland) within a relatively short time, perhaps fifty years in the most favourable sites (Plate 5.6) and 100 years in more severe habitats.[38] Clearly, the vegetation is largely maintained by its management regime and this carries the implication that it has been so since medieval times certainly and probably since the Bronze Age at least. The detailed nature of the vegetation is highly influential in the grazing preferences of, especially, sheep. So the use of the less palatable forage species is influenced by the highly local presence or absence of the more palatable grasses and shrubs; the pool of available invading species would also be crucial especially

TABLE 5.5 Common moorland plant species: tolerance of burning, grazing and drainage

Plant species	Burning	Grazing			Type of grazing	Drainage	Notes
		Light	Medium	Heavy			
Calluna vulgaris	Good	Good if bracken, Nardus absent		Disappears	Sheep; grouse take very small %	Tolerant of all except wettest conditions	Can form pure stands if managed
Vacinium myrtillus	Very good	Not greatly affected by grazing			Occasional grazing by sheep	Absent from wettest habitats	Woodland also: tolerant of shading
Erica tetralix	Very good	Tolerant	Killed by moderate-heavy grazing		Sheep and small mammals	Wet habitats	Only dominant soon after fire
Juncus squarrosus	Tolerant	Young growth grazed if little else to eat; eliminated if grazing ceases			Early growth eaten	Most wet habitats by sheep and cattle	Intolerant of shading
Eriophorum angustifolium	Resists burning	Resistant even to continuous grazing and may thus dominate blanket bog			Sheep	Requires wet soils	Both eliminated by effective drainage
Eriophorum vaginatum	Resists burning and may spread because of it	Tolerant of light – moderate grazing		Weakened if persistently grazed	Sheep	Wet and stagnant habitats, especially in spring	
Scirpus caespitosum	Tolerates burning	Tolerant but expands if grazing pressure is relaxed			Sheep	Peat soils	Eliminated by drainage and increased grazing.
Pteridium aquilinum	Not affected by fire; spreads into adjacent communities weakened by burning	Poisonous to wild and domestic herbivores; capacity to spread lessened by cattle trampling, especially of young plants			None	Well-drained soils	Herbicides current method; cattle and mowing in earlier times kept area lower

Species	Response to burning	Response to grazing		Palatability	Soils/habitat	Notes
Nardus stricta	Withstands burning	Able to withstand light grazing	Weakened by repeated defoliation	Not very palatable but eaten early in year by sheep and cattle	Wide range on acidic soils	Slow growing and does not compete on better soils
Juncus spp (*J. effusus, J. articulatus, J. conglomeratus*)	All tolerant of burning	Invasive under these pressures; *J. conglomeratus* esp resistant to trampling	Resistant	Cattle will eat them but not sheep	Poorly drained habitats	Drainage main method of control
Deschampsia flexuosa	Tolerant	Withstands sustained pressure and will replace species weakened by heavy grazing e.g., *Calluna*			Drier soils	Colonises burned-over heather land
Molinia caerulea	Tolerant of burning	Grazed when young. Was once cut for hay		Preferred by more cattle than sheep	Wet habitats, though not stagnant water	Often fired to remove deciduous material.
Agrostis tenuis, Festuca ovina	Tolerant of some degree of burning	Often closely grazed and recover well from it		All herbivores, such as cattle, sheep, ponies, rabbits and small mammals	Grow best on better drained soils	Present in a range of sward types

Source: Adapted from Ball *et al*. 1982, Table 5–11.

TABLE 5.6 Vegetation change under managed regimes

Vegetation type	Management	Gradual change	Rates of change
Agrostis-Festuca grasslands	Reduced grazing	Invasion by bracken then scrub and woodland	Bracken invasion at 20–40 cm/yr if not trampled Possibly 50+ yr
	Increased grazing on acid soils	If no bracken then more *Vaccinium*, possibly more heather Bryophyte-rich swards develop	
	Increased grazing on base-rich soil	Incursion of species such as *Holcus* and *Lolium*	2–3 yr at 7 sheep/ha 5–10 yr at 7 sheep/ha
Agrostis-Festuca with *Juncus* grassland	Reduced grazing	Taller grassland with *Deschampsia flexuosa*	10–15 yr; then to heather in perhaps 100 yr
	Increased grazing	Prostrate and rosette species increase	3–5 yr with 7 sheep/ha
Nardus grassland	Reduced grazing	to *Vaccinium-Calluna* heath	15–20 yr and then 75–200 yr for scrub woodland
	Increased grazing	Towards *Agrostis-Juncus* if soil moisture levels correct	3–5 yr with 7 sheep/ha
Molinia grassland	Reduced grazing	Invasion by *Calluna* and possibly *Sphagnum*	10+ yr
	Increased grazing	Invasion of *Juncus* or *Festuca-Agrostis*, depending on soils	5–10 yr with 7 sheep/ ha
Calluna heath (well-drained soils)	Reduced grazing & burning	Scrub woodland	20–25 yr or longer
	Increased burning & grazing	(a) Increased *Festuca-Agrostis* grassland (b) Bracken may invade and dominate	5 yr with 5 sheep/ha for initial stages
Calluna heath (wetter soils)	Reduced grazing & burning	Scrub woodland	20–50+ yr
	Increased grazing & burning	*Festuca-Agrostis* or *Festuca-Vaccinium*, thence possibly invaded by *Juncus*	10–25 yr for early stages; rate increased by winter grazing
Eriophorum-Calluna heath	Reduced grazing & burning	Dominance by *Calluna* and perhaps scattered scrub	10+ yr for heather; 50+ yr for scrub
	Increased grazing & burning	Towards dominance by *Molinia* and *Scirpus*	5–10+ yr with 0.5 sheep/ha for initial change in balance.

Source: Ball *et al.* 1982, Table 5–12.

PLATE 5.6 *On the lower moors and where grazing is light, then scrub may develop. The heather is 'leggy', showing it has not been burned recently and the low shrubs are being invaded by gorse and hawthorn. The protective management results in part from the areas's status as a NNR. The Stiperstones, Shropshire.*

in any recolonisation that followed a lessening of grazing densities. There have been strong suggestions that the increases in sheep densities in the uplands have been the cause of declines in bird populations but it is difficult to isolate mechanisms and to disentangle the effects of for example changes in predation pressure. The probable explanation may lie in a combination of the more recent inflation of sheep numbers superimposed on a long-term reduction in carrying capacity. Lapwing and ringed plovers are notable species that have suffered losses in recent decades. If, on the other hand, lambing is outdoors then the carrion feeders benefit from the dead lambs and the placentas: corvids and foxes are the obvious examples.

There are yet other complexities. Nitrogen is deposited on uplands from the atmosphere, as discussed on pages 186–7. Higher levels of nitrogen increase the proportion of grasses; grasses attract grazing animals which then provide a further feedback by urination and dunging. The animals also open up **dwarf shrub heath** and improve the conditions for grass growth. The maintenance of a desired type of vegetation, whether for production or for conservation will therefore require steady and continuous control, as well as local knowledge since for example the levels of nitrogen deposition are not constant across the British uplands. Then, since the early 1990s, sheep dips have changed. Organophosphates were implicated in illnesses of farmers and so are being replaced with synthetic pyrethroids from December 1999. These are 100 times as toxic to aquatic life and so spills are very serious, especially to invertebrates. The Environment Agency has recommendations for disposal that minimise this possibility but with about 11 million sheep in Wales, nearly half of which live in the Wye catchment, and with a single adult sheep holding up to 2 litres of dip after the dipping event and able to wipe out about 11 km of river life, the chances of adverse events are high. The Upper Wharfe (Yorkshire Dales) and Afon Twrch (North Wales) have been contaminated in this way. In the sense that the problem arises in the valleys and not on the moorland itself, this is marginal to our theme, but the sheep will carry the pesticide with them and on their wool wherever they go. The Environment Agency has however been confident that farmers will be vigilant over the problems of disposal of dip materials, with written permission being required for their disposal.

Erosion and fires

The least desired outcomes in the eyes of today's ecologists seem to be:

1. Soil and peat erosion
2. Bracken spread
3. Loss of heather moorland

Each of these will be discussed in turn, though there are some common factors to be brought together at the end.

The extent and causes of upland soil erosion (Plate 5.7) have rarely been considered on a national scale. Yet early results from an England-wide survey showed that of a sample of over 400 upland sites, 43 per cent were eroded and of these, some 85 per cent (156 sites) contained actively eroding scars within 50 m of the National Grid intersection which was the sample node. The extent and severity of soil erosion was strongly influenced by soil type. Peat soils accounted for 60 per cent of all eroded sites although they were less than one-third of all field locations recorded. Stagnopodzols contributed the next largest category: they accounted for 6 per cent of all active erosion. The reasons for the erosion of peats are fairly obvious: their unconsolidated nature makes them vulnerable to many kinds of influence, both natural and human-caused. The high moisture and organic contents of the stagnopodzols may also account for their susceptibility to breakdown and loss.[39] Yet at heart the problem is largely one of use and management, as detailed work in the Peak District and Pennines has shown. Peaty moorland west of Huddersfield (Yorkshire) showed a long-term erosion yield of sediment of about 204 tonnes/km^2/yr, which included an organic fraction of about 39 tonnes/km^2/yr. The last two centuries have been especially important in terms of erosion rates, and incision in blanket peats since 1956 has accelerated erosion rates in a nearby catchment.[40]

PLATE 5.7 *Soil erosion on Levisham Moor, on the North York Moors. This gulley (locally called a 'griff') shows that erosion has taken place in multiple phases, not a single episode of downcutting.*

Disastrous events, like the 1976 fire on Glaisdale Moor (North York Moors), can destabilise the normal processes to a large extent (Plate 5.8). Comparison of air photographs between 1973 and 1983 shows a dramatic increase in drainage densities and hence in erosion (an increase in drainage density in the North Gill of ×9 and in the Bluewath Beck of ×1.3 (Fig. 5.3).[41]

Put into a more general context, the interaction of the factors producing erosion (which is only one of a number of landform-creating processes) suggest that the overall geomorphological setting and thence the strength of linkage between hillslopes and river channels is especially important. Long-term work in the Howgill Fells of Cumbria shows how the various landscape elements contribute to the movement of sediment. Slope failures in the hill-slope zone occur infrequently (six times in thirty years in one catchment) but otherwise there is little sediment supply to the streams. Lower down, basally-induced gulleys accumulate silt in debris cones maybe thirty times per year and stream floods entrain this sediment once every two to five years. A rare flood event (with a return period of >100 years) occurred in 1982, which changed the whole channel system, producing much braiding. The wettest week in thirty years (in October 1998) fed debris flows 400 m downslope which almost connected with the channel system. The effects of a wetter climate can thus be seen to be potentially great.[42]

The surface of the slopes of moorland valleys may be diversified by

PLATE 5.8 *Drastic erosion on Glaisdale Moor (North York Moors), four months after the severe moorland fire of 1976. A few vegetated hummocks remain but most of the burned area (whose boundary is just visible in the background) is bare and clearly open to erosion by wind and rain.*

landslides. Minor occurrences can be found almost anywhere in these uplands and occasionally block transport routes like the Mid-Wales railway line. Where geological instability is caused for example by a sandstone overlying shales at angles of dip of 5–15 degrees then long-term landslides can result. The Edale and Hope valleys in the Peak District (Derbyshire) have more than fifteen such landslides and the Mam Tor slide has been extensively studied since at one time it was crossed by a road which had been a packhorse route but was made into a turnpike in the first decade of the nineteenth century and acquired a blacktop in the 1930s. Such has been the movement of rock slices and debris that the road was permanently closed in 1979 after many attempts at stabilisation had been made. The slide seems to have been initiated about 4,600 BP and has been moving intermittently and differentially since then, especially after periods of intense rainfall. The costs of stabilisation were such that it was more economic to build a new road on a different route especially since total absence of future movement could not be guaranteed.[43]

FIGURE 5.3 (opposite) *The interpretation of aerial photographs allows this portrayal of the changes in the density of drainage network (i.e., the development of gulleys and streams) between 1973 and 1983 on Glaisdale Moor, North York Moors.*
From: M. S. Alam and R. Harris, 'Moorland soil erosion and spectral reflectance', International Journal of Remote Sensing *8, 1987, 593–608, Fig. 2 at p. 596.*
Reproduced with permission of Taylor and Francis, Ltd.: www.tandf.co.uk/journals

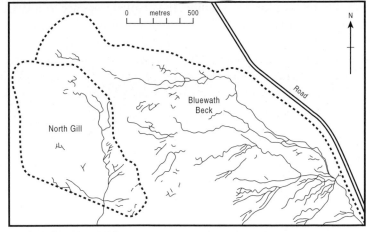

Glaisdale Moor 1973

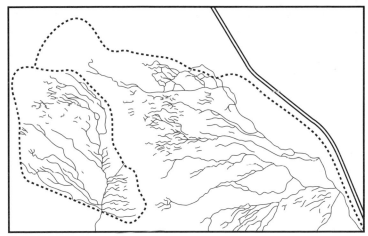

Glaisdale Moor 1978

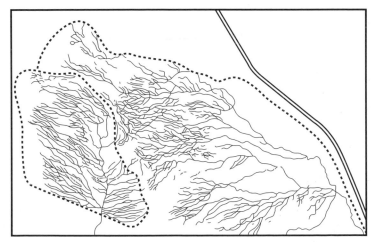

Glaisdale Moor 1983

Back on the zones of low slope where the blanket bogs are found, the 1950s mapping of peat erosion and the speculation about its causes has been amplified into studies of both a historical and experimental nature, with the South Pennines as the area where the processes are now best appreciated.[44] The concept of erosion *and* degradation is central to modern work since the latter is generally a precursor of the former. Degradation consists of a number of related but discernable processes:

1. A reduction in species diversity
2. A reduced cover of *Sphagnum* species compared with 250 years ago
3. An increase in the area of discontinuous plant cover
4. A reduction in the rate of peat accumulation.

The list of degradation agents comprises accidental fires, trampling by walkers, peat-slides and bog-bursts, peat cutting, air pollution and those levels of sheep grazing which are accompanied by deliberate fires and by digging drains. One of the major effects is the erosion of blanket peat in which gulleys 2 m in depth and 1–2 m wide incise themselves at a typical rate of 5 mm/year into the peat. On slopes they form parallel systems but on flat and near-flat summits a reticulation is formed (Plate 5.9). In the South Pennines, about 74 per cent of the blanket bog is affected by such erosive phenomena. This region has seen the most intensive investigations into the origins of peat erosion and a number of conclusions have been formed.

The first is that today's vegetation is often formed on the surfaces of previous erosion phenomena and so does not necessarily give good clues as to the past. Active erosion seems to have been present for at least four to five centuries and has involved a number of distinct episodes. Some of these are associated with periods of dry climate, when the moss *Racomitrium* became a major component of the vegetation, some are due to large-scale fires and yet others to intense storms. The effects of air pollution and intensive grazing by sheep are superimposed on earlier periods of erosion. In the case of Holme Moss in the northern Peak District National Park, the main peat gullies probably originated after clearance of forest from the hillslopes in the eleventh century when well-defined flowlines formed into open drainage channels. Some of them did not become incised until the eighteenth or nineteenth centuries; the surface may have been affected by a severe burn over half the Moss in the eighteenth century and a freak cloudburst in 1777 caused damage to the margins of the peat blanket and possibly accelerated incision. In the Howgill Fells, 800–1100 was again the period of the development of gullies. In the Forest of Bowland (Lancs), the bog-moss *Sphagnum* disappeared during the last 100 years due to a combination of factors which included a very low water-table in the early 1900s (a result of dry years), exceptional summer drought in 1921 and a decline in management standards because of a shortage of game-keepers after WWI. All these led to an accidental but catastrophic burn in

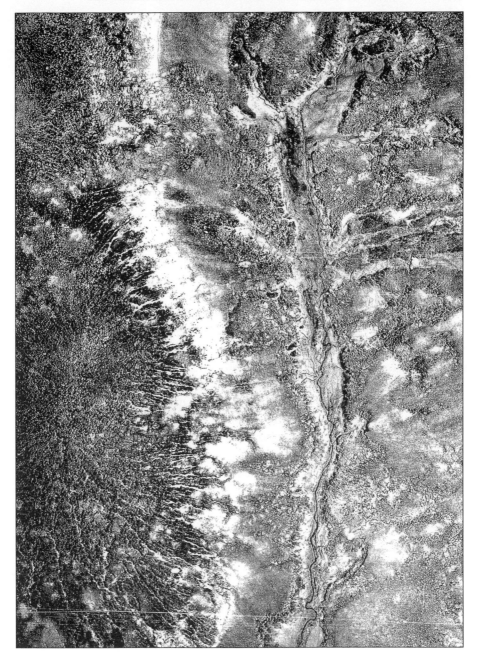

PLATE 5.9 *A vertical aerial photograph of part of the upper East Dart and Black Hill on northern Dartmoor. The peat blanket on the hill has a reticulated erosion pattern, which becomes longitudinal on the slope down to the river. The slope of the hill precludes peat growth and the granite-gravel soil shows through in white. The picture covers an area just over 1 km in width.*
Source: © English Heritage (NHR) RAF Photography.

1921. Now, high levels of sheep stocks appear to prevent the recolonisation of bare peat surfaces.[45]

The importance of some of the episodes of climatic change, such as the inception of erosion during the Little Ice Age (c.AD 1500–1850) must never be overlooked in spite of the temptations to assign erosion to episodes of human-set fire.[46] However, there is an absence of clear evidence that gulleying can develop in an undamaged blanket mire system.[47] Investigations in small lakes with peat-covered catchments have shown enhanced rates of peat erosion between AD 1530 and 1690, putting an emphasis on climate as a cause of erosion, though the levels of charcoal in peat profiles usually rise in the last 500–700 years.[48] The sensitivity of blanket bog to the inception of erosion is clearly related to one set of factors which may not necessarily be those which determine the subsequent course of the breakdown and removal of the peat. In the Cheviots, it seems that topography is the primary influence upon erosion patterns: some of these have stayed stable for the last thirty-two years and in some areas more than seventy years. It is indeed possible that some gulleys are more than 500 years old.[49]

Generalities about blanket bog vegetation and peat erosion stem largely from the management practices that have been catalogued. A *Sphagnum*-rich vegetation (not now common) stems from a long-standing combination of low-intensity grazing and infrequent burning; modest burning frequencies produce dwarf shrubs like *Calluna* and *Empetrum*, and high burning plus high density grazing results in graminoid vegetation dominated by species such as *Eriophorum vaginatum*, *Molinia caerulea* and *Tricophorum caespitosum*. About 82 per cent of the blanket bog in the British Isles has been modified by management, which presumably means that most of the examples in England and Wales have been altered thus. Again for the entire British Isles, one estimate suggests that 350,000 ha (from a total of 22,500 km^2) are affected by erosion.[50] In the Cheviots, an area of 32 km^2 of blanket bog was mapped and 37 per cent of it was eroded, mostly in a linear pattern.[51]

Occasionally, the catastrophic processes can be observed almost as they occur. Two gamekeepers in the Peak District in 1834 saw peat thrown into the air to the height of thirty or forty feet, accompanied by fires and large quantities of water.[52] In 1963, a pair of peat slides on the Moor House NNR in the North Pennines was recorded very soon after they happened. On the northeast slope of Meldon Hill two peat slides occurred on July 6. Their heads were at 640 m ASL and they were 230 m apart, with scars of 230 × 36 m being formed at the widest points and vertical peat faces mostly about 0.6 m high but with a maximum of 1.50 m at the heads. A volume of 4,000 m^3 for one slide and 1,880 m^3 for the other was estimated. They were mostly in grassy areas that also carried *Carex nigra* and *Eriophorum* spp.; at the base of the peat there was a skim of clay which seemed to have provided the interface at which failure took place. In August 1964 a similar slide took place 12 km away on Stainmore Common (at Iron Band) and Meldon Hill itself was the site of a recorded slide in 1870.[53] Certain localities seem to be the sites of clusters of slides but other failures are more solitary (Fig. 5.4). The long-term importance of big bursts and slides in the inception of peat erosion is brought out by findings from the South Pennines that a series of them on the margins of one peat area (Featherbed Moss) initiated the formation of drainage gullies some 1,000–1,200 years ago (Fig. 5.5).[54] Most authors agree that no one period in history has a statistically significant concentration of mass movements in peat.

The cutting of peat for fuel as part of common rights (the right of *turbary*) has existed since medieval times, yet little systematic study of its effects on vegetation and land cover has been carried out. In the South Pennines, one study has concluded that virtually all the peripheral blanket peats have been affected by cutting. Not only do the cuttings show where peat has been removed but dry baulks remain, as do extraction routeways, drains and loading features. Peat-stripping down to the mineral soil over an area of about 2×1 km near Kinder Scout has been recognised and one calculation suggests that the Dark Peak has lost 34 million m³ of peat from land above 370 m ASL.[55] The cut-over areas in Derbyshire seem to be mostly grasslands of *Molinia* and *Nardus*, but on Dartmoor, for instance, there are many areas of rectangular cuttings which are filled with unconsolidated *Sphagnum*. Here too, some areas seem to have had their altitude lowered by cutting and the deep 'peat passes' cut through the blanket peat on the high northern part of the moor are presumably part of the network of pony trackways to carry the peat down to the user settlements. Where mineral soils are exposed then the grassland is likely to carry a higher vole population and thus support the short-eared owl. Figure 5.6 gives an idea of the overall environmental effect of peat cutting on the Howden Moors of the Upper Derwent Valley. In one upland parish in Powys (Mid-Wales), there were seventy-one farms in the early 1950s which could have utilised local peat, extracting about 3,000–8,600 m³/yr of peat (or 3,000–9,500 m³/yr if turf was used), which is close to the growth rate and so the resource can scarcely be comfortably renewable unless used by only a few of the potential exploiters.[56]

The military use of moorlands inevitably results in damage to vegetation and soils; though the soldiers may dig holes for cover, their vehicles are likely to exert the most lasting effects. A study on Dartmoor showed that after a track was abandoned, recovery was quite rapid on heathland and dry grassland but that on the higher moors and on blanket bog, recovery was very slow and in effect the vegetation never recovered its original composition.[57] However, the military have shown themselves to be, usually and within the limits of their missions, much more responsible users of moorland than many recreationists. Though proof is always difficult to maintain, it is often walkers who set off the kind of fires in heather which are not controllable: especially those that happen in a dry summer. Vandalism is of course another source of origin, with a form of class warfare sometimes being the cause of the deliberate and

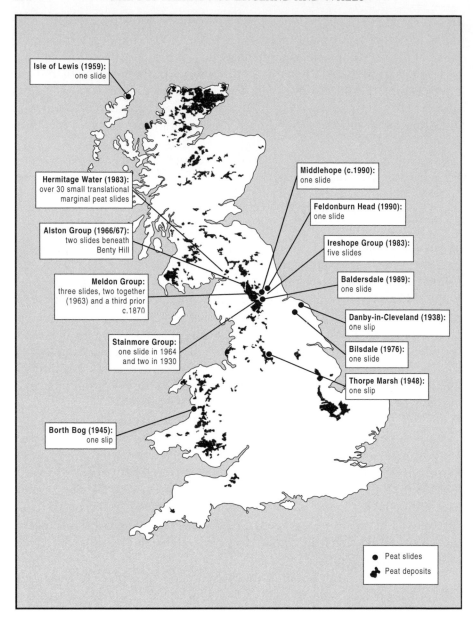

FIGURE 5.4 *The totality of peat deposits in Great Britain with slips and slides in peats identified. There are a few lowland examples e.g., Borth Bog, which is at sea level but most are in the uplands, with a distinct concentration in the North Pennines. The sparsity of records in Scotland might lead us to wonder if the distribution is in some ways an artefact of investigation density.*

Source: © Joint Nature Conservation Committee.

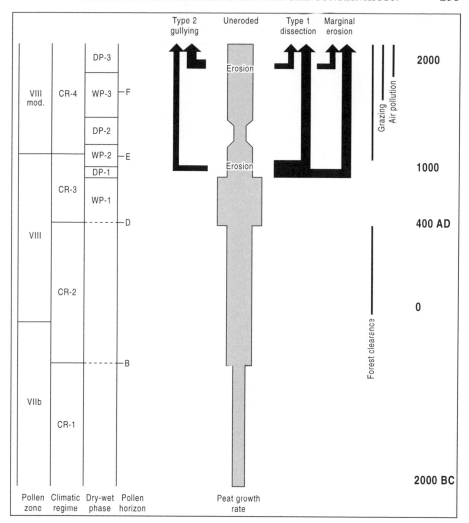

FIGURE 5.5 *The chronology of peat erosion on Featherbed Moss in the Southern Pennines, after J. H. Tallis. The left-hand columns refer to stratigraphic markers and palaeoecological data: the key element here is the DP-WP alternation (Dry-Wet) which is seen as critical in the initiation of most types of erosion of the peat. The emphasis on recent times is apparent, though coincidence with the onset of grazing and pollution is not necessarily exact.*
Source: *J. H. Tallis, 'Mass movement and erosion of a southern Pennine blanket peat',* Journal of Ecology *73, 1985, 283–315, Fig. 13 at p. 307. Reprinted with permission of Blackwell Publishers, Ltd.*

wrong-season firing of grouse moors. Where a large area of heather moor is burned and where the fire then starts to burn the peat as well as the vegetation, which is entirely possible in a dry summer, then recovery of any form of vegetation may be slow, since the effect of the fire (especially if followed by heavy rain) is to allow erosion not only of the burned vegetation and peat but

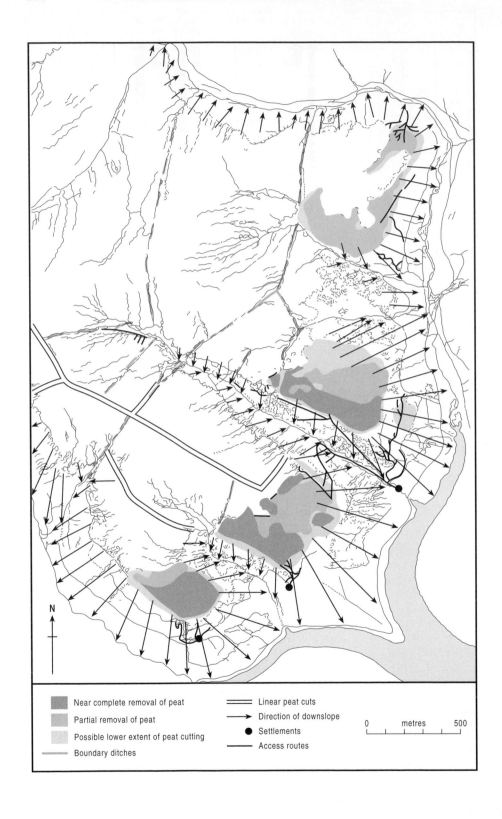

Near complete removal of peat

Partial removal of peat

Possible lower extent of peat cutting

Boundary ditches

Linear peat cuts

Direction of downslope

Settlements

Access routes

0   metres   500

TABLE 5.7 Erosion on the Holme Moss area of the Dark Peak

Commencing about AD 1450: long-standing gulleying on the eastern and north-western parts of the Moss

Early 1700s: a severe fire over much of the Moss produced most of the bare peat visible in that zone today

Around AD 1770: marginal peat slides along the western and north-western sides of the Moss exposed extensive areas of bare soil and rock immediately downslope of the main peat blanket

AD 1800 onwards: air pollution produces the loss of sensitive mosses, especially *Sphagnum.*

Last 60 years: overgrazing by sheep

1976 and 1980: accidental burns which are very slow to recover; *Empetrum* stands especially are affected.

Erection (opened 1951) and replacement (1983) of TV mast causes major disturbance over a limited area.

Source: P. Anderson *et al.*, *Restoring Moorland, Peak District Moorland Management Project, Report Phase III.* Bakewell: Peak District National Park Authority, 1998.

of the mineral substrate as well. In 1970, for example, a 60 ha fire on Sykes Moor in the Peak District brought about bare areas whose overall shape was still visible in 1995; it had even extended somewhat on a narrow slope. In a nearby example (Torside), no significant revegetation had taken place by 1995 on a burned area from 1980 even though sheep had been removed. Studies suggest that '[W]ithout some assistance, subsequent natural colonisation of bare peat, once erosion has begun, seems slow or absent.'[58] The erection and replacement of a TV mast on Holme Moss seems to have been just the latest in a series of erosion-causing events and processes in the Dark Peak (Table 5.7). It stands on a plateau made dark by the scars of erosion channels.

The mapping and recording of moorland fires is a relatively recent activity and so comparative data in time and space are difficult to obtain. For the Derbyshire Peak, Radley asserted that wildfires were rare before 1900. Before 1800, close management by manorial tenants ensured low fire incidence and between 1800 and 1900 the new enclosure of upland grazings and the development of grouse moors gave private landowners an incentive to watch their moors carefully. Particularly dry years presented special hazards and in 1762,

FIGURE 5.6 (opposite) *A landscape much affected by peat-cutting on the Howden Moors plateau of the southern Pennines. Large swathes of the flat terrain are affected, with some areas presumably having all their vegetation removed, though not all at once. Linear peat cuts are also found on Dartmoor near areas of extensive cutting.*
Source: Original map by P. A. Ardron. Reproduced from Blanket Mire Degradation *(1997) by permission of the editors.*

several hundred acres of moors were burned over. In 1826, similar conditions allowed several thousand acres of The Peak to burn, in a national context of extensive fires in Scotland and northern England.[59] Some idea of the recent incidence of uncontrolled fires (i.e., not those set for moor management for grouse or grazing) can be gained from work in the Peak District, where over 300 'accidental' fires have been recorded for 1970–95; in total they burned 8 per cent of the moorland area (Fig. 5.7). Of these, 51 per cent started by roads, paths or the Pennine Way areas and point to recreationists and travellers as the main causes; a further 18 per cent occurred on access land. On Dartmoor, an estimate can be given for deliberate 'swaling' by the graziers in the late 1960s: the average burned each year was 1,104 ha, mostly of *Calluna* moor but *Molinia* and *Ulex* (gorse) were also targeted.[60] Since there are no grouse moors, the control is less well exercised and so some large burns occurred, but they were usually superficial and achieve their aim of removing heather over fifteen years of age. Heathery vegetation on Exmoor seems to have been burned on a rather haphazard and irregular basis; part of this is due to fires set by recreationists and part by the small allotment of moor (an average of 77 ha of rough grazing) to each farmer, for whom co-operative fire management as on the Scottish pattern is not undertaken. A large block of grass moor is also burned each year to remove the previous year's dead grass but this is controlled by a single estate.[61] In the extreme year of 1976, there were sixty-two uncontrolled peat or vegetation fires on the North York Moors between 31 March and 7 September, during the driest summer recorded since 1727, when the relevant climatic records began. The most severe of these was on Glaisdale Moor at an altitude of 380–405 m on slopes below 3° and burned mostly *Calluna* on thin blanket peat or on the peaty horizons of greyed and podsolic soils. It started on August 17 and the smouldering peat was finally extinguished by exceptionally high and periodically intense rainfall in September. In all, 32 ha of vegetation was burned and the fire plus the subsequent rainfall removed 75 per cent of the 19.2 million kg dry weight of peat formerly present. Of the burnt area, only 10 per cent was intact, with charred peat 28 per cent, ash 35 per cent, mineral and humid surfaces 35 per cent, and stone/**regolith** exposures 1 per cent. The revegetation of the site thus started in many instances from surfaces which had lost their seed banks. The recolonisation of bare surfaces was dominated by bryophytes, though these were more successful on ash than on charred and granulating peat. The consolidation of new vegetative growth dates from 1982 but it has been slow and incomplete and by 1984, only 2 per cent of the affected area supported *Calluna*-dominated communities. Some 65 per cent of the area was under the more successful bryophytes (especially mosses of the genus *Polytrichum* like *P. piliferum* and *P. commune*) and 30 per cent still unvegetated.[62] Without doubt, in any area of eroding peat the loss of mineral nutrients from the erosion is high compared with sheep grazing or moor burning. Only the losses in stream water are comparable.[63]

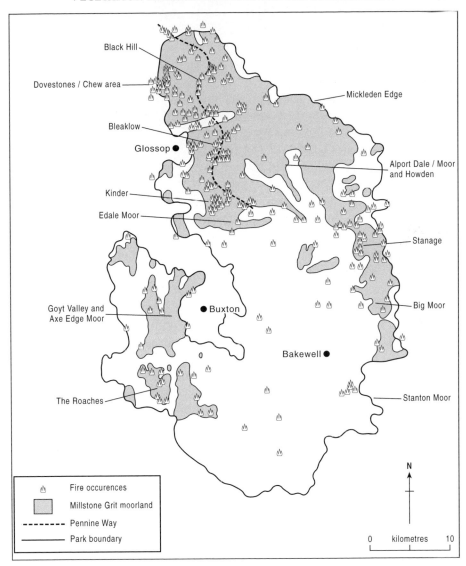

FIGURE 5.7 *The occurrence of fires in and near the Peak District National Park 1970–1995. The relationship between the heather moors on Millstone Grit and fires is obvious though other land cover types are also affected. Proximity to recreation features like the Pennine Way also seems to be significant.*
Source: Original map by P. A. Ardron. Reproduced from Blanket Mire Degradation *(1997) by permission of the editors.*

A more deliberate attempt to reduce the influence of peat on the local eco-system was made by upland managers who took advantage of government and CAP grants (which paid 70 per cent of the cost) to dig drains into peat with the aim of drying them out so as to carry more heather rather than

PLATE 5.10 *Unmelted snow picks out the lines of drains on a moor in the North Pennines. The main lines follow the contours of the hill but are intersected by steep vertical members. The foreground is mostly spoil from nineteenth century lead workings.*

cotton-sedge or deer-sedge. The common term is 'moor-gripping' (Plate 5.10) and the ditches are usually 40–50 cm deep and 15–35 m apart, either run along the contour or dug in a herring-bone pattern. Some twenty-five years have elapsed at Moor House since the first 'gripping' was undertaken (in Great Britain as a whole the practice started in earnest in the 1940s) and studies have shown that *Calluna* peaked about eight years after draining but declined thereafter and it was twenty years before *Sphagnum* showed any statistically significant change in amount. The drains mainly intercept surface water and withdraw the water only along the peat edge. Runoff from above keeps the upper side wet and the drying-out effects were confined to the downslope side of the drain. Intensive studies of the effects which were carried out at 400 m ASL and above showed that the effects were stronger at lower altitudes and in drier places. Nevertheless, government grants stopped in 1985 since the avowed aim of improving the heather had not been achieved and some landowners have adopted a programme of stopping up the drains to keep the blanket bog wet; in Cumbria, English Nature are blocking grips (themselves widening by erosion) with bales of locally harvested heather.[64] It is worth noting that water control on moorland hillsides has a solid nineteenth century history: on Exmoor there are many drainage ditches and attempts to irrigate valley sides so as to create a form of water meadow (Fig. 5.8).[65]

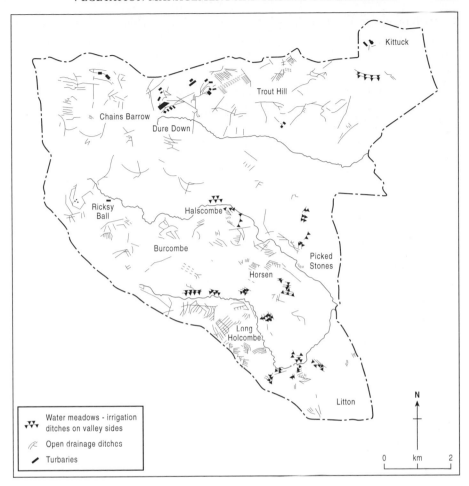

FIGURE 5.8 *Central Exmoor showing modifications of drainage of open lands from peat cutting ('turbaries'), drainage ditches (early attempts at gripping) and at spreading water to irrigate soils at dry times, like lowland water meadows.*
Source: L. F. Curtis, Soils of Exmoor Forest, *Harpenden: Soil Survey of England and Wales Special Survey no. 5, 1971, Fig. 12, facing p. 28.*
Reproduced from National Soil Resources Institute, Soil Survey of England and Wales Special Survey no. 5, 1971. © Cranfield University and for the Controller of HMSO, 1971. No part of this publication may be reproduced without the express written permission of Cranfield University.

Bracken

The cloaking of hill land with bracken (*Pteridium aquilinum*) is one of the main visual elements of upland Britain today. This fern may bring about virtually impenetrable stands after successful competition with other potential dominants. A number of other plant species grow in association with it where the summer shade is not too dense: a common example is the heath bedstraw, *Galium saxatile*. In south-west England the grass *Agrostis curtisii* is found.

Disturbance, fire and biocides allow other species to grow but in dense stands the possession of **allelopathic** compounds assists the dominance of the fern. The spores thrive on fire-enriched ground and the rhizomes allow rapid regrowth after fire, so that it is only a temporary management measure and may indeed weaken competitor species. The main habitat of bracken outside woodlands is on the deeper and well-drained soils of base-poor to neutral status. It will colonise dried-out peat but elsewhere stops where the soils are waterlogged. Colonies are mostly below 450 m and rare above 600 m, avoiding frost hollows and very exposed slopes. The litter decomposes to a nutrient-poor and acidic mor humus.[66]

One reason for its success lies in its toxicity to sheep, so they do not eat it but when more cattle were kept on the hills, they at least trampled the young fronds, though neither did they browse it. The fern was cut for bedding and packing, for its potash content for use in soap and in glass-making and even for silage and fuel, but none of these practices is now common. So bracken is often described as a weed and its rate of increase was estimated to be as fast as 3.77 per cent per year in central Wales in the late 1970s; elsewhere in Great Britain, rates of spread of 1–2 per cent per annum are common. This means that in the 1980s the area of land covered by bracken in Wales was about 1,240 km^2, which is equivalent to the area under tillage, or the whole county of Gwent; and in England, 404 km^2, the same as the Isle of Wight. The total for Great Britain was reckoned by Taylor[67] as 6,361 km^2, the area of Lincolnshire, though smaller figures have been derived from different methods of estimation.

Amenity is also involved since the bracken litter may shelter the ticks (*Ixodes ricinus*, the 'sheep tick') which are the vectors of Lyme disease.[68] Yet again, people do not like to walk through large areas of it since it is very tiring and may also harbour adders. Its spores contain a known carcinogen, ptaquiloside. The main animal beneficiary seems to have been the whinchat, *Saxicola rubetra* but some butterflies are specific to plants which will grow under the fern canopy.[69] On Exmoor, the heath fritillary (*Mellicta athalia*) feeds thus on *Melampyrum pratense*, the cow-wheat; the high brown and pearl-bordered fritillaries (*Argynnis adippe* and *Boloria euphrosyne*) are also favoured by bracken cover. Shelter for fox and badger earths may also increase the diversity of species. In a few places, though, birds such as the willow warbler (*Phylloscopus trochilus*) and the nightjar (*Caprimulgus europaeus*) are more frequent in bracken, though not confined to it. Bracken can harbour some woodland plants even after the trees have gone, as shown on the North York Moors by the occurrence beneath the bracken canopy of dwarf cornel (*Cornus suecica*) and chickweed wintergreen (*Trientalis europaea*). Control measures (mechanical and herbicidal methods are possible) are expensive in relation to the productive value of land released and continuous after-treatment seems essential if recolonisation is to be prevented. In the uplands, chemical control with Asulam herbicide showed a variable rate and direction of revegetation after spraying: the bracken litter offered a poor range of microsites for the germination of plants other than bracken, although an initial cover of the moss *Campylopus introflexus* might be established.[70] In broader contexts, it seems as if any climatic warming at higher altitudes would be likely to reduce the number of the frosts which control its upper limit.[71] A quite large area is thus vulnerable to its spread. Harvesting bracken as a biofuel (Plate 5.11) or as a peat-substitute compost requires more flat land than is normally found under a stand and so is uneconomic. Since much bracken grows on soils which once carried woodland, some ecologists have harboured the thought that such areas are the obvious places to carry broadleaved tree cover once again.

Heather moors

The value of moorland dominated by heather (*Calluna vulgaris*) is seen as high, not least for the aesthetic pleasure of large stretches of purple vegetation in summer and early autumn. The wildlife value is in a few bird species, especially the relatively uncommon merlin, *Falco columbarius*, and the larger predator, the hen harrier (*Circus cyaneus*). The economic value is as sheep pasture and grouse moor, with the latter use being more common on drier areas where something approaching a monoculture of heather can be produced by management. Though pre-eminently a habitat of the Scottish scene, heather moor is also found on most uplands of England and Wales, with the densest stands of *Calluna* on the drier uplands. Biogeographically, the British Isles contain

PLATE 5.11 *An old slide of the author's showing bracken mowing and gathering near Goathland on the North York Moors in 1963, a part of farm practice that has now ceased.*

most of this vegetation formation, though this upland version is also found along the western seaboard of mainland Europe. Five of the upland plant communities are virtually confined to Great Britain and a further six are better represented in Great Britain than anywhere else. The EU Habitats Directive applies to 13 per cent of these communities and the Birds Directive (79/409/EEC) lists eight species of birds out of the forty which are generally reckoned to be characteristic of the upland heather moors.

The main changes to the area of, and habitats within, heather moor are agricultural reclamation and afforestation, bracken invasion and high grazing pressures. In the northern Peak District, there has been a reduction of 36 per cent in the area of heather moor between 1913 and 1980.[72] In England and Wales together, about 20 per cent of the *Calluna* moor present in the 1940s has now been altered significantly and of the remaining, about 70 per cent is estimated to be at risk from further change away from dominance by *Calluna*. Of that proportion, at least half is at risk of degradation of the heather communities (as discussed above) from high sheep densities, for once the number of sheep gets to about two ewes per hectare, the heather starts to give way to other plants. The actual distribution of good heather cover seems to be related to grouse moor management in the north-east of England but in the West Midlands and in South Wales this is less the case. Areas where high sheep densities most affect the *Calluna* density include other parts of South Wales, North Wales and south-west England. Overall, about 47 per cent of the English heather was in good condition and only 24 per cent showed the signs of over-grazing and management neglect; in Wales the figures for suppression by heavy grazing and management neglect were at about 80 per cent.[73] Palaeoecological work on the last 100–150 years of heather moorlands' history suggests that no single agent causes the national decline in heather cover but that intensively grazed heather which is also burned infrequently may allow invasion by grasses such as *Molinia* and *Deschampsia flexuosa*.[74]

The maintenance of healthy heather moor is of vital importance for the maintenance of grouse shooting. As described earlier for the nineteenth century (see pp. 137–41), the basic patterns have not changed and indeed that may well be the attraction to many participants. Commercial involvement has now increased so that landowners may regard the shooting as a marketable part of an enterprise rather than solely a feature of country life (and perhaps inevitably of *Country Life*) after the close of the London Season. The stability of the grouse moor in upland management inventories and social registers has not however been parallelled by evenness in the numbers of grouse available for slaughter every August 12th and onwards through the autumn until December 10th. Ever since Lord Lovat's committee and its report in 1911, this concern has manifested itself from time to time; in recent years there have been two major sets of investigations.[75]

The starting point for grouse studies can be summarised in Fig. 5.9, with its message of fluctuating bags during the period 1900–90. (Other factors, such

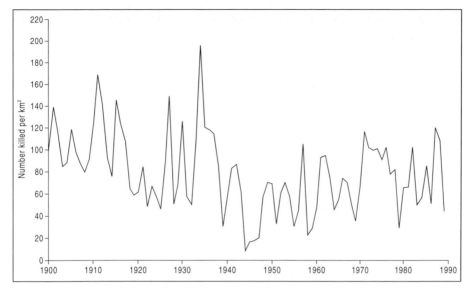

FIGURE 5.9 *The long-term trend in the grouse bag from moors in northern England, now reckoned to be the best sporting moors since their equivalents in Scotland have shown a big decline since 1945. Among the UK's 'wild' fauna, probably only fish popu lations have had such a density of scientific investigation and likewise have a strong set of vested interests.*
Source: *S. Tapper,* Game Heritage, *Fordingbridge: The Game Conservancy Trust, 1992, Fig. 9.3, p. 47. Data courtesy of the Game Conservancy Trust.*

as shooting effort and skill are assumed to have been constant.) There is clearly a cyclic pattern (of about four years' length in northern England) to grouse numbers, which might be expected in a wild population but not, it is assumed, in a managed resource. Lord Lovat's report suggested that better burning was the main need (the ideal was to burn in patches of about 30 m² every 8–15 years) and was rewarded with good inter-war shooting. The ecology-based work of the Institute for Terrestrial Ecology (ITE) concentrated on the terri- torial behaviour of male grouse, with the size of territories and the aggressive- ness of male birds as the main factors in grouse populations. Predation and disease were seen as minor factors. The Game Conservancy's biologists, on the other hand, though that predation (especially by crows, foxes and stoats) was important and that the disease caused by the gut nematode *Trichostrongylus tenuis* was probably the most important of all.[76] It seems also that the nema- tode increases scent emission by grouse and thus increases their vulnerability to predators like the fox, itself encouraged by huge rabbit numbers after the 1970s.[77] High chick mortality is caused by a virus for which the vector is the sheep tick. Joint work by the Game Conservancy and ITE has further emplaced the key role of avian predators, at any rate on one moor in southern Scotland. Here, if hen harriers and peregrine were persecuted instead of protected, post-breeding numbers of grouse should rise by 2.5 times. The long-term decline in grouse numbers is however most likely associated with the loss of heather moorland, itself a consequence of ever higher sheep numbers. Possible management scenarios deriving from this work are rarely in favour of the raptors.[78] Nevertheless, some attention to the long-term can help keep up numbers: the number of gamekeepers who keep down foxes and crows becomes important, as does the spreading of medicated grit. Chick survival and growth can be aided by the construction of artificial flushes if the moor is poor in these insect-bearing habitats. The variability caused by cycles means that a single plot may in one year plummet from 600 birds to 15 the next. Such fluctuations mean that research continues unabated to find their cause: differential territorial behaviour between kin and non-kin (the grouse, not the landowners) is being modelled for its contribution to understanding.[79]

The importance of grouse in the upland economy can be gauged from data which suggest that many estates enjoy an annual average income of £12,000 from their grouse shooting though in Scotland this translated into a net loss of £8,000 per estate. In Great Britain as a whole, there are about 750 estates which shoot red grouse and 62 per cent of them engage in moorland management. In England and Wales this translates to 400,000 ha of heather moorland used as grouse moor, with an average kill of 69 birds per km^2. At high densities, about 46 per cent of the birds are harvested, though some 25 per cent of the birds do not pass over the line of guns.[80] The value is about £50 per bird and so grouse shooting is worth about £10 million p.a. to the British economy, supporting some 2,500 jobs. In the Peak District, the regional income in the 1990s from the shoots and shot birds has been put at £1.45 million per year. (For comparison the total value of abstracted water from moorland reservoirs in the Peak District can be estimated at £100 million per annum.)[81]

Apart from its role in heather cover and the populations of predators, grouse shooting also contributes to nitrogen loss in smoke when burning takes place and 4WD tracks may act as hubs for soil erosion. It is unpopular with other countryside users since keepers are often zealous in keeping walkers away during the nesting season and before shoots; and conservationists believe that predatory birds are killed illegally. If however sheep numbers now decline as a result of lower headage payments, the role of grouse in the income of upland estates is likely to be emphasised; given the success of the 'right to roam' lobby as well, moor managers are likely now to present themselves as keepers of an important piece of 'heritage' and argue for the continuation of their practices as conservationists of a particular type of much-valued upland landscape. They may have to change their view of the hen harrier, since it is strongly implicated in predation on grouse. The species did not breed again in England until 1958, after intensive persecution that began in the nineteenth century. In 1999, twenty-one females had taken up breeding territories but only nine had reared young and in 2001 there were only two successful nests in England. In fact the species only bred on moors with nest protection, confining it to a small proportion of the 230 territories theoretically available on the English uplands. The survival of the harrier may depend upon short-term programmes of diversionary feeding.[82]

The importance of heather in the management of grouse (Plate 5.12) means that the bird is very vulnerable to high stocking densities of sheep which begin to degrade the quantity of *Calluna*. Heavy grazing also reduces the number of the far less common black grouse or blackcock (*Tetrao tetrix*). This bird is present now in quite small numbers (in the North Pennines most territories are between 340–590 m) and shoots commonly record bags of 0.025 birds/km² compared with 0.075–0.10 in the interwar period. The interaction with grazing is postulated in work which suggests that breeding success is 37 per cent poorer in heavily grazed habitats where the plant cover was thinner (the black grouse breeds best where the vegetation is about 40 cm high) and where there were 40 per cent fewer invertebrates. Blanket bog is also important for this bird since the females take the flowers of cotton sedge (*Eriophorum* spp) just before laying and incubating their eggs. It also benefits from moorland fringe habitats of the kind that are lost by reclamation and afforestation, where the taller vegetation is an important habitat for the insects (such as caterpillars and sawfly larvae) needed by the chicks early in life. In general, though, populations are declining across Europe and so the protection of the few remaining populations has a priority air to it.[83]

PLATE 5.12 *A carefully managed grouse moor on the Co Durham section of the North Pennines. The patchwork of heather of different ages is clearly seen; some moors tend to strips rather than these quasi-ovoid patches.*

UK government has recognised the value of heather moorland in its national wildlife and scenery resource as well as its value as fodder and so some conservation policies have been announced, which are discussed in the appropriate later section. More research still apparently needs to be undertaken on *Calluna* since some management proposals are founded on heather which is burned on 10–15 year rotations, which does not apply to extensive areas of upland Britain where the heather plants become layered and respond differently to management processes.[84] This confirms the overall finding that the general ecology of moorland is not a very good guide to measurement at farm scale. The concept of stocking rate, for example, ignores biological variations in space and time; equally the grazing pattern itself is very variable through the year. The correlation of sheep numbers with heather loss is therefore inexact, albeit a tripling of sheep numbers between 1930 and 1976 as has happened in the hill parishes of the Peak District is likely to have exerted some effect.[85] Yet grazing does not itself affect floristic richness since there are usually refugia that are lightly eaten; the difference between very acid and more base-rich soils is much more important. Thus the terms 'under-' and 'over-' grazing have only a limited value for the evaluation of upland vegetation.[86]

Grazing pressures and environmental change

There is however the much broader issue of reducing grazing pressure on moorland vegetation, maintaining plant and animal wildlife, and allowing farmers to make a living. In the 1980s it was recognised that all three seem to be locked into a downward spiral and so attempts on two levels have emerged to try to ameliorate the situation. On one level, there is reform of the Common Agricultural Policy (CAP), which is slow and uncertain in its outcome; at the national level, almost every agency concerned with the uplands has tried to formulate some kind of scheme to try and divert the course of current trends. Some schemes derive from legislation and subsidy payments and others are experimental procedures on a regional basis. They change frequently but a selection of recent redirectives is given in Table 5.8.

Nevertheless, the pressures upon the moorland vegetation and wildlife have been recognised at all levels and many attempts are being made to ameliorate the degradations which are seen as the consequence. The easiest in many ways is the tackling of footpath widening and erosion which results from the feet of large numbers of recreational users. It is not necessarily easy in cost (it has usually to be paid for out of public funds unless the land belongs for example to the National Trust) but is in general uncontroversial. There are, though, some bodies of upland users who regard the process as a form of urbanisation, though it is not clear what other choices they would propose. The control of motor vehicles in the uplands is more difficult: the legal basis for keeping recreational vehicles off some green tracks, for instance, is lacking since the tracks are not legally designated only for walkers, horses and even bikers. So

TABLE 5.8 Schemes for upland management [NB This is *not* intended to be an exhaustive list]

Scheme	Date	Area	Purpose	Agency	Notes
Native woodland in National Parks	1997–	England and Wales: National Parks	Extend and link woods of native species	Forestry Commission	A 'challenge' scheme, with competitive bids
Woodland Management Challenge	1997–	EU Objective 5b areas	Better management of neglected woods	Forestry Commission	Covers costs of all kinds of management, including provision for access
Moorland Regeneration Programme	1997–	North York Moors as part of EU Objective 5b	Marketing initiatives for stock and produce; animal health, bird habitats under SPA	National Park authority	Integrated partnership of over 120 farmers and 26 estates plus statutory agencies
River Regeneration Programme	1996–	River Esk, North York Moors	Habitat improvement for wildlife and anglers	National Park authority	Partnership of many individuals and agencies
ESAs and Stewardship schemes	1987–	LFA status areas: scheme is voluntary	Removal of threat to landscape, nature, heritage from agricultural practices	MAFF/DEFRA	Range of targeted programmes; nearly 400,000 ha by 1999
Moorland Scheme	1995	Any upland	Improvement of heather moor	MAFF/DEFRA	11,000 ha covered and 7,500 sheep removed: now incorporated into 'Stewardship' scheme
Agenda 2000	1998–	LFAs	Agri-environmental schemes to maintain rural employment	EU	MAFF administer Upland Experiments in Bodmin and Bowland
Agenda 2000 with EU Stuctural funds	1998–	Designated regions	Integration of rural policies, protection of regional and local diversity	Countryside Agency	Integration of upland farming systems and the rural economy
EC Habitats Directive; UK Habitat Action Plans	1994–	SSSI, SPA, SAC, NNR, ESA, HAP areas	Wildlife & biodiversity protection	JNCC	Positive management for nature, inc possibility of climatic change

green roads, like some footpaths, can get ever wider and more churned-up.[87] The mountain bike does not mix well with walkers in any environment, let alone in the cover-less uplands. Illegal killing of wildlife meets with almost universal disapproval but is very difficult to police. Groups of enthusiasts together with the approval of a landowner can usually keep watch on the nests of, for example, peregrine falcons (whose young can be sold abroad for hawking) but it is very difficult to do much about an estate whose gamekeepers are systematically killing hen harriers.

Measures of amelioration undertaken by the conservation and countryside management agencies are relatively insignificant in terms of environmental change caused by the heavy grazing of sheep. Policies aimed at supporting marginal agriculture have been targeted at production rather than protection and so every upland farmer has found it essential to increase the number of animals. The effects of this on vegetation and soils (and thence on other life forms) are well understood in general terms though prediction of the impact of a given stocking level on an individual holding is still uncertain. Models which predict the utilisation rate of different vegetation types for particular seasonal stocking systems are being developed. They show the complexity of the shifts in grazing intensity as stocking densities are increased and should enable a land manager to determine the stocking rate which will provide a given mix of vegetation types. The economic context is of course another set of considerations entirely.[88] In areas where for example *Molinia* has replaced *Calluna*, some agencies are trying to re-seed with heather after various forms of treatment, including burning and also harvesting of older heather and the grass. The Stiperstones in Shropshire and Anglezarke Moor in the western Pennines have seen trials of such techniques by public agencies.

An unknown factor in the future of sheep production is the cessation of the legality of hunting with dogs. Though the debate over hunting is largely focused on the lowlands, a number of packs of hounds regularly hunt over moorland and moorland fringes, with the huntsmen on foot or possibly moor ponies rather than large thoroughbreds. About 70 per cent of foxes killed in upland Wales involved foxhounds or terriers though this includes foxes shot when driven by dogs over standing guns. The corresponding figure in the Midlands heartland of foxhunting was 27 per cent. When the evidence is considered coolly, a clear picture of the role of foxes emerges. On average, 17 per cent of lambs born of the Scottish Hills die within 24 hours of birth but of this (admittedly very high) mortality only five per cent can be attributed to predation. Between 0.5–3 per cent of viable lambs (perhaps 1 per cent of those born outdoors) are lost to foxes, so that losses to foxes are small compared with, for example, 30 per cent lost to exposure and starvation. In one survey, 80 per cent of farmers had no evidence that they lost lambs to foxes. Foxes certainly kill healthy lambs but they are also seen around flocks as they scavenge dead and dying lambs and after-births. The cessation of fox control on one estate in Scotland for three years did not produce any increase in predation.[89] It seems therefore that as in the lowlands, the argument should centre on whether this is a form of pleasure that should be legal rather than its role in upland ecosystems.

Moorland restoration

Concern over the condition of moorland has been directed mostly at the loss of heather, and at erosion. The generally poor quality of open grassland with a great deal of *Nardus* is a secondary worry. In a sense, then, the number and regime of sheep grazed is the key to the problems and many references to sheep densities have been made so far. In as much as the simple solution seems to be to find a socio-political route to lower numbers of sheep, there are dangers of over-simplifying matters. What is needed is a model of a particular moor which quantifies the vegetation types so that the annual utilisation rate can be set but broken into monthly units. Thus the seasonal diversification of sward characteristics can be monitored and there is a better chance of matching sheep numbers to sward capacity. It may be, for instance, that taking sheep off at certain times of year, rather than totally, will reduce the pressure on heather. Research suggests that the structural diversity of a heather stand (as brought about by *inter alia* burning and possibly grazing by other vertebrates such as hares and rabbits) is important in influencing offtake by sheep.[90]

In England and Wales the problems of heavy grazing are worst on upland commons, and the situation was exacerbated by an exact quantification of grazing rights that was made under the Commons Registration Act of 1965, with many opportunities for exceeding the capacity of commons to support sheep. Between 1993 and 2000, MAFF investigated 153 cases of overgrazing and erosion, of which 75 per cent related to upland commons.[91] (There are regional differences: on the North York Moors the grazing impact is much the same on commons and on tenanted moors.) Clearly, self-regulation is the best cure (as with the Dartmoor Commoners' Council for example) but this might be more attractive to a low-income group of farmers if a restored set of swards were seen as advantageous. Experiments to restore heather moorland have been carried out in a number of places, with varying success. On Dartmoor, at 400 m ASL on Langridge, both paring and rotovating of *Molinia* sward followed by re-seeding with heather failed to produce a good heather cover since selective grazing by sheep and the invasive capabilities of the *Molinia* were paramount.[92] The replacement of *Molinia* by *Calluna* seems to require high levels of input: in one study of three moorlands, the use of glyphosate was much more effective than combinations of burning and grazing treatments. Even here, *Molinia* litter needed to be removed and heather seed added before success could be measured.[93] One trend for the future was noted in the work by a commercial firm from Yorkshire in re-seeding heather moorland after overgrazing, taking up farmers' grants under stewardship schemes.[94] In most places a public agency has taken the lead: heather cover has been restored on

Anglezarke Moor in west Lancs by the County Council; on the North York Moors by the National Park Authority; in Wales the restoration of habitat for black grouse has been led by the RSPB.

Farming the scenery?

The Foot and Mouth Disease outbreak of late winter-early summer 2001 focused attention on the grazing systems of the nation as never before. Though areas like north Cumbria attracted most media coverage, the announcement of outbreaks in hill lands like the Lake District, Mynydd Eppynt, Dartmoor and Exmoor drew attention to the role of sheep in maintaining the current scenery. It also allowed the agencies concerned with rural land use to point out the Ministry-Client relationship as far as hill sheep are concerned. A Moorland Management Project in the Peak District National Park has attempted to tackle some of the ecological stages involved, using principally the Kinder Estate of the National Trust. Here, the density of sheep numbers could be controlled, and experiments for example in spreading heather seed could be monitored. Local experiments on revegetating burns and on stabilising degrading blanket bog were also carried out. The work was of basic scientific value but also fed into a large ESA (North Peak ESA) which by 1991 occupied nearly 41,000 ha, some 89 per cent of the eligible land. By the adoption of new prescriptions in 1993, the regulations included not only limitations on the numbers and seasonality of sheep and their supplementary feed sites but an agreement to regenerate an area of degraded heather moorland equivalent to 2 per cent of the moorland area. Higher tiers of the agreements reduced the number of sheep (to 0.66 ewes and followers per ha of grazeable vegetation), the number over-wintered by 25 per cent and eventually to designate exclusion areas by erecting temporary fencing. A lasting problem has been the consequences of severe wildfires. Ideally, sheep need to be excluded for two growing seasons but there is no way of achieving this, except perhaps for some form of insurance against the loss of forage.[95]

A brief recapitulation of the situation will recall for us the importance of the Hill Farming Act of 1946, and EC Directive 75/268 which included all the uplands in the category of Less Favoured Areas (LFAs). Hence the number of breeding ewes in LFAs rose by 35 per cent in 1980–2000 and the number of sheep-only farms rose by 17 per cent whereas the proportion of mixed cattle-sheep farms fell by 21 per cent. More cross-bred ewes were introduced: they need more supplementary feed year-round. When increased sheep densities were promulgated as an improvement to the upland economy, no question of the probability of increased erosion entered into the process.[96]

Until the late 1990s all the support payments made on the status of the animals (i.e., not those relating to fencing, grassland improvement, drainage and the like) were headage payments, made on the basis of their number. The advantage to the farmer has obviously been to keep as many stock as possible, using supplementary feed on the hill as necessary. Heavy grazing and erosion, as already noted, are two outcomes. New schemes from 1999 onwards are targeted at the integral management of an area within an LFA, with a whole-farm approach which is supposed to avoid moving the problem down the hillside. The Hill Farm Allowance scheme is fundamental but several payments are still made on a headage basis.[97] Environmental concerns are however central to the new developments and clearly the aim is towards lowering the production and raising the protection. Side-effects keep happening, however: the banning of organophosphorus sheep dips and their replacement by synthetic pyrethrins has allowed pollution incidents to kill native crayfish, fresh-water mussels and salmon.

The management of grazing on common land seems to have been even less profitable to either farmers or environment and recent targeting of ESA agreements at commoners will, it is hoped, produce improvements. The Long Mynd in Shropshire (mostly common land but owned by the National Trust) is an example. The ewe density in the late 1980s was 5.5 ewes/ha, with little heather and a great deal of bracken, poaching of species-rich mires and flushes and trampling of vegetation at sites of supplementary feeding. A ten-year ESA agreement (costing the taxpayers £1.7 million over its lifetime) with the commoners was signed in 2000, which aims at reducing stock levels and taking them off the hill altogether in winter. The immediate aim is to regenerate heather and enhance the populations of upland birds such as red grouse, curlew, wheatear and ring ouzel.

This and other ESA schemes, together with pressures to revise the CAP, are at present nuclei of differences in the management of uplands grazing. They could spread to become the norm but there is nothing inevitable about that and strong counter-pressures no doubt exist.

The statutory protection of upland environments

The revaluation of the uplands that took place under the influence of Romantic ideas has meant that throughout the nineteenth and twentieth centuries there has been, so to speak, a dual economy. Alongside their role as a set of *productive* economies, yielding trees, water, game and grazing, there has been set an important *protective* function in which certain other values have been deemed to merit attention at all levels from the individual person to the national government and beyond. Thus the maintenance of certain types of rural economy, of particular kinds of scenery, of settlement form and function, and of wildlife perpetuation, all together with the setting for a variety of forms of outdoor recreation, have helped to make a set of distinctive environments and landscapes. Such a dualism has naturally created stresses and conflicts, some of which are explored in this section. The general development and impact of outdoor recreation is however dealt with in Chapter 4 and later in this chapter.

Conservation: the appearance of the landscape

No greater icon of the desire to protect upland environments as elements of people's surroundings can be found than William Wordsworth. Though his thinking was sparked by the mountains of the Lake District rather than by moorlands *sensu stricto*, his attitudes became signally pervasive. In his *Guide to the Lakes* of 1810, he ventured the idea that 'persons of pure taste . . . deem the district a sort of national property, in which every man has a right and interest who has an eye to perceive and a heart to enjoy', though this was qualified by the thought that the railway might allow in those incapable of such well-mannered perceptions and pleasures, like labourers, artisans and the humbler class of shopkeepers. The idea was kept alive but there was nothing done until the inter-war period when the pace of urban and industrial change together with demands for relatively large-scale outdoor recreation, evoked in successive governments the desire to protect the 'best' pieces of landscape for the nation and where possible to increase access for walkers.[98] The moving bodies and individuals had in front of them the example of the National Parks of the USA which had been running since 1872, so it is not a surprise to see some proposals which leant in the direction of nature conservation rather than simply of the appearance of and access to the landscape. Bodies such as the Council for the Preservation of Rural England (founded 1926, followed by Wales in 1928) joined with recreation organisations in a Joint Standing Committee on National Parks (JSCNP) in 1936 and thereafter a series of reports, committees and battles was fought over the type and extent of any designations that might be made. The key document of the inter-war period was the Addison Committee, set up by a Labour administration. It reported in 1931 and was heavily influenced by biological considerations; it was useful mostly as a catalyst in promoting bodies such as the JSCNP. Headway amongst landowners on the subject of land use controls was however very slow indeed. As in many other endeavours, it was the national mood of reappraisal during and immediately after WWII that was crucial.

The reports which advocated Parliamentary action in the field of landscape and amenity protection have often been chronicled and discussed. The key stages are those of the Scott report (1942), Dower (1945) and Hobhouse (1947), with the latter leading more or less directly to the National Parks and Access to the Countryside Act of 1949, which also set up the Nature Conservancy. The role of moorlands as candidates for designation as National Parks was central: Dower's 'first list' included Dartmoor, Exmoor, the Craven Pennines, the Peak and the Black Mountains-Brecon Beacons. His 'second list' (these were judgements of worthiness) added the north-east Cheviots, Swaledale Pennines, Berwyns, North York Moors, North Pennines, Howgill Fells, Plynlimmon (sic) and Radnor-Clun Forests. Hobhouse eventually went for twelve National Parks in three groups of four. Of the first four, two were mostly mountainous (The Lakes and North Wales) but the others are moorland: The Peak District and Dartmoor. His second group included the

Yorkshire Dales, the Pembroke Coast (which included the Preseli Mountain) and Exmoor. Group three included The Wall, North York Moors and Black Mountains-Brecon Beacons. In the event, ten parks resulted from the 1949 Act and none of them were lowland areas, like the South Downs, New Forest and Broads.[99] The Act and the subsequent administrative structures have been the subject of almost continuous review and minor amendment ever since, either from official review bodies or quasi-independent groups propelled by the voluntary sector.[100]

In terms of environmental change and management, perhaps three features of the period to about 1970 stand out:

1. The term 'National Park' was never really very helpful. It recognised an aspiration to designate what were perceived as very important scenic areas but it took a great deal of education to persuade people that unlike in the USA for example, the government owned very little land: that in the Yorkshire Dales for instance, 96.2 per cent of the land within the National Park is privately owned.

2. From Dower until the 1970s it was not officially recognised that achievement of the aims of the 1949 Act needed active management, the legislative and financial base (to say nothing of the political will at local level) for which was lacking. In the event, the inter-war impetus towards nature conservation was not carried through.

3. In 1949 (and indeed until the 1970s) it was assumed that farming and forestry were environmentally benign as far as scenery (and wildlife as well, with the possible exception of predator 'control') was concerned. There has never been any obstacle to works for e.g., flood control, though greater landscape sensitivity might be required.[101]

4. It is possible that the emotional concentration on National Parks in the 1950s–80s meant that no more comprehensive countryside planning was attempted.[102]

In the National Parks, the 1949 Act required that scenic values were to be preserved by means of development control of a type stricter than in 'ordinary' areas; that access for recreation was to be facilitated by means of access agreements and the construction of facilities such as car parks; and that a flourishing rural economy was to be perpetuated. That these might not always be compatible was recognised only slowly at an official level and one strand in the discrepancies was recognised by Sandford (1974) when his body called for planning control to be extended to forestry and to farming. This change of attitude was rejected by the government in 1976 and is still on the agenda of the campaigning groups. The Wildlife and Countryside Act of 1981, however, allowed much greater intervention by the National Park authorities in terms of being able to pay land owners not to do certain things and in general to be more friendly to non-traditional uses and visitors.[103] After Circular 22/80

from the Department of the Environment, there was some relaxation of planning restrictions in rural areas even though there were supposed to be no
changes in the National Parks.[104] There is now a new concordat between the
farming community and the National Park authorities as the latter act as facil-
itators in getting funds from EU sources for what emerge in most EU designations as poorer areas.[105] This takes place in the context of a big loss of
farming population and the re-orientation of many of the subsidies that were
directed at production rather than protection. The Treasury has also realised
that sustaining jobs in crafts and tourism, for example, is cheaper than doing
so in farming and forestry.[106]

If we ask what difference National Park status has so far made to moorlands,
then three major strands of impact can be discerned:

1. The authorities recognised the implications for landscapes of the conversion of dry moorlands and acid grasslands into seeded grassland and
   for barley crops. They became involved in negotiations to protect the
   open land for its landscape value and for public access. The highest-
   profile example was Exmoor in the 1970s, which ran to a public enquiry
   chaired by Lord Porchester.[107] He found that about 38 per cent of the
   drier areas had been converted to seeded grassland but only 10 per cent
   of the wetter zone had been subject to the ploughing, liming, fertilizing
   and seeding involved. His conclusion was that only 30 per cent of the
   area denoted as 'Critical Amenity Area' was at risk of conversion but that
   for Exmoor as a whole, nearly 60 per cent of the remaining moorland
   was potentially reclaimable. This study reinforced the efforts in National
   Parks like the North York Moors, also low and dry, to protect key moorland areas. The 1981 WCA required the production and yearly updating
   of a Moor and Heath map for each National Park and the selection of
   critical areas for conservation. This choice was made on the grounds of
   visual character and lack of enclosure rather than botanical or agricultural potential grounds.[108]

2. The National Park authorities have become strongly involved in the protection of heather moorland. This is seen as an important habitat type
   within the European systems of conservation but also as a characteristic
   element of scenery and often a key matrix for access: though walking
   through long heather is not attractive, paths and tracks within it are
   important locations for the trainer-footed as distinct from the booted.
   Heather is usually replaced by acid grassland dominated by *Nardus
   stricta* or by bracken (*Pteridium aquilinum*), depending on grazing and
   burning regimes (see pages 216–50). If sheep congregate regularly at
   supplementary feeding points for example, then soil erosion may result.
   The National Park authorities have had little statutory power to affect
   these outcomes save by buying or leasing the land, entering into compensatory agreements with farmers in which it has been difficult to

compete with EU-based headage payments. New emphases on environment in agricultural support may however gradually turn the tide in favour of fewer sheep and more heather. Heather is of course tied up with the role of grouse: on Exmoor this is of little interest but on the North York Moors the shoots are important parts of both ecology and economy.[109]

3. The National Park authorities have been active in preserving scheduled Ancient Monuments and entering into agreements with owners that prevent further attrition of their archaeological interest.[110]

Overall, therefore, we can recognise a shift in the ability of the National Park managers to intervene in free-market processes on the moorland, propelled both by legislation like the WCA 1981 and the ability to attract funding for bodies like the EU. There was a proposal by some members of the Sandford Committee, that an even higher category of 'national heritage areas' be created within the National Parks, in which no change was to be permitted without the approval of Parliament (Fig. 5.10). The government rejected this and it has, interestingly, never been seriously revived. If we think of the period 1955–80 as one in which the protection of moorland scenery was dominant in the disposition of the nation's attitude to protected scenery, then after 1980 that hegemony can be seen to have been challenged, with for example the conservation of lowland areas (e.g., the New Forest and South Downs) taking centre-stage rather than an elaboration of the existing arrangements for moorland.[111]

The context of the use and planning of the National Parks is changing. Since the 1949 Act, the exercise of planning authority by the National Parks was accepted grudgingly by almost everybody from the Treasury to local inhabitants, with only a few visionary National Park Officers keeping alive the spirit in which the parks were conceived. This has changed to some extent, notably since the parks became the obvious bodies to attract and manage finance under Rural Development Authority schemes from UK sources, and for example Objective 5(b) funds from the EU, as well as being able to attract other European income. Further, the parks were easily assimilable into the logic of the attempts to implement the Rio Accords and the UK version known as Agenda 21. The practical expression has been the National Park plan. The 1972 Local Government Act required the preparation of National Park plans which had to be updated at intervals of at least five years but this obligation was superseded by the 1995 Environment Act which calls for National Park Management Plans (NPMPs) which must be reviewed every five years or less. These plans must set out strategic objectives and policies, introduce an integrated approach across agencies, and set the park within a broader context. This latter includes the post-Rio notion of 'sustainable development' and the numerous obligations and opportunities which result from membership of the EU. As well as the year-to-year activities of development control and rural

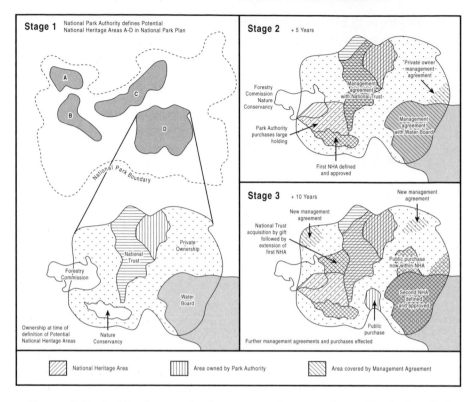

FIGURE 5.10 *A vision for extra landscape protection as put forward in the Sandford Report, with a core of 'National Heritage Area' being largely managed by agreement rather than acquisition by the National Park Authority. Though seemingly radical (and mostly rejected by government) it was in effect an extension of previous policies.*
Source: Department of the Environment, Report of the National Park Policies Review Committee, *[The Sandford Report], London: HMSO, 1974, at p. 104.*

development itself, the NPMP must include a 'Vision' for the park and for the conservation of both nature and culture. For Exmoor, for example, the 'Visions' section of the plan relating directly to environment provides that moorland is not to shrink in area and that heather and grass moorland is to be maintained where possible; that farmed landscapes are to look cared-for; that broad-leaved woodlands ought to be bigger and more frequently seen; and that there should be less reliance on motor vehicles for access, especially in the more remote areas. In the implementation of all these programmes, the 1995 Act allowed that in the event of land use conflicts, conservation was to take precedence over other possibilities. This was advocated in the 1974 Sandford Report but had never before been put into legislation.[112]

Two moorland regions (North Pennines and Bowland Forest) have been the core of designation in the second tier of protected landscape areas in England and Wales. The 1949 Act took some of the 'Conservation Areas' of the

Hobhouse Report and invented the category of Areas of Outstanding Natural Beauty (AONB), which was to receive special care from its county council but for which none of the measures (in terms of development control or finance) aimed at the National Parks were available. Forest of Bowland (802 km^2 was designated in 1964 and the North Pennines (1,983 km^2) in 1978 but neither get special treatment in the only review of AONBs.[113] The report affirmed that the primary purpose of AONB management was to be the protection of the landscape, and that open areas such as moorlands might well receive enhanced public access without the formality of access agreements. In those recommendations, the AONBs were not in many ways different from the National Parks, but their management remained a local affair: the North Pennines AONB belongs to three counties and six districts. Much of the management of the upland has been directed at attracting tourism and maintaining the Pennine Way although Northumberland County Council has attempted to manage the River East Allen for the benefit of its mammalian and fish populations. Both upland AONBs have large areas of grouse moor with the attendant problems of public access and suspicions about the illegal killing of predatory birds. The Durham County Council's plan for its part of the AONB places restrictions on the siting of windfarms and the development of mineral extraction.

The basic ideological structures of the National Parks have not been challenged, needless to say. They have been put forward as a national resource requiring intensive management by professionals and so the 'ownership' has been that of a particular kind of person, notably a private landowner (Fig. 5.11). Landowners have been especially important since they are capable of benevolence in terms of access, provided their basic freedoms are unfettered. While thus seeming to confer benefit on the nation, they have resisted change. The state has retreated since 1949 and the public have by and large acquiesced. So there is an area of 80 sq miles in the Peak District with virtually no public access, and many agri-environmental schemes have attracted landowners who have opted out of the access provisions (with their accompanying payments) of the programmes.[114] The then Minister, Lewis Silkin, introducing the Bill that became the 1949 Act said that it was, '. . . not just a Bill. It is a people's Charter . . .' which has not turned out to be totally true.

Conservation: wildlife

Although moorlands do not have a high diversity of living organisms, they are always of interest to nature conservation, since they form the wildest places in Great Britain, along with mountains and some coasts. Within the overall setting of the legislation pertaining to environmental components, the main phases of wildlife conservation of which we are currently the inheritors have been:

1. Responses to the 1949, 1981 and 1990 Acts of Parliament[115] which permitted the acquisition and management of protected areas such as

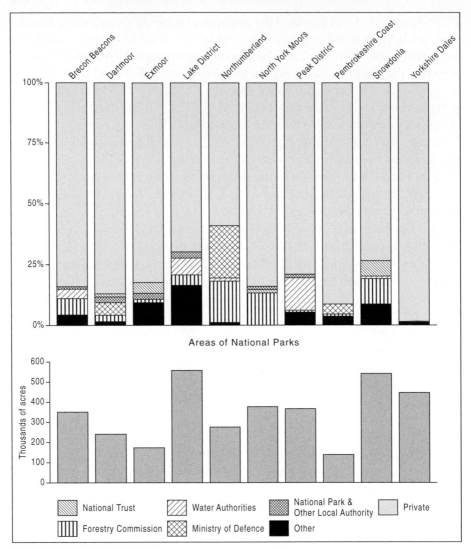

FIGURE 5.11 *Land ownership in the National Parks in the 1970s. (The complete set is given, including therefore the mountain parks.) The predominance of private land is overwhelming and the amount owned directly by the NPAs nugatory. This has increased somewhat since the 1970s but never enough to give much control to the Authorities, although some critical sites may be included, usually recreational areas.*
Source: P. Leonard, 'Agriculture in the National Parks of England and Wales: a conservation viewpoint', Reprinted from Landscape Planning *7, 4 with permission from Elsevier, p. 373, 1980.*

National Nature Reserves (NNRs) and Sites of Special Scientific Interest (SSSIs).

2. European Union legislation in the form of Directives for Habitats and Species which are binding on the UK government;[116] habitats thus protected may be labelled Special Areas of Conservation (SACs); those for

safeguarding birds (especially migratory species) are Special Protection Areas (SPAs).

3. Action following the Rio Convention of 1992 in which the preparation and emplacement of Biodiversity Action Plans at national and local scales are undertaken by national governments.[117]

While many species native to Great Britain are relatively common, between about 10 and 20 per cent of native species are considered threatened. Over a third of the 2,700 native species of mosses, liverworts, and lichens are threatened or nationally scarce. The comparable proportions for seed plants, ferns, and related plants (about 2,300 species in total) and for invertebrates (about 15,000 species in those groups covered by the Red Data Books) are each about a quarter. In the uplands of England and Wales, rare species requiring Biodiversity Action Plans include seven species of plants and animals associated with upland oak woodland. In England and Wales, the yellow marsh saxifrage (*Saxifraga hirculus*) is found only in wet flushes in the North Pennines and is vulnerable to heavy grazing, as are many of the Upper Teesdale rarities (p. 165).[118] The **montane** zone above 600 m ASL contains the dotterell (*Charadrius apicarius*) as its only breeding bird confined to that zone and there were only two pairs per year in the 1990s: it is obviously susceptible to climatic change. Similarly, the summit of Cross Fell in the North Pennines has a small population of the snowfield ground beetle (*Nebria nivalis*) which is unlikely to survive warmer conditions. On a broader scale, the Peak District National Park has above-average populations for certain bird species (it houses 9–11 per cent of the British populations of curlew, golden plover and merlin) of which several (merlin, golden plover, twite, dunlin and short-eared owl) are at the extremes of their population range and therefore vulnerable to all kinds of habitat change.[119]

The involvement of government has been strong in the uplands ever since the 1949 Act, since agricultural subsidies joined the afforestation drive headed by the Forestry Commission, in attempts to fulfil those parts of the 1949 Act which advocated a thriving rural economy. Many uplands, as discussed elsewhere, became the core areas of National Parks and others were candidates for that status, or for AONB prominence, long before obtaining it, as with the North Pennines AONB or never doing so, as with mid-Wales. Designation as a National Park or AONB never carried any axiomatic statutory commitment to nature conservation. The response of the Nature Conservancy to its post-1949 role in the uplands was mostly limited, and maps of designated NNRs and SSSIs in the moorland areas show a scatter of small reserves (like isolated ash woods in the Craven Pennines and the equally isolated pure-oak Wistman's Wood on Dartmoor), with only rare instances of large NNRs. Yarner Wood on the eastern flanks of Dartmoor (150 ha including some lowland heath) was the biggest upland woodland and the Moor House NNR at the head of Teesdale (3,894 ha) (Plate 5.13) were in general exceptions and both were in the first (1952) tranche of NNR designations. Lower in Teesdale, for example, the area containing so many rare species only became an NNR after the Cow Green controversy, although it had been persistently recommended by national committees since 1915. Indeed, a list of NNRs created between 1949 and 1975, which had been proposed between 1915 and 1949, has only Upper Teesdale as a true moorland site from the whole of England and Wales. Maps of reserves managed by County Naturalists' Trusts and the RSPB in 1974 show similar sorts of gaps in higher areas.[120]

The general perception in those years of the moorlands as lacking in diversity and especially in rare species other than birds has carried over into current Biodiversity studies and proposed actions, which mostly depend upon the amendments of grazing regimes. Exceptions are the conservation value of upland deciduous woodland (see below) and the possibilities of retarding the stripping-out of limestone pavement for sale as garden rockery material. Even so, recent conservation work has been carried on in an economic climate in which central government support for commodities has always exceeded that for conservation by a wide margin and any changes have to work at a number of levels. There has first to be a policy which favours conservation and wildlife; this has then to be made operational by appropriate schemes, of which there are often a variety at any one time.[121] These schemes will not work unless

PLATE 5.13 *Part of the Moor House-Upper Teesdale area which has received many conservation designations for its high altitude vegetation and rare flora. This is taken from Cronkley Fell and shows an area of wet flushes within the acid grassland; beyond, the dark colour indicates a large area of juniper (*Juniperus communis*) scrub, an uncommon plant community in England.*

there is a pool of cultural and management skills. Since so many landowners and tenants have had for decades a dominant perception that they were there to produce food, shifts in favour of environmental concerns may meet with management problems even if their purpose and their financing appears acceptable to the local inhabitants.

The evaluation of the uplands for conservation has been undertaken by English Nature and its Welsh equivalent, Countryside Council for Wales (CCW). These have taken the demands of the Biodiversity Action Plan (BAP) and of the relevant Directives from the EU and attempted to prioritise the protection and enhancement of communities and species and indeed their restoration where that action is feasible. The over-riding priority is 'naturalness', where this is seen as the presence of large predators and the absence of alien species, along with some rare species. Diversity as such is not seen as a useful criterion in the uplands. Whether these add up to any degree of being 'natural' in any normal use of the word is debatable in the light of the uplands' history: 'wildness' might be a better word but presumably one which lacks the necessary cultural resonances. The cultural dimension is shown by the broadened criteria developed from the Rio Convention commitments which add to the Convention's identification of biodiversity as a resource and to the need to slow down rates of extinction by setting out reasons for conservation such as moral and aesthetic grounds, stewardship, the benefit to society generally and the economic value of some habitats and species.

English Nature has identified eighteen upland Natural Areas (NAs) and has ranked their conservation significance in terms of the frequent or extensive presence of vegetation communities which are rare or absent outside the UK. In addition it has identified priority NAs for taxonomic groups such as birds, mammals, reptiles and amphibians as well as for plants. 'Upland heath', for example, has seventy-four species of conservation concern. Of the moorland NAs, three (North Pennines, Yorkshire Dales, Dartmoor) rank as 'Outstanding', a further six as 'Considerable' (Exmoor and the Quantocks, Border Uplands, North York Moors, Bowland, South Pennines, Staffs Uplands), and only the White Peak as 'Negligible'. For mammals, the moorland areas are not in general the most important areas of the country, but exceptions include the otter (*Lutra lutra*) in eight moorland areas, the red squirrel (*Sciurus vulgaris*) in the Border Uplands, the Greater Horseshoe Bat (*Rhinolophus ferrumequinum*) which has maternity sites in Exmoor and the Quantocks and hibernation sites on Dartmoor and Bodmin Moor. Its Lesser relative (*R. hipposideros*) is a high priority on Dartmoor and Bodmin Moor and the water vole (*Arvicola terrestris*) is a high priority species in the Peak District and medium in four other moorland areas. The dormouse (*Muscardinus avellanarius*) is found in the woodlands of five moorland NAs. Given the high profile of birds in the UK, attention has been given to their conservation status and a number are singled out for special standing with British views reinforced by designations in European Union Directives. This legislation requires active conservation measures beyond mere protection from killing and is accorded to species for which Britain is a stronghold of international significance.[122] These species are incorporated into the Biodiversity Action Plans. The Plans also cater for other groups of animals, and, naturally for plants as well. The concentration of Arctic-Alpines in Upper Teesdale is an example of concentrated attention to rare species; the BAPs also envisage conservation of upland heather moor (even though it is lacking in diversity) since it is an unusual habitat on the European scale. The National Park authorities are required by the 1981 WCA to make a Moor and Heath map and then to select areas for conservation but the emphasis is on the visual character and unenclosed nature of the terrain rather than on its ecology, or indeed its agricultural potential. Even the National Park authorities cannot control the decline of heather moorland, only encourage landowners and users to adopt schemes which maintain that cover; outside the National Parks the opportunities are even fewer. Bracken eradication programmes are almost entirely confined to National Park areas.[123]

Most schemes for the uplands do not differentiate between moorland and mountain, so that aggregate data for England, for example, may be skewed by the inclusion of the Lake District. Some idea of the penetration of conservation schemes may be gained by the take-up of Moorland Stewardship Scheme (MSS) and Environmentally Sensitive Area (ESA) designations, which between them involve (among other things) moving to lower numbers of sheep and less intensive use of land.[124] By the end of the twentieth century, 108,000 ha of heather moor, 5,520 ha of hay meadows, 6,300 ha of upland limestone grassland and 40,232 ha of species-rich pasture were under management agreements in Great Britain. These need to be set in the context of, for instance, the 565,279 ha of land in the uplands eligible for ESA status of which 380,278 ha were actually under an agreement, of which 13,432 ha of grazing-suppressed upland heather was a core concern. The Moorland scheme has had a slow take-up and in 2000 stood at 11,000 ha of dwarf shrub upland. Capital grants had also enhanced 959 km of hedgerow and 795 km of stone walls. English Nature schemes are focusing on the presence of key indicator species at Natural Area level and then using the lead agencies' skills as identified in the BAP to ensure the future of the species and their habitats. In Wales, a whole-farm agri-environmental scheme called *Tir Gofal* ('Land Care') contains mandatory prescriptions to maintain existing upland heather moor as well as the option of its expansion or re-creation; it is hoped to have 1,000 farms signed up to the scheme by 2002. Most plans recognise that the dominant factors in the trends of wildlife populations in the uplands have been (a) afforestation with coniferous species and (b) agricultural policies which concentrated on production, especially via headage payments on sheep.[125]

The Action Plans published in 1995 for various uplands (as indeed for other regions) focus first on 'habitat statements' which summarise the current status of these areas, the factors affecting the habitat and the action to be taken.[126] Some of these plans were at that stage costed: the protection of limestone pavement (most of which is in Cumbria and North Yorkshire) was given as £130,000 in 1997 and 1998, falling to £100,000 in 2000. Upland oakwood, with a much wider distribution (though with a great deal flanking the Welsh moorlands) and a target of 100,000 ha in conservation management rose from £2 million in 1997 to £6 million in 2000.[127] Even acid grasslands, which are not high in most perceptions of conservation priorities, having a limited biodiversity, are thought to benefit from lower grazing levels, protection from intensification of use and, occasionally, restoration of the habitat. Some conflict with current land management and use is clearly likely. In 2000, English Nature published recommended biodiversity indicators for each region, which are summarised for uplands in Table 5.9.

Another major contention has been between twentieth century afforestation and the preservation of open moorland and moorland-edge landscapes. This is largely treated in the account of afforestation policies (pp. 153–62) but here it will be noted that since the 1980s there has been a new emphasis on the protection of broadleaved woods in the uplands and indeed their extension by fresh planting. The Forestry Commission has a scheme under which National Parks may compete for funds to develop and protect woodland so that such areas, preferably large in size, will develop 'a natural character' and which extend or link existing semi-natural woods. If public access and wildlife

TABLE 5.9 Regional indicators: species and habitats

Region	Habitat type	Species
North-west	Limestone pavement Blanket bog Upland oakwood Upland hay meadows	Otter Curlew
North-east	Blanket bog Upland oakwood Mixed ash wood Juniper wood Upland heathland	Otter Black grouse
Yorkshire	Limestone pavement Blanket bog Upland oakwood and mixed ash wood Upland heathland Upland hay meadows	 Golden plover
South West	Upland oakwood	Otter Horseshoe bats Cirl bunting

Source: English Nature, 'Wildlife yardsticks for the regions', *English Nature Magazine* no 47, January 2000, 9–11.

conservation can be linked to a scheme, so much the better.[128] The first awards were made in 1997 to areas in Northumberland, the Yorkshire Dales and the North York Moors and totalled 330.5 ha. In North Yorkshire, Northumberland and Durham, neglected broadleaved woods of 10 ha or less are also eligible for a grant-aided scheme which pays for selective felling, the removal of invasive species and possibly the restoration of coppice cycles. By 1998, the two north-easternmost counties of England had 125 ha under such strategies. English Nature have also given attention to woodlands in moorland areas since relict deciduous stands are often of 'Ancient' character: in that context, the North York Moors, the Central Marches and the Peak District score highly whereas the central core of Exmoor has very little such tree-clad area.

This broader perspective from the Forestry Commission must be seen in the context of an overall concern for the conservation value of woodlands, especially semi-natural woods dominated by native species. Their value is appraised at national level as being necessary areas to meet the obligations of the UK towards the European Union's Habitat and Species Directive and to its own Biodiversity Action Plan. Other benefits include the maintenance of habitat for native species of animals, the use of woodland extensions as buffer zones for older woods, more commercial timber production, some carbon fixation, and recreational value. If there are no other land use constraints then the area above 250 m has high potential for woodland expansion: in Wales, for example, the area could be tripled by expanding onto lands covered with bracken or onto managed grassland of low conservation value. The English position is very variable: in Northumberland a figure of 4 per cent is realistic, whereas on Dartmoor it might be 34 per cent if there were no other land use constraints which in that case there very probably are many. It has to be remembered that the grant structure has for long meant a loss of income to upland farmers if the number of stock is reduced for any reason.

English Nature have used their classification of Natural Areas as a basis for assessment and the determination of priorities, starting with the area covered by semi-natural ancient woodland. Some moorland regions come at least in the intermediate categories such as 1.5–3.0 per cent coverage rather than the lowest 0–0.5 per cent bracket, though none make the more than 7 per cent grade. None are in the highest ten and Bodmin Moor is in the lowest ten; others in intermediate categories are the Yorkshire Dales, North York Moors, Dartmoor, Exmoor and the Southern Pennines. High levels of woodland fragmentation have been recorded in the Yorkshire Dales, North York Moors, Bodmin Moor and Bowland. Thus areas which would benefit from restoration of ancient planted sites include the North York Moors and the Central Marches as high priorities and the Peak District as worthwhile. All criteria considered, deciduous woodlands have a high significance in the Border Uplands, Northern Pennines, Yorkshire Dales, North York Moors, Shropshire Hills, Exmoor and the Quantocks, and Dartmoor. In most others, the significance level is moderate. Few uplands have large blocks of woodland suitable for minimum intervention and only southern Dartmoor would meet the criteria for a suggested 'woodland district' where semi-natural woodland covers at least two-thirds of the land surface, with a minimum of 750 ha.[129]

For existing upland woods, the major problem that affects conservation values is grazing. Its level has increased greatly in the last 150 years as coppice management has declined; sheep numbers are up whereas cattle numbers are down; there are more deer including the spread of species like the muntjac; and more rabbits are showing resistance to myxomatosis. All these animals have different grazing and browsing preferences. Sheep, for example, are selective grazers but are not selective browsers on trees, especially in bad weather. Ponies are almost entirely non-selective and the impact of deer depends on whether they are resident all year, as roe usually are. Sika deer (*Cervus nippon*) are mostly found in the south of England but their preference for acid-soil vegetation types like heather and Sitka Spruce suggest that it could be a powerful coloniser of uplands; its ability to hybridise with red deer is well-attested. Muntjac (*Muntiacus reevesi*) have spread to the Welsh borders and to Derbyshire and beyond. They are ferocious eaters of ground flora (bluebells and primulas, for instance) and could negate many of the efforts to establish flourishing deciduous woodlands in uplands. Being only fox-sized, they are difficult to control, and have been illegally released into many areas.[130] Upland woods will increasingly require conservation management and a flexible and locally-derived management plan is needed; off-the-shelf general solutions are rarely helpful to the wildlife interest.[131]

The newest context of all for nature conservation is that of climatic change. The scenarios for Great Britain are not uniformly agreed, but most seem to foresee a rise in temperature of 0.8–2.0°C by 2050. Whereas the effects of that kind of change can be easily translated into, for example, river regimes, its effect on living species is less easily modelled. Upland environments such as the Moor House NNR in the North Pennines are seen as warmer and also as drier in the late summer and early autumn. Land use patterns reduce the rate at which species can migrate in response to climatic changes and so disjunct distributions are the norm in any case but are likely to be more fragmented. Upland species can only go 'up the hill' in a warming climate and so may be forced out altogether. A restricted upland species which is the focus of management may be totally lost to the biota and any designation of land as a reserve lose its point. The statutory boundary thus becomes a hindrance to conservation rather than a foundation stone. This situation, we may note, is in some contrast to the position with pesticides, when the uplands acted as a reservoir for species which were becoming greatly diminished in number in the lowlands: the sparrowhawk is an example. The central response was to set up Environmental Change Networks in 1992 which include five upland monitoring sites, two of which are in Scotland and the two in the Pennines and the Cheviots are moorland-based. Monitoring is comprehensive for all the components of the local ecosystem and in the case of the uplands measurement of climate, land management impacts, biodiversity loss, acidification, water quality and the effects of diffuse pollution are all included.[132] Since upland rivers are important in water supply for urban and industrial regions then the prediction of more extremes of flood and of low flow have considerable significance.

Interaction with recreation

The Countryside and Rights of Way Act 2000 is primarily aimed at securing the right of access to 'open country' for walkers. It does however strengthen the protection of all kinds of SSSI; some wildlife offences are now punishable with jail sentences, as can the release of certain non-native species. The management of AONBs is taken further. For existing designations, the Secretary of State can order the setting up of a conservation board, which will have the same kinds of function as a National Park Authority in conserving the 'natural beauty' of the area and enhancing public access and understanding of it, thus adding to the layers of legislation in National Parks and outside them (Table 5.10). If there is no conservation board then the relevant local authority must act.

The 2000 Act has been welcomed by English Nature and CCW but with caution (Plate 5.14). They are both concerned about the possible effects of increased access to uplands upon the wildlife and in particular upon populations of nesting birds. Disturbance of predators is also of concern but is perhaps outweighed by the new powers to try to deter upland managers and others who kill birds illegally.

OVERVIEW

If we add the impacts of recreation, then the vulnerability of the uplands to what are perceived as degradational pressures is obvious. And it is equally clear

TABLE 5.10 Events in the statutory history of the National Parks 1960–90

Year	Event
1968	Countryside Act: establishment of CC to replace National Parks Commission; new powers to NPAs concerning recreation provision, e.g., car parks and ranger services
1974	Sandford Report: Report pointed out that farming and environment not necessarily compatible. Proposals for control of farming and forestry rejected.
1974	All National Parks to have unitary authorities with own staff. Powers over landscape change, farming and forestry and environmental affairs generally weak
1981	Wildlife and Countryside Act: land use changes to be notified and if permission refused then compensation is compulsory
1980s	Establishment of ESAs, of set-aside and tentative movement away from production-driven grant systems.

Plate 5.14 *The uncertainty over public access to moorland is exemplified by this notice in the Peak District National Park, which allows access to the land for walkers except for closure during grouse shooting. The foreground however is not good grouse habitat and the heather moors are out of view. The CROW Act of 2000 will not change the situation much at this locality and in many places the nature conservation bodies will be glad of the possibility of restricting access. The red grouse is however scarcely an endangered species.*

that conservation has to be broadly conceived since the interaction of economy, scenery, nature and recreation is complex. Three negative impacts on wildlife seem to dominate: intensive burning, drainage of blanket bog and illegal predator control. Set in this context, a set of foci for changes in human-environment interactions in moorland areas seems to be emerging. A consensus by the UK conservation agencies identifies the following clusters of attention:

1. Positive management of a network of protected areas for species and habitats.
2. Achieving the targets set under the Biodiversity Action Plans.
3. Influencing government and EU policies so as to reduce grazing pressures by sheep.
4. Promoting best practice for habitat and species management.
5. Increasing public awareness of new attitudes and actions and requiring restraint from various user groups when necessary.
6. Research into, and monitoring of, changes.

Lest it be thought that this is too much of a nature conservationists' list, we can point out that (3) underlies a great deal of landscape concern as well as habitat linkages and that (4) and (5) are common to all responsibilities. However, the Edwards Committee recommended as late as 1991 that nature conservation should be seen as an integral part of all National Park management, with the implication that it had certainly not been so.[133] The position in which nature conservation was generally poorer in National Parks than outside them needs to be an unhappy phase of the past recalled only in histories.[134]

All recreation and conservation is an outworking of a particular theme in our system of values. We enjoy being in rural areas that are 'different' and we wish to see the signs of that in, for example, the sight of heather moor or the sound of the curlew. This is learned behaviour, nevertheless, so we need to delve a little way (though doubtless not far enough) into the ways these regions are represented in our culture.

NOTES AND REFERENCES

1. There is discussion of the idea of 'functional groups' for Scotland in D. D. French and N. Picozzi, 'Functional groups' of bird species, biodiversity and landscapes in Scotland', *Journal of Biogeography* 29, 2000, 231–59.
2. There was a movement to area-based schemes in 2001 under Rural Development Regulation, EU Council Regulation 1257/99 article 13, which includes environmental protection as well as continued agricultural use.
3. For more detail, see S. D. Ward, A. D. Jones and M. Manton, 'The vegetation of Dartmoor', *Field Studies* 4, 1972, 505–33. The Peak District data come from a National Park study of erosion by J. Phillips, D. Yalden, and J. H. Tallis, published by the Park authorities in 1981; see pp. 28–34. They calculated that about 1.2 per cent of the moorland (648 ha) was bare and eroding.

4. G. R. Miller, J. Miles and O. W. Heal, *Moorland Management: a Study of Exmoor*, Cambridge: ITE, 1984.

5. D. B. A. Thompson and J. Miles, 'Heaths and moorlands: some conclusions and questions about environmental change', in D. B. A. Thompson, A. J. Hester and M. B. Usher (eds), *Heaths and Moorland: Cultural Landscapes*, Edinburgh: HMSO, 1995, 362–85. The National Vegetation Classification (NVC) for moors is in J. Rodwell (ed.), *British Plant Communities. Vol. 2. Mires and Heaths*, Cambridge: CUP, 1991. There was an earlier comprehensive survey and map production of the vegetation of upland Wales in 1961–6: see J. A. Taylor, Reconnaissance vegetation survey and maps', in E. G. Bowen, H. Carter and J. A. Taylor (eds), *Geography at Aberystwyth*, Cardiff: University of Wales Press, 1968, 87–110 and the even earlier R. G. Stapledon and W. Davies, *A Survey of the Agricultural and Wastelands of Wales*, London: Faber and Faber, 1936. Vegetation maps in the uplands were attempted early in the twentieth century as in e.g., W. G. Smith and C. E. Moss, 'A geographical distribution of vegetation in Yorkshire. I. Leeds and Halifax district', *Geographical Journal* 21, 1903, 375–401 and for other Pennine uplands in subsequent volumes of that journal.

6. Thompson and Miles, 1995 (Note 5).

7. F. W. Kirkham and J. A. Milne, 'Progress towards defining ecologically-sustainable grazing management: the 'Moorland Biomass' and 'Heather Suppression' projects', *Aspects of Applied Biology* 58, 2000, 151–58. The projects ran from 1995–7 and covered England and Wales from Cumbria to Dartmoor.

8. A. Eddy, D. Welch and M. Rawes, 'The vegetation of the Moor House National Nature Reserve in the Northern Pennines, England', *Vegetatio* 16, 1968/9, 239–84.

9. M. P. Kent and P. Wathern, 'The vegetation of a Dartmoor catchment', *Vegetatio* 43, 1980, 163–72. An earlier vegetation mapping exercise on Dartmoor echoed its time in saying that north-eastern Dartmoor showed, 'a limited degree of adjustment to widespread grazing and other activities'. E. Johns, 'The surveying and mapping of vegetation on some Dartmoor pastures', *Geographical Studies* 4, 1957, 129–37 at p. 129.

10. P. D. Moore, 'The origin of blanket mire, revisited', in F. M. Chambers (ed.), *Climate Change and Human Impact on the Landscape*, London: Chapman and Hall, 1993, 217–24.

11. D. Charman, *Peatlands and Environmental Change*, Chichester: Wiley, 2002, is excellent on all these topics but covers a global range of examples. Much but not all of the treatment of blanket mires is from the British Isles.

12. T. H. Blackstock, D. P. Stevens and E. A. Howe, 'Biological components of Sites of Special Scientific Interest in Wales' *Biodiversity and Conservation* 5, 1996, 897–920.

13. There is a general account in D. A. Ratcliffe, *Bird Life of Mountain and Upland*, Cambridge: CUP, 1990.

14. General accounts of recent research are given in e.g., R. A. Stillman and A. F. Brown, 'Bird associations with heather moorland in the Scottish and English uplands' in D. B. A. Thompson, A. J. Hester and M. B. Usher (eds), *Heaths and Moorland: Cultural Landscapes*, Edinburgh: HMSO, 1995, 67–73; D. B. A. Thompson and J. Miles, 'Heaths and moorland: some conclusions and questions about environmental change' in Thompson *et al.* op. cit. 1995, 362–85; M. B. Usher and S. M. Gardner, 'Animal communities in the uplands: how is naturalness affected by management?', in M. B. Usher and D. B. A. Thompson (eds), *Ecological Change in the Uplands*, Oxford: Blackwell Scientific Publications, 1988. British Ecological Society Special Publications Series no. 7, 75–92. More specialised research material is found in journals such as *British Birds, Biological Conservation* and the research publications of the RSPB.

15. P. J. Hudson, 'Spatial variations, patterns and management options in upland bird communities', in M. B. Usher and D. B. A. Thompson (eds), *Ecological Change in the Uplands*, Oxford: Blackwell Scientific Publications, 1988. British Ecological Society Special Publications Series no. 7, 381–97.

16. Such as the Wild Birds Directive 79/409/EEC, and the Habitats Directive 92/43/EEC which allows the creation of SACs (Special Areas of Conservation).

17. I. Newton, P. E. Davies and D. Moss, 'Distribution and breeding of red kites in relation to land use in Wales', *Journal of Applied Ecology* 18, 1996, 210–24.

18. S. J. Parr, 'Changes in the population-size and nest sites of merlins *Falco columbarius* in Wales between 1970 and 1991', *Bird Study* 41, 1994, 42–7.

19. I. Newton, P. E. Davis and J. E. Davis, 'Ravens and buzzards in relation to sheep-farming and forestry in Wales', *Journal of Applied Ecology* 19, 1982, 681–706.

20. W. H. Pearsall, *Mountains and Moorlands*, London: Collins New Naturalist, 1950.

21. J. Coulson, L. Bauer, J. Butterfield, I. Downie, L. Cranna and C. Smith, 'The invertebrates of the northern Scottish Flows and a comparison with other peat-land habitats', in Thompson *et al.*, 1995 (Note 14), 74–94. See also J. Coulson and J. B. Whittaker, 'The ecology of moorland animals' in O. W. Heal and D. F. Perkins (eds), *The Production Ecology of some British Moors and Montane Grasslands*, Berlin: Springer-Verlag, 1978, 52–93; J. Coulson and J. E. L. Butterfield, 'The invertebrate communities of peat and upland grasslands in the north of England and some conservation implications', *Biological Conservation* 34, 1985, 197–225.

22. J. C. Coulson and J. E. L. Butterfield, 'The invertebrate communities of peat and upland grasslands in the North of England and some conservation implications', *Biological Conservation* 34, 1985, 197–225.

23. See for example R. D. Bargett, J. C. Frankland and J. B. Whittaker, 'The effects of agricultural management on the soil biota of some upland grasslands', *Agriculture, Ecosystems and Environment* 45, 1993, 25–45; P. R. Holmes, D. C. Boyce and D. K. Reed, 'The ground beetle (Coleoptera: Carabidae) fauna of Welsh peat biotopes – factors influencing the distribution of ground beetles and conservation implications', *Biological Conservation* 63, 1993, 153–61; M. B. Usher, 'Management and diversity of arthropods in *Calluna* heathland', *Biodiversity and Conservation* 1, 1992, 63–79.

24. The material on goats comes from G. K. Whitehead, *The Wild Goats of Great Britain and Ireland*, Newton Abbot: David and Charles, 1972, and the 29 January 2001 Rural News page of the National Farmers' Union website (seen at 17/05/01), www.nfucountryside.org.uk/news. It is curious that during WWII dogfish were marketed as 'rock salmon'.

25. R. J. Fuller and S. J. Gough, 'Changes in sheep numbers in Britain: implications for bird populations', *Biological Conservation* 91, 1999, 73–89.

26. M. Rawes and O. W. Heal, 'The blanket bog as part of a Pennine moorland', in O. W. Heal and D. F. Perkins, 1978 (Note 21), 224–43.

27. S. A. Grant, G. R. Bolton and L. Torvell, 'The responses of blanket bog vegetation to controlled grazing by hill sheep', *Journal of Applied Ecology* 22, 1985, 739–51.

28. P. D. Hulme, R. J. Pakeman, L. Torvell, J. M. Fisher and I. J. Gordon, 'The effects of controlled sheep grazing on the dynamics of upland *Agrostis-Festuca* grass-land', *Journal of Applied Ecology* 36, 1999, 886–900. (This work is from southern Scotland but is very likely applicable even further south.)

29. S. A. Grant, D. E. Suckling, H. K. Smith, L. Torvell, T. D. A. Forbes and J. Hodgson, 'Comparative studies of diet selection by sheep and cattle: the hill

grasslands', *Journal of Ecology* **73**, 1985, 987–1004. (Also from southern Scotland.)

30. F. M. Chambers, D. Mauquoy and P. A. Todd, 'Recent rise to dominance of *Molinia caerulea* in environmentally sensitive areas: new perspectives from palaeoecological data', *Journal of Applied Ecology* **36**, 1999, 719–33.

31. R. Tipping, 'Towards an environmental history of the Bowmont Valley and the northern Cheviot Hills', *Landscape History* **20**, 1998, 41–50.

32. R. H. Hobbs and C. H. Gimingham, 'Vegetation, fire and herbivore interactions in heathland', *Advances in Ecological Research* **16**, 1987, 87–173.

33. S. A. Grant *et al.*, 1985 (Note 29); D. E. Suckling, H. K. Smith, L. Torvell, T. D. Forbes and J. Hodgson, 'Comparative studies of diet selection by sheep and cattle: the hill grasslands', *Journal of Ecology* **73**, 1985, 987–1004; S. A. Grant, L. Torvell, H. K. Smith, D. E. Suckling, T. D. Forbes and J. Hodgson, 'Comparative studies of diet selection by sheep and cattle: blanket bog and heather moor', *Journal of Ecology* **75**, 1987, 947–60. Most of these studies derive from the Southern Uplands of Scotland but seem to be germane to England and Wales as well. A summary of grazing (and burning) effects on blanket bog and wet upland heath can be found in English Nature Research Reports no. 172, 1996: *Literature Review of the Historical Effects of Burning and Grazing of Blanket Bog and Wet Heath*, Peterborough: English Nature and Countryside Council for Wales – there is not much history beyond about thirty years, however.

34. P. Anderson, J. H. Tallis and D. W. Yalden, *Restoring Moorland. Peak District Moorland Management Project Report Phase III*, Bakewell: Peak District National Park Planning Board, 1998.

35. R. H. Marrs, M. Bravington and M. Rawes, 'Long-term vegetation change in the *Juncus squarrosus* grassland at Moor House, Northern England', *Vegetatio* **76**, 1988, 179–87; M. Rawes, 'Changes in two high-altitude blanket bogs after the cessation of sheep grazing', *Journal of Ecology* **71**, 1983, 219–35; M. Rawes, 'Further results of excluding sheep from high-level grasslands in the North Pennines', *Journal of Ecology* **69**, 1981, 651–69.

36. P. Anderson and E. Radford, 'Changes in vegetation following reduction in grazing pressure on the National Trust Kinder Estate, Peak District, Derbyshire, England', *Biological Conservation* **69**, 1994, 55–63.

37. D. Welch, 'Response of bilberry *Vaccinium myrtillus* L. stands in the Derbyshire Peak District to sheep grazing, and the implications for moorland conservation', *Biological Conservation* **83**, 1998, 155–64.

38. Hawthorn has been shown to colonise grassland in Snowdonia in periods of low grazing pressure from sheep such as the 1920s; present animal numbers prevent recruitment to stands of woody species. See J. E. G. Good, R. Bryant and P. Cargill, 'Distribution, longevity and survival of upland hawthorn (*Crataegus monogyna*) scrub in North Wales in relation to sheep grazing', *Journal of Applied Ecology* **27**, 1990, 272–83.

39. M. McHugh and T. Harrod, 'Upland soil erosion in England and Wales: the influence of soil', in T. P. Burt *et al.* (eds), *The British Uplands: Dynamics of Change*, Peterborough: JNCC Report no. 319, 2002, 147–8.

40. J. C. Labadz, T. P. Burt and A. W. R. Potter, 'Sediment yield and delivery in the blanket peat moorlands of the southern Pennines', *Earth Surface Processes and Landforms* **16**, 1991, 255–71; P. White, D. P. Butcher and J. C. Labadz, 'Reservoir sedimentation and catchment sediment yield in the Strines catchment, UK', *Physics and Chemistry of the Earth* **22**, 1997, 321–8.

41. M. S. Alam and R. Harris, 'Moorland soil erosion and spectral reflectance', *International Journal of Remote Sensing* **8**, 1987, 593–608.

42. A. M. Harvey, 'Coupling between hillslopes and channels in upland fluvial systems: implications for landscape sensitivity, illustrated from the Howgill Fells, northwest England', *Catena* 42, 2001, 225–50.

43. A. C. Waltham and N. Dixon, 'Movement of the Mam Tor landslide, Derbyshire, UK', *Quarterly Journal of Engineering Geology and Hydrology* 33, 2000, 105–23.

44. M. M. Bower, 'The distribution of erosion in blanket peat bogs in the Pennines', *Transactions of the Institute of British Geographers* 29, 1961, 17–30; M. M. Bower, 'The cause of erosion in blanket peat bogs', *Scottish Geographical Magazine* 78, 1962, 33–43; more recently: J. H. Tallis, 'The Southern Pennine experience: an overview of blanket mire degradation', in J. H. Tallis *et al.* (eds), *Blanket Mire Degradation. Causes, Consequences and Challenges*, Aberdeen: Macaulay Land Use Research Institute, 1997, 7–15.

45. J. H. Tallis, 'Fire and flood at Holme Moss: erosion processes in an upland blanket mire', *Journal of Ecology* 75, 1987, 1099–1129; A. W. Mackay and J. H. Tallis, 'Summit-type blanket mire erosion in the Forest of Bowland, Lancashire, UK: predisposing factors and implications for conservation', *Biological Conservation* 76, 1996, 31–44; J. H. Tallis, 1997 (Note 44), 7–15.

46. C. K. Ballantyne, 'Late Holocene erosion in upland Britain: climatic deterioration or human influence?', *The Holocene* 1, 1991, 81–5.

47. J. H. Tallis and R. Meade, 'Blanket mire degradation and management', in J. H. Tallis *et al.* (eds), *Blanket Mire Degradation. Causes, Consequences and Challenges*, Aberdeen: Macaulay Land Use Research Institute, 1997, 212–16.

48. J. H. Tallis and E. A. Livett, 'Pool-and-hummock patterning in a southern Pennine blanket mire: I. Stratigraphic profiles for the last 2800 years', *Journal of Ecology* 82, 1994, 775–88; N. Rhodes and A. C. Stevenson, 'Palaeoenvironmental evidence for the importance of fire as a cause of erosion of British and Irish blanket peats', in J. H. Tallis *et al.* (eds), *Blanket Mire Degradation. Causes, Consequences and Challenges*, Aberdeen: Macaulay Land Use Research Institute, 1997, 64–78.

49. D. Wishart and J. Warburton, 'An assessment of blanket mire degradation and peatland gulley development in the Cheviot Hills, Northumberland', *Scottish Geographical Journal* 117, 2001, 185–206.

50. R. J. Hobbs, 'Length of burning rotation and community composition in high-level *Calluna-Eriophorum* bog in N. England', *Vegetatio* 57, 1984, 129–36; O. M. Bragg and J. H. Tallis, 'The sensitivity of peat-covered upland landscapes', *Catena* 42, 2001, 345–60.

51. Wishart and Warburton, 2001 (Note 49), pp. 185–6. There is a map of peat mass-movements in Great Britain in A. Mills, J. Warburton and D. Higgitt, 'Peat mass movements: a world and regional database', in T. P. Burt *et al.* (eds), *The British Uplands: Dynamics of Change*, Peterborough: JNCC Report no 319, 2002, 167.

52. P. Anderson *et al.*, 1998 (Note 34) p. 41.

53. D. T. Crisp, M. Rawes and D. Welch, 'A Pennine peat slide', *Geographical Journal* 130, 1964, 519–24; P. A. Carling, 'Peat slides in Teesdale and Weardale, Northern Pennines, July 1983: description and failure mechanism', *Earth Surface Processes and Landforms* 11, 1986, 193–206. See also K. Beven, A. Lawson and A. McDonald, 'A landslip/debris flow in Bilsdale, North York Moors, September 1976', *Earth Surface Processes* 3, 1978, 407–19.

54. J. H. Tallis, 'Mass movement and erosion of a southern Pennine blanket peat', *Journal of Ecology* 73, 1985, 283–315.

55. P. A. Ardron, I. D. Rotherham and O. L. Gilbert, 'The influence of peat-cutting on upland landscapes – case studies from the South Pennines', *Landscape Archaeology and Ecology* 2, 1996, 56; P. A. Ardron, O. L. Gilbert and I. D.

Rotherham, 'Factors determining contemporary upland landscapes: a re-evaluation of the importance of peat-cutting and associated drainage, and the implications for mire restoration and remediation', in J. H. Tallis, R. Meade and P. D. Hulme (eds), *Blanket Mire Degradation. Causes, Consequences and Challenges*, Aberdeen: Macaulay Land Use Research Institute, 1997, 38–41.

56. P. J. Wisniewsky, L. M. Paull and F. M. Slater, 'The extractive potential of peats in mid-Wales with particular reference to the county of Powys', *Biological Conservation* 22, 1982, 239–49.

57. D. J. Charman and A. J. Pollard, 'Long-term vegetation recovery after vehicle track abandonment on Dartmoor, SW England, UK', *Journal of Environmental Management* 45, 1995, 73–85.

58. P. Anderson *et al.*, 1998 (Note 34), p. 15.

59. J. Radley, 'Significance of major moorland fires', *Nature* 205, 1965, 1254–9. On moors west of Sheffield a 2,000-acre fire in 1868 was attributed to bilberry-pickers who had been ejected from their accustomed picking-grounds.

60. Ward *et al.*, 1972 (Note 3) for Dartmoor; P. Anderson, 'Fire damage on blanket mires', in J. H. Tallis, R. Meade and P. D. Hulme (eds), *Blanket Mire Degradation. Causes, Consequences and Challenges*, Aberdeen: Macaulay Land Use Research Institute, 1997, 16–28, deals mostly with the Peak District National Park. Controlled fire is allowed only between 1 November and 31 March and is subject to the Heather and Grass Burning (England and Wales) Regulations (Statutory Instrument 386, 1949, amended 1987). MAFF (now DEFRA) has a burning code of 1992.

61. G. R. Miller, J. Miles and O. W. Heal, *Moorland Management: a Study of Exmoor*, Cambridge: ITE, 1984,

62. E. Maltby, 'The impact of severe fire on *Calluna* moorland in the North York Moors', *Bulletin d'Écologie* 11, 1980, 683–708; E. Maltby, C. J. Legg and M. C. F. Proctor, 'The ecology of severe moorland fire on the North York Moors: effects of the 1976 fires and subsequent surface vegetation development', *Journal of Ecology* 78, 1990, 490–518.

63. D. T. Crisp, 'Input and output of minerals from an area of Pennine moorland: the importance of precipitation, drainage, peat erosion and animals', *Journal of Applied Ecology* 3, 1966, 327–48. The area of Moor House NNR studied comprised about 11–20 per cent eroding peat and the rest was blanket bog.

64. J. A. Stewart and A. N. Lance, 'Effects of moor-draining on the hydrology and vegetation of northern Pennine blanket bog', *Journal of Applied Ecology* 28, 1991, 1105–17; A. M. Robinson, 'The hydrological effects of moorland gripping – a reappraisal of the Moor House research', *Journal of Environmental Management* 21, 1985, 205–11. There appears to be no estimate of the area to which it has been applied in England and Wales; in Scotland it was about 20,000 ha/yr in the 1960s and 1970s. Restoration plans for gripped areas are highlighted in *Cumbrian Nature* (the newsletter of the regional team of English Nature) Issue 12, Summer 2002, front page.

65. See Fig. 12 in L. F. Curtis, *Soils of Exmoor Forest*, Harpenden: Soil Survey of England and Wales Special Survey no 5, 1971.

66. J. S. Rodwell (ed.), *British Plant Communities. Vol. 3. Grasslands and Montane Communities*, Cambridge: CUP 1992, 487–98.

67. J. A. Taylor, 'The bracken problem: a local hazard and global issue', in R. T. Smith and J. A. Taylor (eds), *Bracken. Ecology, Land Use and Control Technology*, Lancaster: Parthenon Press, 1986, 21–42.

68. B. J. Sheaves and R. W. Brown, 'A zoonosis as a health hazard in UK moorland recreational areas: a case study of Lyme Disease', *Journal of Environmental*

Planning and Management **38**, 1995, 201–14. Although known as the sheep tick, *Ixodes* will obtain blood meals from most terrestrial species including humans.

69. D. S. Allen, 'Habitat selection by whinchat: a case for bracken in the uplands?', in D. B. A. Thompson, A. J. Hester and M. B. Usher (eds), *Heaths and Moorland: Cultural Landscapes*, Edinburgh: HMSO for Scottish Natural Heritage, 1995, 200–5.

70. R. J. Pakeman and R. H. Marrs, 'Vegetation development on moorland after control of *Pteridium aquilinum* with asulam', *Journal of Vegetation Science* **3**, 1992, 707–10; R. J. Pakeman, M. G. LeDuc and R. H. Marrs, 'Moorland vegetation succession after the control of bracken with asulam', *Agriculture, Ecosystems and Environment* **62**, 1997, 41–52. Asulam is methyl-N-(aminobenzenesulphonyl)-carbamate. Some idea of the detailed knowledge needed for successful chemical control can be gained from e.g., S. J. Whitehead and J. Digby, 'The morphology of bracken (*Pteridium aquilinum* (L.) Kuhn) in the North York Moors – a comparison of the mature stand and the interface with heather (*Calluna vulgaris* (L.) Hull). 1. The fronds', *Annals of Applied Biology* **131**, 1997, 103–16.

71. J. A. Taylor, 'Bracken: global weed and environmental issue', *Geography Review* **11** (5), 1998, 12–15; R. Brown, 'Bracken on the North York Moors: its ecological and amenity implications in national parks', in R. T. Smith and J. A. Taylor op. cit. 1986 (Note 67), 77–86; R. H. Marrs and P. J. Pakeman, 'Bracken invasion – lessons from the past and prospects for the future', in D. B. A. Thompson, A. J. Hester and M. B. Usher (eds), 1995 (Note 69), 180–93; D. S. Allen, 'Habitat selection by whinchats: a case for bracken in the uplands?', in D. B.A. Thompson, A. J. Hester and M. B. Usher (eds), 1995 (Note 69), 200–5; R. J. Pakeman and R. H. Marrs, 'The conservation value of bracken *Pteridium aquilinum* (L.) Kuhn-dominated communities and an assessment of the ecological impact of bracken expansion or its removal', *Biological Conservation* **62**, 1992, 101–14; R. J. Pakeman, M. G. Le Duc and R. H. Marrs, 'Bracken distribution in Great Britain: strategies for its control and the sustainable management of marginal land', *Annals of Botany* **85** (Supplement **B**), 2000, 37–46.

72. P. Anderson and D. W. Yalden, 'Increased sheep numbers and the loss of heather moorland in the Peak District, England', *Biological Conservation* **20**, 1981, 195–213.

73. D. B. A. Thompson, A. J. MacDonald, J. H. Marsden and C. A. Galbraith, 'Upland heather moorland in Great Britain – a review of international importance, vegetation change and some objectives for nature conservation', *Biological Conservation* **71**, 1995, 163–78; R. D. Bargett, J. H. Marsden and D. C. Howard, 'The extent and condition of heather on moorland in the uplands of England and Wales', *Biological Conservation* **71**, 1995, 155–61. Methods used in the investigation of trends in heather moorland on Dartmoor are discussed in G. J. Clark, 'Mapping past dwarf-shrub heath extent on Dartmoor', *Aspects of Applied Biology* **58**, 2000, 179–84.

74. A. C. Stevenson and A. N. Rhodes, 'Palaeoenvironmental evaluation of the importance of fire as a cause for *Calluna* loss in the British Isles', *Palaeogeography, Palaeoclimatology, Palaeoecology* **164**, 2000, 211–22.

75. There is an outline description of the programmes of the Institute of Terrestrial Ecology and of the Game Conservancy in J. R. Lawton (ed.), *Red Grouse Populations and Moorland Management*, London: British Ecological Society Ecological Issues no. 2, 1990, 31–2, as there is in J. R. Krebs and R. M. May, 'The moorland owners' grouse', *Nature* **343**, 1990, 310–11.

76. R. Moss and A. Watson, 'Adaptive value of spacing behaviour in population cycles of red grouse and other animals', in R. M. Sibley and R. H. Smith (eds), *Behavioural Ecology. Ecological Consequences of Adaptive Behaviour*, Oxford: Blackwell Scientific Publications, 1985, 275–94. P. Hudson, *Red Grouse. The Biology and Management of a Wild Gamebird*, Fordingbridge: The Game Conservancy Trust, 1986; S. Tapper, *Game Heritage*, Fordingbridge: The Game Conservancy Trust, 1992, 46–9. The chief direct predators are foxes and crows but mustelids such as stoats will 'spill over' onto grouse even if their chief food, the rabbit, is also plentiful. Mink may also spread up onto the hills from rivers. See the down-to-earth instructions in P. Hudson and D. Newborn, *A Moorland Management Manual*, Fordingbridge: The Game Conservancy Trust, 1995.

77. A. Dobson and P. Hudson, 'The interaction between the parasites and predators of Red Grouse *Lagopus lagopus scoticus*', *Ibis* **137** (**suppl.**), 1995, S87–S96.

78. S. J. Thirgood, S. M. Redpath, P. Rothery and N. J. Aebischer, 'Raptor predation and population limitation in red grouse', *Journal of Animal Ecology* **69**, 2000, 504–16; S. J. Thirgood, S. M. Redpath, D. T. Haydon, P. Rothery, I. Newton and P. J. Hudson, 'Habitat loss and raptor predation: disentangling long- and short-term causes of red grouse declines', *Proceedings of the Royal Society of London Series B*, 2000, 651–6. The moor is at Langholm, just north of the Scottish border on the west side of the extension of the northern Pennines into the Southern Uplands. Moors either side were subject to illegal persecution of hen harriers, for which the above authors note there has never been a successful prosecution.

79. J. Matthiopoulos, R. Moss and X. Lambin, 'Models of red grouse cycles. A family affair?', *Oikos* **82**, 1998, 574–90.

80. Assuming a modern scenario of 6–12 guns, 16 beaters for each drive and 4–5 drives deriving from 80–240 ha on each drive.

81. P. Anderson *et al.*, 1998 (Note 34), p. 26.

82. English Nature, *State of Nature. The Upland Challenge*, Peterborough: English Nature, 2001; 'Fighting against extinction', *English Nature Magazine* no. **59**, January 2002, 16. Their language is tactful but illegal persecution by grouse moor managers seems to be central; S. Thirgood, S. Redpath, I. Newton and P. Hudson, 'Raptors and red grouse: conservation conflicts and management solutions', *Conservation Biology* **14**, 2000, 95–104.

83. D. Baines, I. A. Wilson and G. Beeley, 'Timing of breeding in black grouse *Tetrao tetrix* and capercallie *Tetrao urogallus* and distribution of insect food for the chicks', *Ibis* **138**, 1996, 181–7; S. Tapper, 1992 (Note 76), 50–3.

84. A. J. MacDonald, A. H. Kirkpatrick, A. J. Hester and C. Sydes, 'Regeneration by natural layering of heather (*Calluna vulgaris*) – frequency and characteristics in upland Britain', *Journal of Applied Ecology* **32**, 1995, 85–99.

85. Anderson and Yalden, 1981 (Note 72).

86. S. A. Grant and T. J. Maxwell, 'Hill vegetation and grazing by domesticated herbivores: the biology and definition of management options', in M. B. Usher and D. B. A. Thompson (eds), 1988 (Note 14), 201–14.

87. The southern parts of the Yorkshire Dales have probably the biggest problem and the NPA tries hard to find powers to exercise control.

88. S. A. Grant and H. M. Armstrong, 'Grazing ecology and the conservation of heather moorland: the development of models as aids to management', *Biodiversity and Conservation* **2**, 1993, 79–94.

89. These data come from a report to the Burns Enquiry on hunting with dogs: R. McDonald, P. Baker and S. Harris, *The Ecological and Economic Impact of Foxes in Britain*, University of Bristol School of Biological Sciences, *c*.1999; seen at http://www.ntws.org.uk/fox_pests.htm on 13 January 2001. The Burns report

itself is *Report of the Committee of Enquiry into Hunting with Dogs*, London: The Stationery Office, 2000. In the summer of 2002, the League Against Cruel Sports alleged that a fell hunt in northern England was artificially encouraging foxes by maintaining artificial earths and dumping sheep carcasses as food. See www.league.uk.com/news/media_briefings/2002/

90. S. A. Grant and H. M. Armstrong, 'Grazing ecology and the conservation of heather moorland: the development of models as aids to management', *Biodiversity and Conservation* 2, 1993, 79–94. The work is mostly Scottish and is also concerned with red deer but otherwise seems applicable.

91. DETR, *Protection and Better Management of Common Land in England and Wales*, London: Department of Environment, Transport and the Regions, 2000.

92. R. Weaver, M. Kent and S. Goodfellow, 'Dynamics and management of semi-natural vegetation on Dartmoor', in M. Blacksell, J. Matthews and P. Sims (eds), *Environmental Management and Change in Plymouth and the South West*, Plymouth: University of Plymouth, 1998, 21–51.

93. P. A. Todd, J. D. P. Phillips, P. D. Putwain and R. H. Marrs, 'Control of *Molinia caerulea* on moorland', *Grass and Forage Science* 55, 2000, 181–91.

94. 'Sowing the seeds of a new venture', *Countryside Focus* 19, April/May 2002, p. 7.

95. P. Anderson *et al.*, 1998 (Note 34).

96. R. Evans, 'Soil erosion in the UK initiated by grazing animals', *Applied Geography* 17, 1997, 127–41.

97. Details are summarised in *English Nature, State of Nature. The Upland Challenge*, Peterborough, English Nature, 2001, 59. Three premiums – one for ewes and two for cattle are still paid on a headage basis – and in the words of the above document, 'Without reform of these, any revision of the LFA payments are likely to be extremely limited and upland biodiversity will continue to decline.'

98. The element of snobbery was still alive in some interwar discussions when it was mooted that access to National Parks should be restricted to the right sort of people – members of the YHA, for example.

99. G. E. Cherry, *Environmental Planning 1939–1969. Volume II. National Parks and Recreation in the Countryside*, London: HMSO, 1975.

100. The main example of the former is Department of the Environment, *Report of the National Park Policies Review Committee*, London: HMSO, 1974 [the Sandford Report], with the government response in Circular 4/76; of the latter, the Council for National Parks' sponsored (but with secretariat provided by the Countryside Commission) report, *Fit for the Future: Report of the National Parks Review Panel*, Cheltenham: Countryside Commission, 1991, CCP 334 [the Edwards Report].

101. See for example R. D. Hey, 'River engineering in the National Parks: the case of the River Wharfe, UK', *Regulated Rivers: Research and Management* 5, 1990, 35–44.

102. G. E. Cherry, 1975 (Note 99).

103. N. Stedman, 'Conservation in National Parks', in F. B. Goldsmith and A. Warren (eds), Conservation in Progress, Chichester: Wiley, 1993, 209–39. J. Blunden and N. Curry (eds), *A People's Charter? Forty Years of the National Parks and Access to the Countryside Act* 1949, London: HMSO, 1990.

104. N. Curry, 'Controlling development in the National Parks of England and Wales', *Town Planning Review* 63, 1992, 107–21.

105. There is a plethora of schemes and titles, bound to go out of date between writing and publishing. In 1999, for example, EU Structural Funds were at the heart of four schemes in the North York Moors National Park, one of which involved an upland river and another was a Moorland Regeneration Scheme.

Plans to secure other EU funds of an agri-environmental nature were proceeding as well.

106. P. Leonard, 'Agriculture in the National Parks of England and Wales: a conservation viewpoint', *Landscape Planning* 7, 1980, 369–86, gave an estimate for 1976 in which the costs per job per year were crafts £660, tourism £900, hill farming £3,800 and forestry £28,700.

107. Lord Porchester, *A Study of Exmoor*, London: DoE/MAFF, 1977.

108. P. Anderson, 'Habitat and landscape conservation: current strategies in the National Parks', *Ecos* 6, 1985, 18–24.

109. G. R. Miller, J. Miles and O. W. Heal, *Moorland Management: a Study of Exmoor*, Cambridge: Institute of Terrestrial Ecology, 1984; Anon, *Moorland Management*, Helmsley: North York Moors National Park, nd [1986].

110. J. Greville (ed.), *Managing the Historic Rural Landscape*, London and New York: Routledge, 1999.

111. M. Shoard, 'The lure of the moors', in J. R. Gold and J. Burgess (eds), *Valued Environments*, London: Allen and Unwin, 1982, 55–73.

112. Countryside Commission and Countryside Council for Wales, *National Park Management Plans Guidance*, Cheltenham: Countryside Commission, CCP 525.

113. K. H. Himsworth, *A Review of Areas of Outstanding Natural Beauty*, Cheltenham: Countryside Commission, 1980, CCP 140. The Commission clearly decided to spend very little money on its production: it is typed with an unjustified right-hand margin and no pictures nor even a map. But compared with most such reports it is interestingly written and even quirkily individual in places.

114. G. Parker and N. Ravenscroft, 'Benevolence, nationalism and hegemony: fifty years of the National Parks and Access to the Countryside Act', *Leisure Studies* 18, 1999, 297–323.

115. The National Parks and Access to the Countryside Act 1949; the Wildlife and Countryside Act 1981 (as amended 1985; it often appears in the literature as 'the WCA') and the Environmental Protection Act 1990.

116. The key document is The Habitats and Species Directive 92/43/EC, as amended by 97/62/EC. The Directive on Wild Birds is 79/409/EC. Basically, birds were first and the later Directive covers other groups of organisms. These Directives are implemented in the UK by Conservation (Natural Habitats, &c.) Regulations 1992.

117. See the UK Government, *Biodiversity. The UK Action Plan*, London: HMSO, 1994, Cmd 2428; UK Government, *Biodiversity. The UK Steering Group Report. Vol. 2 Action Plans*, London: HMSO 1995. The official response is the *Government Response to the Steering Group Report*, London: HMSO, 1996. Target dates for action go as far as 2010.

118. English Nature, *State of Nature. The Upland Challenge*, Peterborough: English Nature, 2001.

119. R. A. Stillman and A. F. Brown, 'Population sizes and habitat associations of upland breeding birds in the South Pennines, England', *Biological Conservation* 69, 1994, 307–14.

120. J. Sheail, *Nature in Trust. The History of Nature Conservation in Britain*, Glasgow and London: Blackie, 1976. It is probably fair to say that County Wildlife Trusts are less heavily engaged in conservation within moorland blocks than in lowlands, though there are exceptions.

121. There are schemes which are national, some purely local and some regional. Some have been in operation for several years, like Environmentally Sensitive Areas (ESAs) since 1991. Other programmes swing in to replace one or more

previous procedures – in Wales in 1999, *Tir Gofal* replaced *Tir Cymen* and ESAs, for example. The websites of English Nature, CCW and DEFRA are good places to search for recent developments but do not usually have much historical material on the preceding stages. Many of the schemes are listed and evaluated (but the accounts are inevitably out of date) in e.g., W. M. Adams, *Future Nature. A Vision for Conservation*, London: Earthscan, 1996, especially chapters 8 and 9; D. Evans, *A History of Nature Conservation in Britain*, London and New York: Routledge, 2nd edn, 1997.

122. D. A. Ratcliffe and D. B. A. Thompson, 'The British uplands: their ecological character and international significance', in M. B. Usher and D. B. A. Thompson (eds), *Ecological Change in the Uplands*, Oxford: Blackwell Scientific Publications, 1988, 9–36, Special Publication no. 7 of the British Ecological Society. (Of necessity, large parts of this chapter deal with Scotland.)

123. P. Anderson, 1985 (Note 108); D. Statham, 'Managing the wilder countryside', in S. Glyptis (ed.), *Leisure and Environment: Essays in Honour of Professor J. A. Patmore*, London: Belhaven Press, 1993, 236–52.

124. Both of these are MAFF schemes, introduced under pressure from conservation bodies. The MSS became the Upland Stewardship Scheme in 1999. MAFF also introduced regional management experiments in the Forest of Bowland and on Bodmin Moor in the late 1990s. The initial (1988) tranche of ESAs included the North Peak, Clun and the Pennine Dales; further areas of the Peak and Exmoor followed in 1993, with Dartmoor and the Shropshire Hills in 1994.

125. C. J. Bibby, 'Impacts of agriculture on upland birds', in M. B. Usher and D. B. A. Thompson (eds), *Ecological Change in the Uplands*, Oxford: Blackwell Scientific Publications, 1988, 223–36, Special Publication no. 7 of the British Ecological Society; M. Yeo and C. Keenlyside, 'Integrating nature conservation and agriculture in the uplands of Wales: long-standing problems and new opportunities', in T. P. Burt *et al.* (eds), *The British Uplands: Dynamics of Change*, Peterborough: JNCC Report no **319**, 2002, 79–82.

126. *Biodiversity: The UK Steering Group Report vol 2: Action Plans*, London: HMSO, 1995.

127. Ibid. (Note 126). These are estimates of the cost of the Action Plans, not the actual expenditure.

128. Forestry Commission, *A New Focus for England's Woodlands*, Cambridge: Forestry Commission, 2001, with an earlier version in 1998; S. Maidment and R. Britton, 'The challenge of protecting and expanding England's upland woodland resource', in T. P. Burt *et al.* (eds), *The British Uplands: Dynamics of Change*, Peterborough: JNCC Report no. **319**, 2002, 95–9.

129. See English Nature Research Reports nos 186, 1996 and 239, 1997.

130. N. Chapman and S. Harris, *Muntjac*, Fordingbridge, Hants: The Mammal Society, 1996; R. Putnam, *Sika Deer*, Fordingbridge, Hants: The Mammal Society, 2000.

131. English Nature, *Grazing in Upland Woods: Managing the Impacts*, Peterborough: English Nature, 1996.

132. Data and descriptions are available (at 02/2000) at www.nmw.ac.uk/ecn/

133. *Fit for the Future: Report of the National Parks Review Panel*, Cheltenham: Countryside Commission, 1991, CCP 334 [the Edwards Report], p 19.

134. This was articulated by D. Thompson in the report of a 1990 workshop (Cheltenham: Countryside Commission, CCP 323, 1990) and discussed in Stedman, 1993 (Note 103), p. 223.

CHAPTER SIX

More detailed histories

In this section, space is given to some more detailed accounts of particular aspects of particular places. They are essentially unique to their locality and isolate phases of special interest in the evolution of the environmental character of that particular upland, though they do not give complete accounts of scenic or biotic attributes. Three regions have been chosen:

1. **The North Pennines.** The biggest block of remote moorland in England, with a vigorous nineteenth century industrial history but now dependent upon sheep, grouse and some tourism; never designated a National Park and only as an AONB after some agonising. A region dry enough on the east to be dominated by grouse moors.
2. **Dartmoor.** Now a considerable recreational focus, Dartmoor has had a great deal of pre-nineteenth century mining but has avoided a number of large-scale water and forestry projects. The National Park authority has not however managed to eliminate the very strong military use of the upland.
3. **South Wales.** A core of rural uplands is designated as a National Park but there are fingers running south between valleys formerly dominated by coal and iron. The upland economies are now dominated by sheep and tourism but with all the results of the industrial economy in terms of afforestation and water catchments.

Within these regions, certain periods and topics are selected for discussion:

1. **North Pennines**
    i. The role of human history in the survival of rare species and their present-day perpetuation; environmental conflicts with grazing, shooting and water development (all periods).
    ii. Lead mining in the nineteenth century (industrial era)
    iii. Today's rural economy and its environmental connotations (twentieth century)
2. **Dartmoor**
    i. Hunter-gatherer vegetation change, and peat growth (hunter-gatherers)

    ii. Pre-nineteenth-century mining (agricultural era)

    iii. National Park management (twentieth century)

3. **South Wales**

    i. The rural economy in the face of the disappearance of coal and iron from the southern valleys; the tide of tourism washing in from the north (twentieth century)

THE NORTH PENNINES

The counties of Durham, Cumbria and Northumberland meet in a kind of heart-shaped block of upland, whose boundaries appear naturally to be on the watershed between the Greta and Arkengarthdale to the south, the Tyne and Irthing to the north, the Pennine escarpment that overlooks the Vale of Eden to the west and the last outcrops of unenclosed moor and heath to the east. The area designated in 1988 as an AONB is a little different, for all kinds of reasons, but traces more or less the hill-foot of that block of territory (Fig. 6.1). Its topography is typical of upland England, even down to the presence of a sill of volcanic rock; its vegetation history echoes that of most other uplands south of the Scottish Highlands and its human tenure shares the same major phases and occupations as Dartmoor, for example. Today's environmental status, however, is quite markedly affected by ways in which certain parts of the main stages have provided an intensified climax of change or have left behind phenomena which have rather later in time given rise to a peak of attention which has stimulated environmental management.

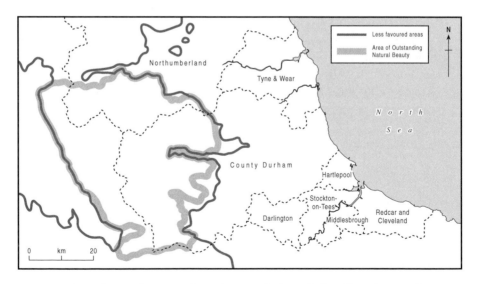

FIGURE 6.1 *The Northern Pennines, showing the boundaries of Less Favoured Areas (LFAs) and the AONB. As the LFA boundary shows, the AONB is to some extent a chunk of terrain with similar areas to north and south, though the Tyne gap is an obvious natural break. Note also the tri-county split of the AONB.*

PLATE 6.1 *The 'sugar limestone' (where Carboniferous Limestone has been baked by contact with hot basalt) is very crumbly and provides a continuously open habitat, especially when grazed by sheep. It is one of the set of habitats (here seen on Cronkley Fell) which offer refugia for plants last commonly found in England in the late-glacial period before deciduous forests were widespread.*

Conservation of rare species

Much of the area would be rated dull by any collector of plant records or compiler of bird lists. High moorland subject to fire, sheep and bad weather is not full of obvious biodiversity. One small region however (as mentioned in previous chapters) has attracted considerable attention and that cognisance has in more recent years spread out to the North Pennines more widely as a refuge for a diminishing species of bird, the black grouse.

Ever since the eighteenth century the area of Upper Teesdale around Widdybank and Cronkley Fells has attracted attention because of its rare flora, although recognised only belatedly in terms of protected status. Once the Cow Green reservoir was a certainty, there was an acceleration of interest in the plants themselves (and any animals that might also have such status), and also in the mechanisms by which they came to be there. The standard explanation in terms of local climate, the crumbly soils on the Sugar Limestone and the gaps in the blanket bog were recognised eventually to be inadequate since they did not account for the dynamics of vegetation change during the years since the last abundance of an Arctic-Alpine flora, presumably during the last cold phase of the late-glacial sequence, ending in about 10,000 BP.

The plants that comprise the Teesdale flora fall into a number of habitat groups (Plate 6.1). For example, a diverse assemblage of rare and local arctic-alpine plants such as bird's-eye primrose (*Primula farinosa*) and Scottish asphodel (*Tofieldia pusilla*) are found in a base-rich flush fed by springs, within a gentle depression at an altitude of 495 m. This fills up with drifted snow in the winter to a depth of up to 10 m and is a locality for the very rare round-mouthed whorl snail, *Vertigo genesii*. Several of the flushes are the most important sites for marsh saxifrage (*Saxifraga hirculus*) in the UK; they contain 80–90 per cent of the UK population of the species. Upper Teesdale is also a southern outpost for many of the rarer arctic-alpine plants characteristic of a kind of alpine habitat and has a unique relict mountain flora. Teesdale sandwort (*Minuartia stricta*) is restricted to Upper Teesdale, and other rare species found in this habitat type include false sedge (*Kobresia simpliciuscula*), hair sedge (*Carex capillaris*) and Scottish asphodel (*Tofieldia pusilla*). Extensive stands of *Sesleria albicans-Galium sterneri* grassland also contain a rich assemblage of relict arctic-alpine species, such as spring gentian (*Gentiana verna*) and alpine forget-me-not (*Myosotis alpestris*), confirming Upper Teesdale as one of the most important arctic-alpine refugium sites in the UK. Near High Force there is the second most extensive area of juniper (*Juniperus communis*) formation in the UK and the largest south of Scotland. The main area of juniper scrub grows on the igneous Whin Sill, at moderately high altitude. In Upper Teesdale the juniper has developed mainly on heath but small patches of juniper scrub also occur on calcareous soils, including the sugar limestone grassland. The hay meadows bear several rare species of lady's-mantle (*Alchemilla acutiloba*, *A. monticola* and *A. subcrenata*) and abundant globe-flower (*Trollius europaeus*). They are also important for their floral diversity since they commonly carry mean numbers of species of 25 per square metre but going as high as 41 spp/m^2. Lastly, the lead spoil heaps carry a number of plants genetically adapted to tolerate the high levels of toxic metals.

There are then three types of biological rarity in Teesdale: the plant and animal communities of the unenclosed land, especially of Cronkley Fell and Widdybank Fell; the riverside meadows which are home to a high diversity of flowering plants; and the high moors that contain populations of black grouse. In many accounts of these objects of conservation, prominence was given to the 'natural' factors of climate and open ground and very little to any features of human land management that helped their perpetuation. Though the level of detail varies a great deal, the interactive nature of human manipulation of ecosystems and their non-human components is now generally recognised, both at the intellectual level and as an essential ingredient of today's efforts at conservation and 'sustainable' resource use.

The lamentable history of the disruption of habitats by the short-termism of the Cow Green reservoir development has been discussed in Chapter 4. Many accounts of the vegetation produced in the course of the controversy emphasise the combination of limestones, soils with a high base status but an open and friable nature, and a treeless environment brought about largely by a harsh climate. The work carried out by palaeoecologists and historians in response to the development proposals, however, constructed a history in which human activity had played a role. Accounts of land use history now given in assessing management success in ESAs, for example, point to the role of prehistoric communities in deforestation of Upper Teesdale, for example, and the subsequent opening of the landscape for relatively intensive land uses.[1] The effects of hunter-gatherer occupation of the land are treated in a somewhat ambiguous set of statements:[2]

> 2.3 Man's ability to modify the environment inevitably increased, initially by the use of fire to drive animals for hunting purposes and, perhaps, to create and expand clearings attractive to browsing mammals. For most of the Mesolithic period, the overall effect of such clearances would have been counterbalanced by climatic improvement. This meant that, by 5,000 BC, when the temperature was some 2°C higher than today, the older birch/juniper woodland and grassland had been succeeded by a predominantly pine, elm, oak, lime woodland with hazel.

This seems to confuse the possible effects of upward movement of deciduous woodland with the creation or maintenance of openings within such woodland or indeed discount the possibility enunciated for Dartmoor and the South Pennines that humans tried to prevent the upward expansion of trees during the mid-Holocene. For Upper Teesdale, pollen analytical evidence at Dufton Moss (368 m ASL) produced by Squires suggests an intense burning episode during the Later Mesolithic (probably about 8,000 BP but no radiocarbon dates are available) which could be interpreted as a rapid interruption of the process of colonisation by a birch wood.[3] In general, the conclusion for the Mesolithic is that humans, using fire as a main tool

> made small temporary clearances in the mixed deciduous valley woodlands and to have cleared extensive areas of the birch woodlands on Cronkley Fell.[4]

Later interpretations would be more sparing in their use of the concept of 'clearing' but the tone is one of much more manipulation of the vegetation than recent MAFF reports allow. It seems likely that the effect of humans in maintaining open areas with unstable soils has been essential for the survival of those Teesdale rarities of Arctic-Alpine affinity that require unstable soils with a high base status. Given the 'climatic improvement' of the mid-Holocene then the Mesolithic may well have been the period which presented the greatest chance of the open-ground flora being competed out: literally overshadowed.

Later prehistory is well documented by palaeoecological work and it is of little consequence whether any one period creates more treeless land than another, provided that the areas of suitable soils are not totally tree-covered and that (probably more importantly) areas remained free of the blanket bog which started to grow in the mid-Holocene on areas formerly woodland or scrub.

> 2.4 During the Neolithic period, as man began to farm the land he started to have a more significant effect on the appearance of the landscape. Analysis of pollen samples from peat deposits recorded the beginnings of an extensive clearance of forest cover. By the end of the Bronze Age (*c.*2,000–850 BC) much of the uplands and valley sides had been cleared for both arable and pastoral farming. Increasing precipitation led to a deterioration in the fragile upland soils. This led to the development of peat deposits and moorland vegetation on the higher ground, with extensive soil erosion leading to silting in the valley bottoms. During the late Iron Age and Romano-British periods, the climate again became slightly warmer and drier but the process of change on the uplands was irreversible. The higher ground continued to be used for grazing, with arable cultivation and settlements restricted to the dales.[5]

This MAFF statement is an acceptable summary of the pollen analytical work, though work from the higher sites would suggest it is some time into the Iron Age before moorland vegetation (as traced in the proportion of *Calluna vulgaris* pollen) becomes very widespread. This means more remnant woodland and hence the continued importance of human economies in keeping open areas. The possession of domesticated animals would have made this an easy and unconscious activity.

The designation of some of the meadows in Upper Teesdale as SSSIs and their inclusion in an ESA in which particular forms of management are encouraged (and subsidised) leads to curiosity about their environmental history. A romantic narrative can see them as more or less direct descendants of the Norse occupation of the Pennine Dales, with a land management system that has close links with those of, for example, Iceland or upland Norway even today. It is thought likely that the meadows have had a more complex environmental history in which a number of cultural groups have been implicated since the time when the river valley woods were removed and enough animals were kept to maintain either grassland or, more likely, arable land on the lower lands below the fells. Until the nineteenth century, many such valleys had a fair degree of self-sufficiency for basic cereals. Oats in particular (and possibly rye as well) were grown for human food. Thus the nutrient status of the meadows may well derive in part from phases as cultivated fields at any time from prehistory to the nineteenth century, though to what extent any such phases can lose their identity to a later grassland regime is not known. Areas of arable fields, marked by lynchets and rigg and furrow, are evident as field monuments, though their age is uncertain.

There is no Domesday record for the Upper Tees valley but large areas may well have been granted to feudal landlords, with hunting in mind: the term 'Forest' is used of part of the dale. Better established is the record of monastic establishments, who used a system of granges, the best-known of which was Rievaulx's horse stud at Kaveset, (now Friar House near Low Force); in the grant in 1131 of five acres of land, enclosed with hedges and ditches for winter pasture for horses and colts, went with it. After the dissolution, large estates were the main beneficiaries, with a customary pattern of leased farms with access to common grazing land on moors. Their control in Upper Teesdale seems to have been complete enough to prevent much piecemeal encroachment on the commons. The doubling of sheep numbers between 1600 and 1700 may also have helped to counter pressures for intaking from the rough grazings. Isolated settlements may have come with the development of lead mining but they too seem more controlled than for example in Weardale. Increasing enclosure pressures were common throughout the seventeenth to nineteenth centuries and the eventual pattern of small fields with stone walls and dispersed hay barns (which in Teesdale are less likely to be stone-built than in the Yorkshire Dales) may well be not much older than the nineteenth century. In Upper Teesdale, the number of farms virtually doubled in the years 1769–1847. Some of this was by fission of existing units and might well have added to the number of intensively managed hay meadows. Interaction with the mining history of the region is also likely since most miners were part-time farmers as well. In 1848 there were four farms in Harwood (above Langdon Beck) and all the male occupiers were described as lead miners. So where the protection of uncommon or rare species is concerned, a knowledge of the nineteenth and early-mid twentieth-century regime of grazing, fertilising and mowing is important.[6] For example, in 1803 Lord Barnard's surveyor was recommending that tenants increased the amount of meadow to guard against bad winters. This also turned out to be a time when many field boundaries were altered.[7]

The management system as it existed down to the 1950–70 period revolved around the few fields that were suitable for cutting and these were mown between July 1st and the end of August. Before mowing, they were grazed by the sheep after lambing and then closed up between May and cutting-time; after the hay crop was taken there was a further period of grazing. Only manure from the barns was spread on the fields. Cattle were often winter-foddered from haystacks that had been built in the middle of the fields and this may account for the high densities of species of lady's-mantle (*Alchemilla* spp) found irregularly distributed within some of the hay meadows.[8]

In the post-war period, and particularly since the 1970s, these flower-rich meadows came increasingly under threat from agricultural intensification. This might involve some or all of drainage, re-seeding, applications of nitrogenous fertiliser, and cutting meadows earlier for silage. Big bale silage is of course far less dependent upon the weather than hay-making. Intensification was also associated with neglect of the traditional features, such as drystone walls or stone built field barns. Such changes also had an impact on the populations of characteristic birds like the yellow wagtail (*Motacilla flava*) and on breeding habitat for some waders.

The recognition at national level of the values of these meadows has been to make them part of an Environmentally Sensitive Area (along with other pastures and grasslands in the North Pennines) and to offer a two-stage management agreement to farmers, along with varying levels of subsidy. The management regime approximates that of the 1950s, with special attention paid to the continuation of grazing, the role of bagged fertilisers and the sequence in which the fields are cut.[9] The rare flora has been protected by standard methods: of declaring the designation of SSSIs and NNRs, now amplified by SAC designation under EU Directives, but not by much active management. The two NNRS of Moor House and Upper Teesdale were designated by UNESCO as a single biosphere reserve in 1976. The NNRs are adjacent to each other and share many attributes, and, as of 1999, their management (until then split between the Cumbria and Northumbria English Nature teams) has been integrated. While ownership of Moor House is in the hands of English Nature, Upper Teesdale is owned principally by the Strathmore and Raby Estates. The landowning bodies are in general more interested in grouse than sheep and so grazing densities on the unenclosed land have never reached some of the levels of other uplands. Visitor pressure has remained modest and uninformed damage is mostly restricted to picking gentians. Gardeners may well have removed more rarities than anybody else. The sheep probably help in keeping some of the habitats open and so their removal would not support conservation of the flora: almost certainly some of the rare plants would be shaded out by growth of grasses on soils which became more stable, for instance.

Although the species occurs throughout much of northern Eurasia, the Black Grouse (*Tetrao tetrix*) is in decline in virtually all areas where changes in population size have been investigated. In Britain, they were once widespread but are now confined to the uplands, inhabiting open woodland and the moorland fringe. In England, Black Grouse are now restricted to the North Pennines and adjacent parts of Northumberland with a very small and vulnerable group of birds in the Peak District. A survey of lekking Black Grouse in 1998 showed that there are now only around 700–800 blackcock in England, and, presumably, a similar number of greyhens.

In the North Pennines, the Black Grouse is very much a bird of the moorland edge. Its decline here is thought to be attributable to a number of causes including increased grazing intensity of the hill margins, in the Pennines predominantly by sheep, but with rabbits having important local effects too. This has removed key food plants, nesting cover and invertebrate food required by growing chicks. Ideal grazing conditions must maintain sufficient food plants (heather, cotton sedge, herbs and grass seed), adequate cover for nesting (heather, grasses and rushes over 30 cm high) and a plentiful supply of insects for chicks (especially caterpillars of sawflies and moths). The loss of scrubby woods which are important food sources through the winter has added to the decline as has the drainage and heavy grazing of mires and wet flushes, destroying food sources and cover for nesting and chick rearing. There is some indication that Black Grouse breeding attempts may fail in poor weather more readily in heavily grazed areas.

In late 1996, the North Pennines Black Grouse Recovery Project was established, following the launch of a UK Biodiversity Action Plan for this species. The Project is a collaborative venture between the Game Conservancy Trust, Ministry of Defence, Royal Society for the Protection of Birds, English Nature and National Wind Power. There is a Project Officer operating within parts of Northumberland, County Durham, Cumbria and North Yorkshire, who works closely with land owners and managers developing management plans to improve conditions for Black Grouse on their land. In the North Pennines, the main courses of action are reductions of grazing on moor edges and adjacent rough pastures, small-scale planting of shrubby woodlands and encouragement of diligent legal control of predators. Thirty-eight management plans had been produced by the end of 1998, which range from shrub planting on smallholdings to a combined set of recommendations which cover areas in excess of 1,000 hectares of large estates. Combined, these recommendations cover over 11,000 hectares of suitable, or potentially so, Black Grouse habitat in the North Pennines. Over 65 per cent will be initiating some positive management for Black Grouse. The project's immediate aim is to reduce grazing on moorland edges and adjacent rough pastures, and to plant shrubs and trees on a small scale. Together, the sites hold about 30 per cent of the black grouse population in the region.

Looked at overall, the effects of environmental history on today's environment in Upper Teesdale are strong. The combinations of natural features, land use history and public interest have lead to a very strong presence of statutory denominations for conservation purposes, with one feature or another being nominated for the latest designations available, for example, from the EU. Just how much this would have happened without the spotlight of the Cow Green reservoir battle is impossible now to say.

The impact of lead mining

As brief mentions earlier in the book have pointed out (pp. 122–7), the mining of lead in all the northern Pennine Dales has a long history. In the Teesdale-Weardale-Allendale region, this is notably so, with Roman evidence preceding the extraction of lead in medieval times. Hexham and Blanchland abbeys owned land that later produced lead but there is no evidence that they ever worked it, but in Teesdale there are a number of remains and suggestive place-names that pre-date the mid-fourteenth century.[10] The miners' target was veins and replacement bodies in the hard rocks (limestones, sandstones and shales) of the Carboniferous formation and the total extraction out of this block from Roman times to the recent closures was of the order of 390,000 tonnes of lead concentrates containing 60–75 per cent of lead. The period from the seventeenth to nineteenth centuries became the time of greatest activity, with lead extraction peaking from 1815 to 1880 and zinc from 1880 to 1920. Only lead and silver were worked until the late nineteenth century when the associated minerals (sphalerite, fluorite, witherite and barytes) became more important than lead and prolonged the life of some mines. Fluorspar mining in Weardale persisted until 1981 and in the Derwent basin until 1987. In the Tees basin, the last mine closed in 1955.[11]

These dales also yielded bulk minerals from quarrying, like limestone and gannister (a high-silicon sandstone used for lining furnaces in the steel industry) from Weardale, and hard basalts from Upper Teesdale. Together, these developments accelerated the penetration of railways in the nineteenth century, which brought in cheap coal to lower the costs of smelting as well as transport the products. The effects of the lead mines have persisted in the form of lead-laden silts on river floodplains for 60–70 km downstream, in many tips, shafts and levels, and in the reconstruction of some of the processing plants for the tourist industry.

The direct environmental effects of lead mines are, like most other mining, seen as holes and heaps. The holes might be shafts which went down at a variety of angles, or levels which were nearer the horizontal, or scars where overburden had been washed away to expose a vein. The heaps were short-lived if they were of ore on the way to be processed (but might be contributing all kinds of silty materials to runoff), or longer-lived if they were waste materials with no obvious re-use on site. The mouth of a level would have a heap of rejected material from the dressing of the ore that took place immediately it was brought to the surface. It was also common practice to re-smelt some of the tip material at smelting plants. So there was a dispersed landscape of extraction sites, each with holes and heaps, contributing to a smaller number of smelting plants, where the heaps predominated. However, as examples at Nenthead show, refining was usually initiated on the site of a mine, so that there was an on-hand supply of ore. The exigencies of the winter months presumably made it useful to have a weather-independent source of materials to keep the smelter occupied.

The North Pennines' topography lent itself to the practice of damming water at the head of a stream and then allowing the head to flood downhill so as to wash as much as possible of the peat, soil and glacial drift from the top of a lead-bearing vein. Examples can still be seen today, with sharp notches breaking the skyline, as on the north side of Teesdale between Middleton and Newbiggin. Occasionally, more permanent dams were built and it seems as if a few of these 'hushes' were re-used, though the accepted picture is of a single event. They were however used as well for reworking wastes and dressing-floors, not usually on the site of a vein in these cases. Below the hush, a fan of debris was built up and this cloaked the pre-existing soils, proving a new open habitat for plants able to colonise such surfaces. They seem to have fallen from favour at about the end of the eighteenth century and so are not usually a dominant feature of the main period of lead mining in the nineteenth century; in Upper Weardale, the last remaining working hush is dated at 1846–47.[12] A hush which exposed a deep vein might then be taken over for opencut working and contribute more waste to the surface, to be washed off and to provide an open habitat.

Where tips were long-lived then a special suite of plants able to tolerate lead-rich soils evolved. Some were species able simply to grow on the toxic surfaces, others genetically adapted races of species of other habitats. On the open debris of the dumps, many common plants of swards of unaltered limestone are found, but there is an abundance of three northern plants which seem to have adapted especially well: spring sandwort (*Minuartia verna*), alpine penny-cress (*Thlaspi alpestre*) and alpine scurvy-grass (*Cochlearia pyrenaica*). If a continuous grass cover has developed, then the mountain pansy (*Viola lutea*) is often abundant and in a few localities there are high densities of thrift (*Armeria maritima*); in this enclosed turf the tiny fern *Botrychium lunaria* (moonwort) is found. There are also plants that seem to require an additional soil element as well as lead.[13]

The 'hole' aspect affected the environment in the simple way of contributing to the quantity of heaps and road-beds but also in the more complicated fashion of adding drainage water to surface watercourses. From Nenthead to Alston there ran underground the Nent Force Level, which was a drainage tunnel of five miles' (8 km) length built in the late eighteenth-early nineteenth centuries that had a much-visited waterfall at one end of a canal-like section, along with a platform known as Jennie's Dancing Loft.[14] Some of the 'levels' which drain great complexes of underground tunnels must have produced high volumes of silt-laden water which carried fines which were themselves lead-rich. Values for lead, zinc and cadmium in floodplain sediments of the mining areas are up to ten times greater than baseline values for pre-mining sediments but show an overall decrease downstream, undergoing various forms of chemical and physical transformation.[15] Much of the metal-contaminated sediment of the Tees appears to be in alluvial stores which undergo limited reworking and so are sequestered for the time being.[16] In the Tyne valley, lead values are attenuated more rapidly downstream because they are diluted by uncontaminated sediments derived from rapid riverbank erosion (Fig. 6.2). In the Tyne, metals seem to move downstream as a series of sediment pulses or 'slugs'. So significant remobilisation is currently active and metals are reinjected into the river water from time to time. In a nearby region, the concentrations of lead and zinc in surface floodplain soils at Reeth in Swaledale are all raised and some are above the danger levels identified by regulatory agencies.[17] The toxic lead fumes from smelters were a constant cause of complaint since they poisoned the land and the cattle. A mining agent in

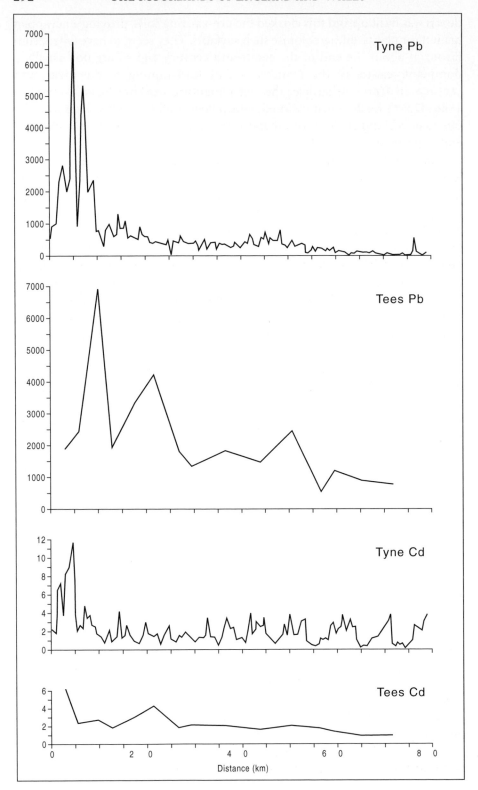

the late eighteenth century reported of Langley Mill (just south of Haydon Bridge on the Tyne) that:

> the Reek of Smoak of a Lead Mill is of such a Nature not very readily to mix with the air, and is frequently driven in an unbroken Current upon the Earth's Surface to a Miles Distance, nay we have ourselves seen the Smoak of Lead Mills at the Distance of 6 or 7 Miles.[18]

One response was to build a wall to enclose the smelter in the belief that the fumes would be trapped; another was to build long flues which dispersed the fumes and also precipitated some more lead. At Allenheads the Beaumont Company built a flue of 3,424 yards and at Rookhope one of 2,548 yards.[19]

One secondary effect of smelters was the gaining of fuel; in ore furnaces the main source was peat. A year's supply was generally cut in May or June in an operation involving carts and a great number of men, women and children. Coal might also be mined in local open pits (as on Tan Hill) and both it and ore, and the finished lead, were transported along 'Galloway roads' by packhorses.[20] Eventually, coal, the turnpike road and the railway made this older system obsolete. Such developments in transport confirmed the dales as the sites of iron mining, for ironstones were found in the Rookhope and Middlehope valleys. There was an iron smelter in Stanhope Dene from 1845–1918, which used coal brought in by railway and it spawned daughter plants at Wolsingham from 1864 (active until recently) and at Tow Law, 1850–82.[21]

Today's remains do not always give a clear impression of the environmental make-over (Plate 6.2) which attended a lead-processing centre such as the head of a mine like the Park Level at Killhope in upper Weardale or the complex at Nenthead at the head of a tributary of the South Tyne. At Nenthead, even recent tidying work has failed to remove the lunar aspects of the valley in which the operations took place. Given the toxicity of lead, it is not surprising that even today there has been little revegetation of the land surface and at its height in the nineteenth century, the valleys of the burns to the south of what the 1:25,000 map dismisses as LEAD MINE (rems) must have been virtually devoid of life other than that of humans, their parasites and their pack animals.

FIGURE 6.2 (opposite) *The fate of (mostly) nineteenth-century heavy metals (Pb = lead, Cd = cadmium) in the sediments of the Tyne and Tees valleys measured along a downstream gradient. The inference is that the metals in the Tees basin are generally sequestered in the sediments and undergo limited reworking whereas those in the Tyne are more often shifted in a series of pulses as slugs of material are reworked.*
Source: Reprinted from The Science of the Total Environment *194/195, M. G. Macklin, K. A. Hudson-Edwards and E. J. Dawson, 'The significance of pollution from historic metal mining in the Pennine orefields on river sediment contaminant fluxes to the North Sea', p. 393, 1997, with permission from Elsevier.*

PLATE 6.2 *A common sight in the North Pennines (here at Rookhope: see the discussion of W. H. Auden in Chapter 7) is a set of heaps of spoil from former lead mining operations. The whole slope in the background has been affected by mining as well. In use, the waste heaps were unvegetated, thus allowing precipitation to wash fines into the run-off, increasing the heavy metal levels in solution and suspension.*

Forces in the rural economy

The northern part of the North Pennines, between the Swale-Tees watershed and the Tyne valley, consists of about 2,000 km^2 of land with a population of some 12,000 people, which makes it the most sparsely populated region of England. Agriculture still contributes 26 per cent of all employment and pockets of high registered unemployment (Alston at 9–12 per cent) conceal a great deal of underemployment and deprivation. The relative prosperity of the nineteenth century, when the population was twice as large and the dual economy pushed farming as high as 550 m, has left a terrain in which hill farming is under constant threat of total unprofitability but where the scope for diversification is limited, where mineral extraction has ceased except for a few quarries, where tourism is confined to a few favoured sites and where the attitudes to change are much influenced by a few large estates. The region houses twelve reservoirs, 650 ha of MoD training land, 537 ha of NNR-SSSI (about 27 per cent of the AONB) and conservation areas belonging to at least five other organisations. Most of these facts have environmental linkages.

The major environmental influence is farming. The area between the Yorkshire-Durham border and the Tyne valley which has been designated as an AONB has about 1700 farm units, in which the number of very large and very small units alike has been increasing through the post-1970 period. Some 222 of those units are rented from one of the four large estates in the region. The husbandry practised is almost entirely dependent on animals to the level of about 90,000 sheep and 100,000 cattle. These are supported on enclosed grass, grassland managed for hay or silage and moorland of various types. There are thirty-nine blocks of *Calluna* moor which comprise about 38 per cent of the AONB (77,094 ha), which is about one quarter of the total heather moorland in England and Wales. Nearly all of this moor is used for grouse shooting; it could also be used for sheep grazing and some of it bears a mixed regime.

The basic agriculture is a mixture of hill sheep farming, where 90 per cent plus of the grazing is on open moorland, and upland farming where 30 per cent plus is on enclosed pastures. In the latter there has been more income from cattle than sheep, and the farms have often been able to raise their income by improving pasture quality and livestock housing.[22] This has been much more difficult for the hill sheep units, for whom the only way forward, encouraged by subsidies, was to increase the size of the flock. In 1969–85, for instance, the sheep numbers in the AONB area went up by 40 per cent and the overall stocking density by 12 per cent as dairy herds steadily disappeared. The proportion of rough grazing (i.e., open moor) is about 38.5 per cent of the AONB and the proportion of the population employed in forestry and agriculture in 1984 was 35 per cent which is very high even for uplands in England and Wales.

As discussed in the section on vegetation and its management, the density of sheep is a strong influence on moorland vegetation and soils, including their erosion. In this region, there has come to be a heavy reliance on headage payments under various national and EU schemes. Without them, farm viability would be very low indeed, in spite of the fact that some of the estates try to keep rents to a level which reflects the income of farmers. Headage payments have meant that there was an increase in sheep numbers by about 27 per cent in 1981–8, for example, a relatively modest figure which reflects the influence of the estates in resisting high levels of grazing on heather moors. Nevertheless, the replacement of heather by coarse grasses such as *Nardus* can be readily seen, and the encroachment of bracken is little different from other uplands. In these circumstances, the take-up of agri-environmental schemes has been quite high, with about 300 agreements under the ESA schemes, for example. The various schemes offer a steady and predictable cash income in return for changing the balance between production and protection. Small woodlands can also attract grants and there are also schemes for encouraging developments to cater for visitors, but this is not an area with the pull of, say, Dartmoor or Wensleydale.[23]

It is an area with a special attraction for sportsmen, however. The high land which is either common land or private moorland now has the best grouse moors in Great Britain, in which (in 1990) a driven brace of birds is worth about £100. Thus a day's shooting for ten people which kills 100 brace is worth £10,000 to the landowner. There may be up to 15 days per year of such sport and every extra brace of grouse is worth about £2000 on the value of the moor. The shoots may provide 10–75 per cent of an estate's income, depending on the density of the birds in a given year. Since a certain density of sheep starts to shift the vegetation away from heather (see the discussion in Chapter 5), the estates have sought to reduce or at any rate contain sheep numbers: on private moors this is done directly; where common lands exist then estates have for instance bought up farms so as to control the number of stints. The estates have also seen the presence of visitors as inimical to the shooting and so have discouraged developments on their land and by their tenants (Plate 6.3).

In particular the North Pennines estates saw the creation of bodies like the North Pennines Rural Development Board (1969–71) and the proposal for an AONB made in 1976 by the Countryside Commission as threats to their position on the land. Both were seen as steps to greater environmental controls and higher visitor numbers. The RDB had powers which it never used to control the transfer of land between farms and from farming into forestry and so it could have encouraged, for instance, the creation of larger farm units. The AONB has virtually no powers and since it has very little central organisation it cannot fight for EU 5b and Objective 2 funds in the manner of National Parks. It came into being in 1988 after a Public Enquiry (the first ever for National Park or AONB designation) but has been of little significance in

PLATE 6.3 *The trackway is indeed not a public right of way but no opportunity is lost to point this out. We might feel that the owner of this grouse moor (near Blanchland in Northumberland) would prefer the shooting butt to point at the road rather than the birds.*

rural changes, which seem to be dictated by forces of extreme locality on the one hand and those from beyond our shores on the other.[24]

DARTMOOR

I once met the Governor of Dartmoor Prison (which is at Princetown, 450 m ASL) in his office. His previous job had been the equivalent post in Cyprus and I asked him about the contrasts. He did not hesitate: 'it rains every bloody day here'. As John Hooker put it in the sixteenth century,

> And this one thing is to be observed that all the yere through out commonly it raineth or it is fowle wether in that more or desert.[25]

So then as now, many an Atlantic depression experiences its first orographic uplift over Dartmoor but that has not prevented it being an attractive area for prehistoric settlement at any rate until the Iron Age, nor precluded attempts at colonisation during medieval warm periods, or diminished its recreational importance in the twentieth century. The climate has not stood in the way of mineral extraction: indeed the reliability of water supply may have enhanced it. Bad weather is presumably good for toughening up soldiers and it hides shell-holes. The flavour of Dartmoor is presented here in three slices: the environmental relations of the last hunter-gatherers, who seem to have been responsible for much of the openness of the high moor, the continuing environmental and landscape presence of extractive industry whose importance preceded that of nineteenth-century industrialisation, namely tin, and the active role of the National Park authority who try to contain today's pressures for change along with an active concern for those inhabitants who still make a living from environmental resources (Fig. 6.3).

The last hunter-gatherers on the high moor

It is not now controversial to suggest that the moorlands of England and Wales were once largely covered in deciduous woodland, nor that pre-agricultural human groups engaged in the manipulation of that woodland in the direction of greater openness. Yet in 1960, the first notion was mostly confined to a small scientific community; the second was radical and on the whole contrary to the conventional wisdom. The two ideas came together first on the North York Moors in work by Geoffrey Dimbleby but this was based on pollen stratification in mineral soils[26] and thus decried by some as suspect. The first published paper which showed that late Mesolithic (i.e., hunter-gatherer communities) affected woodland vegetation but used conventional peat pollen as evidence came from Dartmoor.[27] This was sited at Blacklane Brook, on the high watershed of the southern half of the moor. Excellent later work from other parts of the Moor has mostly been directed by Professor Caseldine of the University of Exeter.

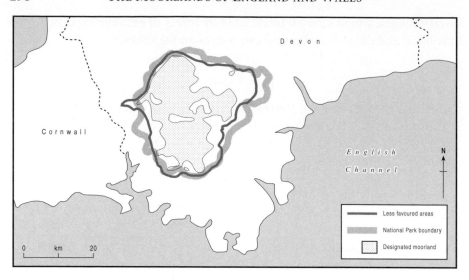

FIGURE 6.3 *The coincidence of Less Favoured Area (LFA) and the National Park boundary on Dartmoor. Note the virtual coincidence of both and the unitary occurrence within a single county.*

The presence of peat over much of the high ground of Dartmoor has furnished material for palaeoecological analysis but it has also very likely obscured many of the remains of Mesolithic date. Compared with uplands in the Pennines or the North York Moors, for example, the find records of Mesolithic flint material are very sparse. The evidence for human presence and effects during the early and middle Holocene (c.10,000–5,000 BP) is therefore mostly the presence of charcoal and the pollen of plants of open ground in peats of appropriate age, as dated by radiocarbon assay.

Analysis of early Holocene peats at Black Ridge Brook on northern Dartmoor (447 m ASL) is interesting in this context since there is a band of charcoal about 1 cm thick along at least 4 m of the peat face, occurring in peats dated as older than 8,785 BP. The fire depressed the values of birch pollen but increased those of grasses; it succeeds a band of very wet bog-moss (*Sphagnum*) accumulation. It is attributed by the authors to natural causes, an explanation favoured even in lowland areas for late Devensian and early Holocene charcoal accumulations.[28]

By contrast, charcoal occurring in peats from the mid-Holocene at a number of sites on the high moor is attributed to human presence. Most such material occurs between 7,700 and 6,100 BP on the northern moor and 7,660–6,010 BP on the southern moor. The deposits are accompanied by falls in the proportion of tree pollen and increases in indicators of open ground. At Pinswell (461 m) relatively undisturbed hazel woodland was changed to a plant community more characteristic of the woodland edge, where heather (*Calluna*) was present along with an indicator of opening in woodlands such as *Melampyrum pratense*, cow-wheat. At Blacklane Brook (457 m) there is a triple sequence in which there is an initial phase with a rapid onset of distur- bance phenomena (~8,000–6,010 BP), followed by a 'maintenance' period (6,010–~5,200 BP) and an intensification just before the decline of *Ulmus* which happens in most European pollen diagrams around the time of the transition to a Neolithic culture, though rarely completely coincident with it. In the second and third phases the acidification of the locality is shown by rising values for *Calluna*. These indications of an active presence of a Mesolithic population are found on slightly lower ground at Tor Royal (390 m) and but less so at Bellever (*c.*330 m) where the charcoal seems more likely to be domestic in origin and to coincide with no change in the vegeta- tion.[29] The inference, therefore, is that the effect of the Mesolithic groups was strongest on the higher ground and that one of their effects (possibly one of their aims) was to perpetuate open ground and scrub rather than closed woodland. In consequence, they tried to hold in place a kind of tonsure effect against the climatic trend of climbing forests (Plate 6.4).

The detailed work at Pinswell has elucidated the transition from a *Corylus*- dominated woodland at 7,000 BP to the establishment of true blanket bog (with *Calluna*, *Narthecium ossifragum*, *Sphagnum* and *Potentilla*) about 1,000 years later. Intermediate stages include a secondary upland woodland with open stands of *Corylus* along with *Calluna*, *Sphagnum* and *Melampyrum*, an

PLATE 6.4 *North-west Dartmoor: Kitty Tor is surrounded by moorland where once it would have been submerged in deciduous forest. The tor itself might have been an island of open rocks or might even have retained some woodland when all the rest had been removed.*

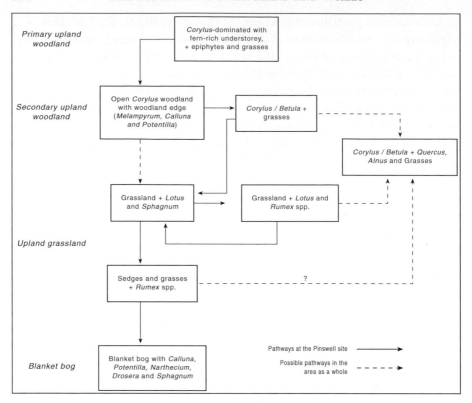

FIGURE 6.4 *Some vegetation pathways on the higher part of northern Dartmoor under Mesolithic influence. The solid lines are from data at a site at Pinswell (461 m ASL) and the pecked lines amplify the information by using data from other localities in the general area. The general sequence is from hazel (Corylus) woodland through grassland which is converted through the addition of sedges to blanket bog.*
Source: C. Caseldine and J. Hatton, 'The development of high moorland on Dartmoor: fire and the influence of Mesolithic activity on vegetation change', in F. M. Chambers (ed.), Climate Change and Human Impact on the Landscape, *London: T. M. Chambers, 1993, Fig. 11.5 at p. 129. With kind permission of Kluwer Academic Publishers.*

acid grassland with *Sphagnum* and *Lotus* which passes into a sedge-grass sward with *Rumex* species. This last is the immediate precursor of the blanket bog, signalled by the presence of the acid-tolerant insectivore *Drosera* (Fig. 6.4). The acid grassland seems to have no equivalent in the vegetation of today or the recent past. The driving forces in the transition, which might have taken as little as 600–700 years, were burning and grazing (by wild and not domestic animals) and the withdrawal of the human groups suggested by Caseldine and Hatton did not result in the regeneration of woodland cover. In view of the critical importance of this phase in the development of the environments of upland Dartmoor, it is worth quoting Caseldine and Hatton in full. Their results:

demonstrate the nature of the irreversible changes that Mesolithic communities initiated on the highest moorland slopes, changes which paved
the way for later extension of peat and diminution of woodland as a
result of climate and later use of the moorland for grazing.[30]

Pre-nineteenth-century mining

Documents allowing the tin miners to dig for tin anywhere on Dartmoor have
been found dated 1201, 1305 and 1510. But in 1574 the consent of the owner
was required for the extraction of ore from meadows, manured agricultural
land, orchards and gardens. This suggests that the most easily worked ores
were running out, causing the miners to encroach upon enclosed land rather
than simply use the moor. There is another variable: the tide of settlement and
enclosure creeping up the moorland valleys. Thus it seems likely that a dual
economy developed in which a parcel of land might be devoted to tin or to
animals as the situation proved most profitable. It seems probable that by the
thirteenth century, colonists were occupying the valleys of the upland and
were involved as well in the tin industry. Similar confirmation from the
fifteenth–seventeenth centuries is recorded.[31] Stratigraphic investigations
confirm that there is no enhancement of river silts by tin earlier than the
twelfth century and that the latter half of the thirteenth century (River Erme)
and the sixteenth century (River Avon) were periods of high production.[32]
The tinners extracted tin ore from the alluvial deposits of the rivers and the
environmental consequences of this 'tin-streaming' can still be seen: the
channel is turned over as the gravels are dug. The ore-rich gravels are washed
by the diverted stream and the heavier cassiterite is separated from the gravel
matrix. Most of the silt and sand fractions are washed out and are free to clog
the river channel whenever the energy level drops. Problems in the estuaries
of the Dart and the Plym were instigators of Acts of Parliament in the fifteenth
century. Today, the floodplains contain straight boulder-banks with variable
amounts of fine-sediment fill. Fine sediments may contain enhanced levels of
tin since the streaming process was less effective for fine sediments than for
the gravel fraction. Concentrating water flow for the streaming process
reduced the volume of the 'natural' channel by an average 18 per cent and the
quantity of fine material released onto floodplains meant that downstream
silting happened long after tin working by this method had ceased.[33]

In the early phases for which there is good documentary and field evidence,
the total production from the Moor seems to have been 47–82 tons/yr from
1400–1600, with a peak of 252 tons in 1524. Since production in 1300 was only
33 tons, a shift in techniques is generally postulated as the reason for the later
growth. Somewhere about 1400, the alluvial stream gravels were no longer the
chief source of ore. Between about 1400 and 1650 the method of working
changed, to what is usually called 'openwork'. This is no less than following a
lode away from a stream and taking out the ore by open-cast digging. The ore
was removed with pick and shovel and then rinsed, with the hope that the

water exposed more of the lode. An open gulley 250 m long by 30 m wide and 10 m deep might be left.[34] Environmentally, the difference with tin-streaming was not great, especially since waste heaps were often created over the worked-out stream gravels. There was an additional manipulation needed, however, for there was no longer an automatic supply of water. So water was led in channels ('leats') about one metre wide and 60 cm deep to the lower lodes. If the ore-bearing site was too high then a reservoir might be constructed behind a stone and earth dam, with gutters leading into it from either side. (Neither method could guarantee water supply in very dry or very cold weather.) Taken together with tin-streaming, the environmental impact in river valleys would have been considerable even if mining had stopped in, say, 1650 (Plate 6.5).[35]

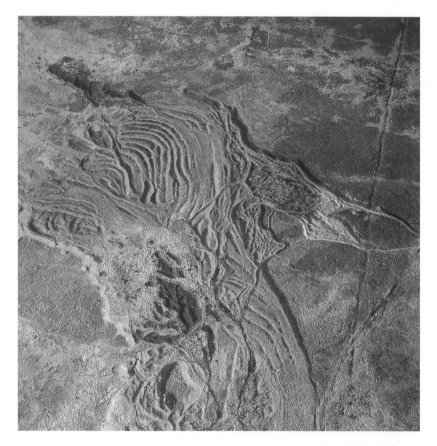

PLATE 6.5 *The remains of tin-streaming on Dartmoor. The curving ridges of gravel are left behind as the tin ore is separated from waste by running water often brought in by leats. In medieval times the smelting took place on the spot but later 'blowing-houses' used bellows to blast air into charcoal or peat to produce the metal. This example is supposed to be from South Tawton on the north edge of Dartmoor but has probably all disappeared beneath the A30.*
Source: Crown Copyright/MOD. Reproduced with permission of the Controller of Her Majesty's Stationery Office.

The eighteenth century was a time of variable fortune for Dartmoor tin: there was a revival in the early eighteenth century, failure by 1750 but then prosperity until the 1830s.[36] Resistance to extinction was offered by way of moving to shaft and adit mining. This enabled deeper lodes to be exploited and also necessitated the removal of less 'dead ground' (unproductive over-burden). So large mines of the late eighteenth century and early nineteenth century used this method and for a time could be profitable, even in remote places like the Steeperton Tor mine at 457 m ASL. It opened in 1799 on the site of earlier workings, using turf as its main fuel, then installing a large water-wheel to power the bellows. Its last tin was produced in May 1879, with the remnants of the operation, its leats, reservoir, adits and shafts adding to the alterations of water-courses and landforms by earlier openworks and alluvial streaming.[37] Only one other mine existed in such a remote place (Wheal Prosper on the Walkham) and another late survivor was at Ailsborough with both blast and reverbatory furnaces receiving ores from mines ten miles away until its demise in 1831. They demonstrate the entrepreneurial penetration of mining and its environmental impacts into the fastnesses of remote areas when the price is right.

The role of the National Park Authority (NPA)

Earlier discussion has reminded us that the National Parks of England and Wales are rarely public land: that even the moorland is either in the direct ownership of a manorial lord or is common land with positively assigned rights, usually of medieval origin. In the case of Dartmoor, 37 per cent (about 35,825 ha) of the National Park is common land, with 1,500 commoners exer-cising registered rights. The rights of turbary, estovers, pannage and extrac-tion of sand and gravel still exist.[38] The right of common of pasture is without doubt the most important of the rights, which exist for 145,000 sheep, 33,000 cattle, 5,450 ponies and 12,330 unspecified grazing units. The numbers cur-rently grazed are much smaller. In 1994, Dartmoor was designated as an ESA and sheep numbers peaked in the early 1990s at about 240,000 for the whole National Park area (i.e. bigger than the commoners' zone) and have fallen slightly since then. The high numbers of visitors has given some scope for innovative ventures but they tend not to be full-time enterprises.[39]

In this context, the applicable legislation gives the NPA only limited scope to impose a resolution of conflicts over land use, or even to combat obvious environmental degradation. But by making the best information available to the rural community, by acting as a conduit for external funds and identify-ing with the local rural community as much as with the national scene, the Dartmoor NPA has been a significant player on the environmental scene. A listing of the most important actions is given as Table 6.1, in which they are classified into two groups: those which essentially restrain actions deemed to be environmentally degrading, and those which promote a diverse and stable environment.[40] (Agriculture and forestry are not subject to normal planning

TABLE 6.1 Environmental processes manipulated by the Dartmoor NPA

Restraining actions	Promoting actions
Resistance to extension of military use of the moor	Dartmoor Commons Act 1985 management of livestock and of recreation
Resistance to further reservoirs in open country, e.g., proposals for Swincombe in 1970	Endorsement of MAFF proposals for ESA status for Dartmoor region, 1994–
Deflection of china clay waste dumping proposals	Signposting of roads according to most appropriate use
Introduction of 40 mph speed limit on moorland roads	Traffic Management Strategy, with Devon CC 1994–
Negotiation of Access Agreements	Guidance on suitable planting schemes for appropriate areas
Monitoring landscape change e.g., the spread of bracken	Surveys for, and inter-agency co-ordination of, biodiversity action
Concordat with Forestry Commission and other bodies against afforestation of bare land.	Moorland fire plans
Bye-laws to control off-road mountain biking 1998–	Management agreements with farms for environmentally sensitive practices Moorland restoration by clearing rushes and adventitious Sitka Spruce. Repair of eroded trackways and paths

control either inside or outside the National Parks.) There is not space here to discuss all the measures but it can be seen that the positive column outweighs the negative, though some of them have an element of both restraint and inducement, as with moor fire plans, which aim to control swaling (purposeful/deliberate burning), which had become somewhat anarchic and not directed at old-growth *Calluna* and *Molinia* where its effect is most beneficial in encouraging new shoots.

The foundation stones of the National Park Authority's involvement with the land cover and use are without doubt the Dartmoor Commons Act 1985, and the Authority's role in the Biodiversity Action Plans for the region. The 1985 Act recognised that the medieval structures controlling grazing and other common rights were no longer working, producing over- and under-grazing, poor animal health and indeed abuse of rights (Plate 6.6). All this was in the context of increasing recreation pressure, especially for access. The Act as passed is essentially a large-scale management and access agreement between the commoners and the National Park Authority. Regulation of grazing is the foremost activity controlled by a Commoners' Council but the National Park Authority is empowered to restrict access in order to protect

PLATE 6.6 *Although Dartmoor has two large blocks of moorland, its eastern side has interspersions of moor, woodland and enclosed pasture over which the NPA has much more difficulty formulating management policies due to the fragmented nature of the land cover. This side, too, has the heaviest recreation pressures.*

ancient monuments, SSSIs or tree growth. Erosion problems caused by commercial riding operations are also within their purview since this was one of the four major problems which led to the Act and the only one not ensuing from grazing operations. The Act also stipulates that foals must be removed from the commons during the months of January–April and that nobody may play radios or musical instruments to the annoyance of others. The Biodiversity work includes management plans for rare species such as the blue ground beetle and the bog hoverfly. As well as the protection of habitats such as blanket bog, the raised bog at Tor Royal (near Princetown) and semi-natural deciduous woodland, management is proposed for unusual habitats such as Rhôs pasture, which is a species-rich *Molinia* and rush pasture, with 20 per cent of its English area on Dartmoor. Maintenance involves controlled grazing by cattle and/or ponies in the summer and autumn months. The plant community supports particular species of hawkmoths, hoverflies and dragon-flies, some of which fall under the EU Habitats and Species Directive. The scrubby fringes of the Rhôs pasture may also be the home of dormice. The Action Plan addresses the maintenance of the areas of the pasture type, the annual maintenance regime and arrangements for access and interpretation. What is perhaps special about Dartmoor is the incorporation into SACs of the very large SSSIs which comprise the whole of the blanket peat areas of the

North and South moorland blocks (about 120 km²): an unusual size for England.[41] There are even 50–70 pairs of red grouse on Dartmoor, at the southern edge of their range.

The Dartmoor National Park Authority provides a good example of the ways in which inadequate national legislation can be supplemented by regional action and ways in which partnerships between statutory bodies and communities can be built up, dispelling a great deal of the animosity that has been the more usual relationship between the park planning committees and their local residents.

SOUTH WALES

The topographic map (Fig. 6.5) sets out clearly the well-known blocks of moorland in South Wales which frame the valley of the Usk. At the valley head is the block known as Mynydd Eppynt, part of mid-Wales and then along the southern side come the Black Mountain (*Y Myndydd Du*), Fforest Fawr and the Brecon Beacons (*Bannau Brycheniog*); to the north are the generally softer outlines of the Black Mountains (*Mynyddoedd Duon*), along whose eastern-most ridge runs the border with England. The map also delineates the terrain

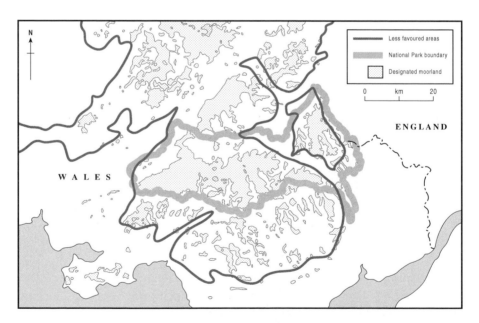

FIGURE 6.5 *The boundaries of LFA designation and the Brecon Beacons National Park in South Wales. Note that the LFA area here far exceeds the National Park's boundaries so that a large proportion of the hill land separating the formerly industrial valleys of South Wales is designated as LFA. Only the more severe of the two LFA categories is plotted here.*
Source: Reprinted with permission of Welsh Assembly GIS Services, 'Less Favoured Areas, Moorland and National Parks', Aberystwyth, 2002.

(less well known in this context) which lies between the Beacons and the Vale of Glamorgan. Here, the best recognised features are the valleys which have been famous in British industrial history but which, we may note, are separated by interfluves of upland all of which once carried continuous moorland vegetation. Near the heads of these valleys, towns grew which are among the highest of their kind: Merthyr Tydfil at 230 m, together with Beaufort and Brynmawr at more than 400 m ASL, for example. These ridges were later fragmented after 1780 by industrial uses and are now unified to some extent by the largest commercial forests in Wales, so extensive that they and the other plantations ensure that there are more exotic than native trees in the Principality.

A major theme of a closer look at environmental change in South Wales, therefore, has to concentrate on the contrasts between the wild and consistently rural uplands – which form the core of the Brecon Beacons National Park – and those which form southward-pointing fingers of higher ground that are in effect part of a highly industrialised region, even though its time as one of the nation's principal coal- and metal-working regions seems to have gone. Both zones have sheep, both have 'new' forests, both have water impoundments, though not in equal measure; the wilder north offers a variety of environment-based recreations, though the south absorbs a great deal of local use. The Beacons are wild enough to be important for military training and the Black Mountains near enough to Hereford to afford the occasional sighting of the SAS being manly.

The rural north

This region is dominated by large blocks of upland: the Black Mountains for example are the foremost continuous area of high ground in Wales except Snowdonia, with 54 km^2 of land above 550 m and 249 km^2 above 300 m, with a total of 157 km^2 of unenclosed land in the 1930s. Here as in the other blocks, the uplands are high enough to create a rain shadow to the east so that while Fforest Fawr has some areas with over 2,300 mm of rainfall per year, the Black Mountains are deluged with only about 1,500 mm at most.

The human element in the environmental history follows reasonably predictable paths. All the uplands were occupied during prehistory from the Mesolithic onwards, though the impact of the human communities is only variably known.[42] The detail of the investigations at Waun-Fignen-Felen on Fforest Fawr is scarcely repeated in the few pollen diagrams from the Black Mountains, for example. Put together with some palaeoecological work from the Brecon Beacons, a pattern of early maintenance of forest gaps, followed by upland agriculture in the Neolithic and Bronze Age periods is apparent, with some withdrawal from higher ground in the Middle-Late Bronze Age, very likely under the influence of a deteriorating climate.[43] That there was a Roman presence is evident from roads and camps on some very high interfluves, especially north into central Wales. Some of the roads are lined with gravel pits, suggesting another environmental impact of their construction and use.[44]

Hence, the medieval land cover of South Wales was in large measure a humanised artefact, with post-clearance blanket bog on the flat summits, grasslands whose composition reflected the number and types of stock grazed at different seasons, hillside woodlands dominated by oak but managed for underwood yield, and valley arable lands focussed on cereals (oats in particular) and pulses. Irish immigration and the activity of the pre-Norman church may have caused clearance and intensified the agro-pastoral use of the uplands from the sixth century onwards. We can envisage the additional pressures on timber (as well as underwood) brought about by castle-building when Edward I and his successors decided to bring Wales within their control.[45] Other social differences with environmental consequences may have been (the evidence is unexplored) the holding of rights of common under Welsh law (which persisted in east Carmarthenshire until the early sixteenth century), and the general lack of Cistercian activity compared with mid-Wales' houses like Strata Florida. Abbey Dore, just east of the Black Mountains, produced a great deal of wool but its holdings appear to be fragmented and the abbey's relation to upland pastures is not clear.[46]

Between 1000–1300 the medieval warm epoch made the uplands of South Wales accessible to a variety of grazing regimes. A study at the western end of the Black Mountain shows a variable dynamic of outstations for summer grazing, which seems to range from traditional transhumance to relatively close outstations of the permanent settlement, all between 250–450 m in altitude; the dating control is however poor. After about 1300 the uplands became summer pastures only and the boundary between the enclosed land and the common in South Wales seems to have had a constant level by the sixteenth century.[47] As in England, the aristocracy maintained hunting parks and there is a good example on the eastern flank of the Sugar Loaf mountain near Abergavenny, where its boundary ditch is still a strong element in the landscape, in places separating a wood-pasture from open moor. Another connection was the cultivation of rabbits using artificial 'pillow mounds' of loosened soil similar to those found on Dartmoor and the North York Moors; three large developments followed the enclosure and sale of part of the Great Forest of Brecon (in this case the Crown Allotment) in 1819. The warrens operated between 1827 and 1860 but thereafter reverted to open moorland. At their height however they punctuated the moorland with structures like the pillow mounds as well as pit traps, warreners' houses and enclosed fields for growing the root crops which were the rabbits' supplementary feed.

The eighteenth century marked the intensifying commercial connections between this part of South Wales and the outside world. In particular, this took the form of cattle exports for example from Brecon to London, a feature remarked upon by Daniel Defoe in his travels. The output of cattle seems to have been restricted by insufficient winter feed and this may have been a factor in the progress of enclosure of hill land in the nineteenth century: in the Black Mountains the moorland edge was pushed to 300–400 m ASL in the years after

1844. The cause of 'improvement' is typified by the enclosure of parts of Fforest Fawr after 1819, when the Crown Allotment was acquired by a London businessman, John Christie. He moved to Brecon in 1820 and developed limestone quarries at Pen-wyllt (south of the main moorland block) and Pwll Byfre (about 2 km to the north-east and at 480 m ASL) from them extended a tramway south to the Swansea canal and north to Sennybridge, with the aim of making enough lime available to supply his farms and convert the upland pastures. He also saw the advantages of linking regional workings of coal and iron. The moorland relevance of the development is best seen in aerial photographs which show the main line at nearly 400 m ASL on the watershed and the many branches penetrating to different mineral deposits which Christie tried to develop. The intrusion of the tramroads affords examples of cuttings, embankments, drainage channels, culverts and depots; these led to the pock-marking of some moorlands with many small pits and quarries: the ridge of Cribarth above Abercraf is one example.[48] The whole is reminiscent of the way in which railways came to remote valleys like Rookhope in the North Pennines.

The changes in the moorland area between the middle of the nineteenth century and the 1940s are chronicled by the work of Parry and by the Land Use Survey (LUS) of the 1930s.[49] Parts of Figure 4.10 have shown the difference over about 100 years in south-east Brecon, suggesting a retreat of moorland in the case of the main blocks and a 'tidying up' of patches of lower terrain (perhaps better described as heath rather than moor) in the Usk valley. To some extent it obscures the outcome of poor harvests in 1875–85 in many upland parts of Wales when hill farms were abandoned and the amalgamation of holding accelerated the trend to keeping sheep at the expense of cattle. By the 1930s renewed processes of abandonment and retreat were apparent and the LUS for Brecon, for example, makes much of them. It talks of how in the last 100 years (i.e., to the mid-1930s) 50 farms plus 152 small holdings, small farms and cottages were vacated in the Black Mountains alone, and produced a photograph of an abandoned farm north of Llanbedr with unkempt hedges, and invasions of bracken and scrub into former enclosures. Where large farms were merged, then the overall intensity of use was uneven and large bracken zones resulted.[50] The progress of 'enduring conversion' is shown in Figure 6.6.

The Romantic re-appraisal of the hill lands came to the Brecon Beacons no less than many other uplands. One famous result is the watercolour of Llanthony Abbey ruins by J. M. W. Turner, in which Loxidge Tump to the north is depicted as a peaky mountain in order to add drama to the generally stormy scene. The site's fascination (rather akin to that of Fountains in Yorkshire) drew the poet Walter Savage Landor to buy the abbey and adjacent farm lands with the idea of establishing a haven of social bliss within an inspiring landscape. He seems to have had little gift for getting on with people and the project failed: all we can now see are a few sweet chestnut trees in poor condition (Plate 6.7) and the roofless remains of a hillside 'house' which was a

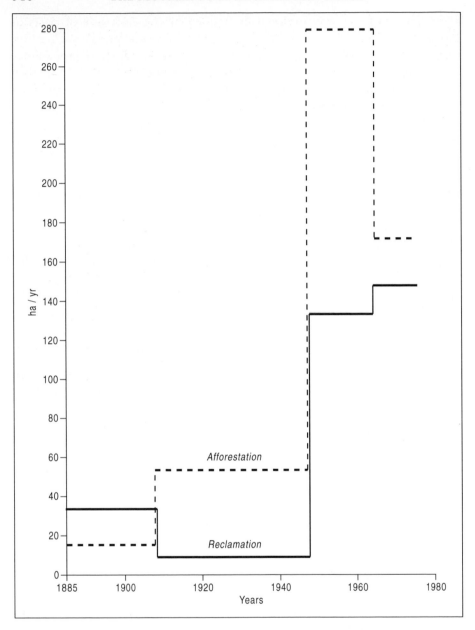

FIGURE 6.6 *The annual rate of enduring conversion of land from moorland in the Brecon Beacons National Park 1885–1980. Both reclamation for more intensive agricultural use and afforestation are recognised. Before the establishment of the Forestry Commission, there was a definite process of conversion to woodland, though this increased markedly during and after WWI. The effect of WWII on reclamation can be seen as can the enduring effects of the subsidy structures established in the 1940s.*
Source: M. L. Parry, A. Bruce and C. E. Harkness, Changes in the Extent of Moorland and Roughland in the Brecon Beacons National Park, *Birmingham: Department of Geography, University of Birmingham, 1982. (Surveys of Moorland and Roughland no. 7), Fig. 6.1 at p. 36.*

PLATE 6.7 *Remains of Landor's landscape scheme at Llanthony in Gwent. Between the chestnut trees the ruins of the former monastery can just be seen.*

kind of superior belvedere.[51] The remote feeling of this valley attracted a Roman Catholic monastery to Capel-y-ffin in 1870 and thereafter the sculptor and typographer Eric Gill set up his studio and domestic ménage there from 1924–8. Pleasure came to the uplands of Fforest Fawr during the late nineteenth century and early twentieth century in the form of grouse moors but they were not kept up in the manner of the northern hills of England.[52]

The LUS records the poor state of sheep farming and could not have seen the resurgence of that activity under the various subsidies available in post-1946 times, culminating with the EU's underpinning of marginal areas as sites of meat production. One outcome in the period 1948–75 was the loss of 8,195 ha of moorland in the Brecon Beacons National Park area, so that the park area was 49 per cent moorland in 1975 compared with 55 per cent in 1885. Sheep densities rose rapidly after about 1950 and they were out-wintered, with hay being brought in to support the higher densities (Fig. 6.7). The results do not differ from elsewhere in England and Wales, with shifts in forage species being the most obvious outcome; soil erosion is however limited and on the Beacons distinctly secondary to that caused by recreation. Commercial peat extraction on Fforest Fawr as late as the 1950s has also left some scars. Deforested areas also yielded rough pasture in the years 1909–48 when wartime demands left open land that was not reforested: Fan Frynych (south-west of Brecon) and the southern slopes of the Sugar Loaf near Abergavenny are examples.

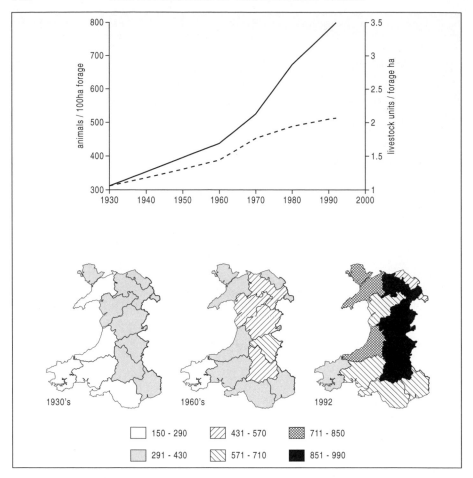

FIGURE 6.7 *The density of grazing animals in Wales during much of the twentieth century. Horses, cattle and sheep are included (though not horses after 1958) and the continuous line is the absolute number per 100 ha of forage; the dotted line represents livestock units (where sheep count less than cattle) per hectare of forage. This makes the rise look less spectacular. The maps show the distribution of the increases in animals per 100 ha and emphasise the contribution of the upland core of mid-Wales, with the south contributing a little less to the increased densities.*
Source: M. Shrubb, L. T. Williams and R. R. Lovegrove, 'The impact of changes in farming and other land uses on bird populations in Wales', Welsh Birds 1(5), 4–26, 1997, Figs 3 and 4. Reproduced by permission of the Welsh Ornithological Society.

A summary of the moorland edges' movements in the past 100 years would emphasise that over half of all the reversions were in the 1948–64 years yet two-thirds of that land was then reclaimed in 1964–75 (Fig. 6.6). In one sense the recent pattern of reclamation was dictated by the reversions of the 1930s or even the land that was allowed to fall out of more intensive usages in the 1880s. After 1964, afforestation and grazing land were approximately equal destinations of reclaimed terrain until the 1970s when the losses of open land

TABLE 6.2 Land cover in the Brecon Beacons National Park 1970s–80s: km^2

Land cover type	1970s	1980s	Net change	Gross change
Upland heath	78.14	69.63	−8.51	13.83
Upland grass moor	387.99	381.71	−6.28	26.84
Blanket peat grass moor	1.58	1.57	−0.01	0.01
Bracken	67.97	75.00	7.03	12.25
Upland heath/grass mosaic	47.67	49.57	1.90	11.04
Upland heath/bracken mosaic	4.45	4.24	−0.21	0.75
Upland heath/blanket peat mosaic	2.36	2.35	−0.01	0.01
Eroded bare peat	0.33	0.44	0.11	0.17
Total moor and heathland	590.49	584.51	−5.98	64.90
Quarries and mineral workings	2.76	2.26	−0.50	0.76
Derelict land	4.04	3.23	−0.81	1.09
Total area of National Park	1,351.4[a]	1,351.4	0.0	277.1[a] (20.5%)

[a] Some land uses are omitted so that this column is not the total of its cells.
'Upland heath' is heather-dominated; 'blanket peat grass moor' is mostly *Molinia*.
Source: J. C. Taylor, *Landscape Change in the National Parks of England and Wales, Final Report Vol. III The Brecon Beacons*, Silsoe: Cranfield Institute of Technology, 1991, Table 5.5.

shown in Table 6.2 represent conversions to improved pasture well in excess of afforestations, though the overall figure for the park is not necessarily all related to the moorland margin.[53]

The kinds of evaluations given form by Father Ignatius and Eric Gill are relevant to today's environment in the sense that valleys like that of the Honddu are now highly rated as recreational environments and this adds to the pressures to protect the area against certain forms of change. The attraction of the Offa's Dyke path as it follows the ridge overlooking the Golden Valley of Herefordshire would be rather diminished if it failed to contrast with the forested parts of the trail further south, just as hang-gliding off Hay Bluff requires an open landscape devoid of trees. Pony-trekking, on the other hand, is enhanced by a climb through woodland to open heights beyond.

There are few surprises, then, in the environmental history of this part of South Wales. The progress of differentiating a set of land cover systems from the wildwood is not very different from most other uplands in Wales in spite of, or indeed because of, a colonial relationship with England. The difference between South Wales and most other uplands is further south and has a very basic origin in a solid geology which contained coal, ironstone and limestone.

The industrial south

If we open out the Ordnance Survey's 1:50,000 sheet 170 (*Vale of Glamorgan*), then the block of territory 40×20 km on the northern part of the sheet looks like a forest. If it were not for the narrow valleys with a density of settlement that scarcely seems to relate to other symbols on this map, then the presence of moorland ecosystems on the intervening ridges and plateaux could be almost overlooked. Sheet 171 to the east shows far less afforestation, with the southward-pointing fingers of high ground much more apparent.

The moorland environments between the coast at Port Talbot and near Neath as far east as the escarpment from Abergavenny down through Ponty-pool to Cwmbrân were very much the same as those further north until the eighteenth century. The Cistercians had abbeys at Margam, Neath and Llantarnam and these had upland holdings as well as interaction with upland environments. Neath had a few upland patches of land but Margam had a coal-working grange in 1519 in 'the Gorse Moor' and their grange at Resolven worked an extensive mountain area with a chapel; Llantarnam had a coal-mine at Landerfel near Cwmbrân.[54] These point forward to the industrialisation of the valleys, in the early phases of which (after about 1750) there was a great demand for timber, which resulted in the clearance of woodland, and for fuel, which brought about conversion to coppice. (Such woodland as remained was also under pressure whenever there was a coal strike.) The woodlands were largely found on the valley sides and so were the resultant coppices: extensive areas were noted along the Ebbw and Sirhowy rivers in the early nineteenth century. Today there is a considerable contrast between the valleys tributary to the Usk, where a belt of woodland and enclosed pasture separates the moor from the valley floor, and the Valleys, where bare slopes fall directly into urban-ised and industrialised ground without so much as a thorn bush on the way: the upper Rhondda Fawr provides a good example (Plate 6.8).

Above the woodlands, the moors were, until the advent of intensive indus-trialisation in the nineteenth century, grazed by cattle, sheep and goats. The highest and flattest areas were covered in *Molinia* and *Nardus* grassland, with *Eriophorum* on the wettest parts, where there was some peat accumulation. The greatest area was, however, acid *Agrostis-Festuca* grassland. There was some heather and gorse and a few well-drained slopes were dominated by *Vaccinium*. Variable quantities of bracken would under favourable conditions spread into the fescue grasslands, at a rate which increased as cattle were taken off the hill and the sheep no longer controlled carefully, in other words as the twentieth century progressed. Industrial pollution is held to account for the expansion of *Molinia* and *Eriophorum*, the latter at the expense of *Sphagnum* mosses. Nonetheless, before the afforestation of the inter-war period, the LUS talked of moorland as the greatest uninterrupted land type in Glamorgan, with its map showing no evidence of the changes due to afforestation.

The industrialisation of this region was dominated at first by iron at the heads of the valleys, using local outcrops of coal as fuel. The area of moorland between Blaenavon and the Usk was an early site for this industry and the combination of mines, ironworks and canals has led to its designation as a World Heritage Site: the only one on the moorlands discussed in this book. After 1830 it was coal mining above all in deep pits that characterised it envi-ronmentally. The main impacts on moorland can be divided into three cate-gories:

PLATE 6.8 *Cwmparc from Craig Ogwr, east of Rhondda Fawr in mid-Glamorgan. The former coal-mining settlement was planted during the nineteenth century onto the moorland which is characteristically bare of any deciduous woodland. As is now common in the eastern part of South Wales, large blocks of Forestry Commission plantations are also present. The patterns within the grassland suggest former occupation sites, probably farms from a pre-industrial era of grazing based high up the hillsides.*

1. Extraction of iron from chains of pits that ran round the hillside and the nearby dumping of spoil
2. The use of the intake land in valleys and lower valley sides for collieries and other works, housing and communications.
3. The conveyance of spoil from collieries by aerial conveyors or tramways to tips on the hillsides or on summit plateaux.

The total effect was that of obliterating the moorland by covering it with works or with wastes (Fig. 6.8) and the fragmentation of agricultural holdings, so that the land value was low enough to allow the Forestry Commission to acquire large holdings, with the aim also of creating some alternative employment. A few small enclosures at the moorland edge near Brynamman were re-used during the coal strike of 1926.

Aerial photographs are good sources of synoptic views of these processes, though few exist for the inter-war period when the land cover effects were at their height; most come from times when iron had gone and coal was contracting, and when older tips had grown some plant cover, especially if they were stable. A picture of Clydach Terrace at Brynmawr in 1993 shows the effect of extracting one seam of coal, producing a contoured groove, plus the

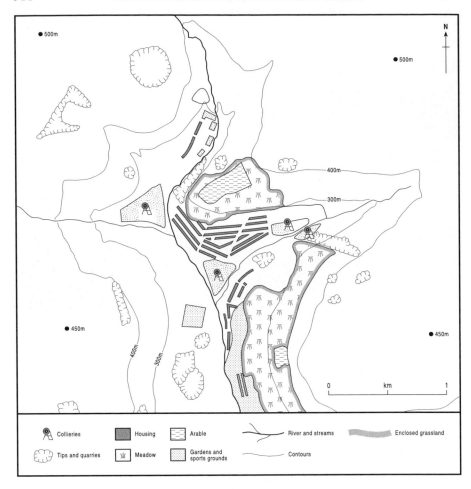

FIGURE 6.8 *A simplification of the Land Use Survey map of Blaengarw in the 1930s. The town (at 190 m ASL) sat in a moorland frame, and the history of iron and coal extraction is one of the colonisation of moorland. The sites of tips show that wastes had to be transported out of the narrow valleys. In the original publication, the patch of arable is labelled as 'oats', the common cereal of upland Britain in the nineteenth century. But no less than Harrogate, this urbanisation was mostly of open land.*
Source: A. M. Thomas, The Land of Britain, *pt 31:* Glamorgan, *London: Geographical Publications, 1938.*

spoil from iron extraction showing almost as a set of latter-day pillow mounds that cover the whole area of the photograph.[55] A picture of the two Rhondda valleys and the intervening ridge of Mynydd Ty'n Tyle is, by contrast, little dominated by industrial remains, though a line of iron pits streaks the hill above Ferndale; the housing directly abuts the moor, with no enclosed grazing land at all on the west side of the Rhondda Fach.[56] Collections of paintings of collieries inevitably show the mines against a backdrop of hills, usually flat-topped.[57] The decline of the coal industry has meant the disappearance of

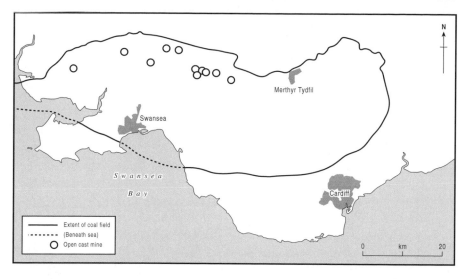

FIGURE 6.9 *The distribution of open-cast coal mines along the northern area of the South Wales coalfield in the 1990s. This area, like West Durham, was one of the earliest to be exploited and now is being worked over again, so to speak. Many of the pits are in a moorland framework.*
Source: IPR/40-8C British Geological Survey. ©NERC. All rights reserved.

many installations and the re-modelling of tips under a number of centrally-financed reclamation programmes. Even so, the original moorland does not reappear and where possible some use other than grazing is found for the reclaimed land. The site of the tip at Aberfan, which slid into a school in 1966, killing 124 people (including 116 children) is a grassy tongue extending from the moorland through the enclosed grass and bracken lands to the former site of the school. The Merthyr Vale colliery alongside the River Taff has gone, as have all the other deep collieries as work-places.

In a kind of reversion to the patterns of the early nineteenth century, the main concentration of open-cast coal mines in the 1990s was in the northern part of the valleys, west of Merthyr Tydfil, with a few outside the upland areas both to the west and the south (Fig. 6.9). The latest manifestation of an industry with some environmental implications is the windfarm, of which there is an example on Mynydd Maendy, but the scale of mid-Wales is nowhere yet in view, so to speak.

Thus the effects of industry on moorland are without doubt greatest in the north of the valleys region of South Wales, especially where early iron and coal were both the focus of industry and where towns were built directly onto moorland sites as at, for instance, Blaenavon, Beaufort and Merthyr Tydfil. (For a comparison we have to think of Harrogate in Yorkshire, developed after an enclosure award.) A similar process is still at work in this area since the upper parts of late twentieth century industrial estates near the Heads of the

PLATE 6.9 *The new 'head-dyke' at Rassau in mid-Glamorgan, South Wales. In earlier times many moors were separated from enclosed pastures by a ditch and bank and this is repeated in the upper edge of an industrial estate at the head of the Valleys. Today's demarcation is distinguished with more verticality than was once the case.*

Valleys road enclose relict patches of moorland and at Rassau near Beaufort, a ditch and fence separate the industrial plants from the moorland in an echo of a medieval head-dyke (Plate 6.9). A minor road east of Merthyr Tydfil, on the way to Ffos y Fran, runs through a piece of moorland dissected by former railway routes and plastered with industrial waste tips; it also seems to act as a fly tip for the whole neighbourhood. West of Merthyr, afforestation has covered some of the wastes just as it has the moorland and so a third phase of land cover since 1750 has become established. A few of the plantations are spreading spontaneously onto adjacent slopes where these are little grazed if at all.

North and south

Though there is an economic interdependence of north and south here (pony-trekkers come from Cardiff, but that is where the specialised hospitals for the whole region are found), there is a definite difference to the environments. The north is defiantly rural and the moorlands are central to the image as well as to the patterns of biological production and of the runoff of water into major systems. Both have water reservoirs to supply towns and industries though only in the north is a natural lake (Lyn y Fan Fach) 'improved' by damming as distinct from drowning a former valley. The south is fragmented by the relics of industry and is home to large areas of coniferous afforestation. Yet even if farming were to change radically in the wake of redirected subsidy patterns, the likelihood of rapid economic and environmental change feels greatest in the valleys of the region.[58] One element in this is the high density of local population who (if their job prospects were better) might seek more local environmentally-based recreation, and the other is the development of tourism based on historical materials. This has happened at Blaenavon, for example, with the Big Pit (a colliery devoted to visitors) and the Iron Works (early blast furnaces) and might be extended, though there is presumably some limit to this type of industrial attraction and in any case the environmental linkages of such developments are not always well explored. But the coalescence of environmental perception of the north and the south around pleasure and protection rather than production seems a likely future.

INTER-REGIONAL COMPARISONS

To compare these three regions in environmental terms is not perhaps very profitable: all have high sheep densities, all have had mining periods in their history, two are now largely National Parks but the North Pennines are an AONB and the Valleys of South Wales do not attract such designations. All are vulnerable to changes in the support systems for hill farming such as are coming from the EU and which may be infused with any different thinking about agro-environment schemes which emerges after the 2001 Foot and Mouth Disease outbreak. So the idea of the 'eternal hills', which has been undermined by earlier chapters is further destabilised and instead we have the notion of a set of alternative futures for the moorlands instead of a single trajectory.

NOTES AND REFERENCES

1. See for example *Historical Monitoring in the Pennine Dales ESA 1987–1995*, ADAS Report to the Ministry of Agriculture, Fisheries & Food, April 1996. London: MAFF.
2. MAFF (Note 1) p. 12.
3. R. H. Squires, A Contribution to the Vegetational History of Upper Teesdale, PhD thesis, University of Durham, 1970, 2 vols. Dufton Moss is examined in more detail in I. G. Simmons, *The Environmental Impact of Later Mesolithic Cultures*, Edinburgh: Edinburgh University Press, 1996, 71–2.
4. Squires (Note 3) p. 214.
5. MAFF (Note 1) p. 12.
6. A. Younger and R. S. Smith, 'Hay meadow management in the Pennine Dales, Northern England', in R. J. Haggar and S. Peel (eds), *Grassland Management and Nature Conservation*, Aberystwyth: British Grassland Society Occasional Symposium No. 28, 1994, 137–43.
7. B. K. Roberts, 'Man and land in Upper Teesdale', in A. R. Clapham (ed.), *Upper Teesdale. The Area and its Natural History*, London: Collins, 1978, 141–59.

8. M. E. Bradshaw, 'The distribution and status of five species of of the *Alchemilla vulgaris* L. aggregate in Upper Teesdale', *Journal of Ecology* **50**, 1962, 681–706.

9. R. S. Smith and L. Jones, 'The phenology of mesotrophic grassland in the Pennine Dales, Northern England: historic hay cutting dates, vegetation variation and plant phenologies', *Journal of Applied Ecology* **28**, 1991, 42–59; R. S. Smith and S. P. Rushton, 'The effects of grazing management on the vegetation of mesotrophic (meadow) grassland in northern England', *Journal of Applied Ecology* **31**, 1994, 13–24.

10. J. Pickin, 'Early lead smelting in Teesdale', in L. Willies and D. Cranstone (eds), *Boles and Smeltmills*, Matlock: The Historical Metallurgy Society, 1992, 25–7.

11. K. C. Dunham, *Geology of the Northern Pennines Orefield. Vol. 1. Tyne to Stainmore.* Memoirs of the Geological Survey of the UK, London: HMSO, 1948, 2nd edn 1990; R. A. Fairbairn, *Weardale Mines*, Keighley: Northern Mine Research Society Monograph no. 56, 1996.

12. D. Cranstone, 'To hush or not to hush: when, where and how?', in B. Chambers (ed.), *Men, Mines and Minerals of the North Pennines*, Friends of Killhope, 1992, 41–8. [no place of publication given]

13. D. A. Ratcliffe, 'The plant communities' in A. R. Clapham (ed.), *Upper Teesdale The Area and its Natural History*, London: Collins, 1987, 64–87; P. T. Buchanan, 'Metalliferous plant communities: the flora of lead smelting in the Upper Nent Valley', in L. Willies and D. Cranstone (eds), *Boles and Smeltmills*, Matlock: The Historical Metallurgy Society, 1992, 58–61.

14. P. N. Wilson, 'The Nent Force Level', *Transactions of the Cumberland and Westmorland Antiquarian and Archaeological Society* **NS LXIII**, 1963, 253–80.

15. K. A. Hudson-Edwards, M. G. Macklin, C. D. Curtis and D. J. Vaughan, 'Processes of formation and distribution of Pb-, Zn-, Cd-, and Cu-bearing minerals in the Tyne basin, North-east England: implications for metal-contaminated river systems', *Environmental Science and Technology* **30**, 1996, 72–80.

16. K. Hudson-Edwards, M. Macklin and M. Taylor, 'Historic metal mining inputs to Tees river sediment', *The Science of the Total Environment* **194/195**, 1997, 437–45.

17. M. G. Macklin, K. A. Hudson-Edwards and E. J. Dawson, 'The significance of pollution from historic metal mining in the Pennine orefields on river sediment contaminant fluxes to the North Sea', *The Science of the Total Environment* **194/195**, 1997, 391–97; C. M. Sedgwick, 'Historic metal contamination at Reeth, Upper Swaledale', in A. J. Howard and M. G. Macklin, *The Quaternary of the Eastern Yorkshire Dales*, London: Quaternary Research Association, 198, 67–75.

18. A. Raistrick and B. Jennings, *A History of Lead Mining in the Pennines*, Newcastle: Davis Books/Littleborough: Kelsall Publishing, 1983, 239–40. [A 1983 edition of a book of the same title first published in 1965 by Longmans, Green in London.]

19. Ibid. (Note 18) p. 244

20. Ibid. (Note 18) pp. 268–70

21. R. A. Fairbairn, 1996 (Note 11) pp. 40–3.

22. M. C. Whitby (ed.), *Agriculture in the North Pennines*, Newcastle-upon-Tyne: University of Newcastle Agricultural Research Group, 1986. (This was a rapidly prepared report for the Countryside Commission ahead of the Public Inquiry into the establishment of the AONB.)

23. About thirty years ago there was a big scheme to develop a winter sports centre in Weardale. Subsequent winters have shown how marginal this would have been.

24. Most of this information comes from North Pennines AONB Steering Group, *The North Pennines Area of Outstanding Natural Beauty,* Durham: Durham County Council 1995; O. Wilson, 'Landownership and rural development in the North Pennines: a case study', *Journal of Rural Studies* **8**, 1992, 145–58.

25. See W. Blake, 'Hooker's Synopsis Chorographical of Devonshire', *Transactions of the Devon Association* **47**, 1915, 345. Annual average rainfall at Princetown is 2,150 mm/85 inches.

26. Key publications include: G. W. Dimbleby, 'Pollen analysis of terrestrial soils', *New Phytologist* **56**, 1957, 12–28; G. W. Dimbleby, 'The ancient forest of Blackamore', *Antiquity* **35**, 1961, 123–8. Criticism came in e.g., H. Godwin, 'Pollen analysis in mineral soil: an interpretation of a podzol pollen-analysis by Dr G. W. Dimbleby', *Flora* **146**, 1958, 321–7.

27. I. G. Simmons, 'Pollen diagrams from Dartmoor', *New Phytologist* **53**, 1964, 164–80. The advances in the field have been such that this would not now be acceptable in a front-line journal. In particular the phenomena assigned to inter-ference with the vegetation were not then radiocarbon-dated. The key site, at Blacklane Brook, was re-investigated in the 1970s and happily the same phenom-ena were found and this time a ^{14}C date was obtained. See I. G. Simmons, J. I. Rand and K. Crabtree, 'Further pollen analytical investigations at Blacklane Brook, Dartmoor', *New Phytologist* **94**, 1983, 655–67.

28. C. J. Caseldine and D. J. Maguire, 'Lateglacial/early Flandrian vegetation change on northern Dartmoor, south-west England', *Journal of Biogeography* **13**, 1986, 255–64.

29. C. Caseldine and J. Hatton, 'The development of high moorland on Dartmoor: fire and the influence of Mesolithic activity on vegetation change', in F. M. Chambers (ed.), *Climate Change and Human Impact on the Landscape*, London: Chapman and Hall, 1993, 119–31; C. J. Caseldine, 'Archaeological and environ-mental change on prehistoric Dartmoor – current understanding and future directions', *Quaternary Proceedings* 7, 1999, 575–83; I. G. Simmons, J. I. Rand and K. Crabtree, 'Further pollen analytical investigations at Blacklane Brook, Dartmoor', *New Phytologist* **94**, 1983, 655–67. C. Caseldine and J. Hatton, 'Vegetation history of Dartmoor – Holocene development and the impact of human activity', in D. J. Charman, R. M. Newnham and D. G. Croot (eds), *The Quaternary of Devon and East Cornwall: Field Guide,* London: QRA 1996, 48–61; S. West, D. Charman and J. Grattan, 'Palaeoenvironmental investigations at Tor Royal, central Dartmoor', in D. J. Charman *et al.* 1996, 62–80. For a broad review see C. Caseldine and J. Hatton, 'Into the mists? Thoughts on the prehistoric and historic environmental history of Dartmoor', *Proceedings of the Devon Archaeological Society* **52**, 1994, 35–48.

30. Caseldine and Hatton, 1993 (Note 29) p. 130.

31. P. Newman, 'Tinners and tenants on South-West Dartmoor: a case study in land-scape history', *Transactions of the Devon Association* **126**, 1994, 199–238.

32. V. R. Thorndycraft, D. Pirrie and A. G. Brown, 'Tracing the record of early allu-vial tin mining on Dartmoor, UK', in A. M. Pollard (ed.), *Geoarchaeology: explo-ration, environments, resources,* London: Geological Society of London Special Publications **165**, 1999, 91–102.

33. S. B. Bradley, 'Characteristics of tin-streaming channels on Dartmoor, UK', *Geoarchaeology* **5**, 1990, 29–41; C. C. Park, 'Tin streaming and channel changes. Some preliminary observations from Dartmoor, England', *Catena* 6, 1979, 235–44.

34. P. Newman, 'The moorland Meavy – a tinners' landscape', *Transactions of the Devon Association* **119**, 1987, 223–40.

35. C. Gerrard, *The Early British Tin Industry*, Stroud: Tempus Publishing, 2000. This deals with the period to AD 1700.

36. T. A. P. Greeves, 'Tin smelting in Devon in the eighteenth and nineteenth cen-turies', *Mining History* **13**, 1996, 84–90. Appendix A contains a list of documented tin smelting houses in Devon from 1700–1890.

37. T. A. P. Greeves, 'Steeperton Tor tin mine, Dartmoor, Devon', *Transactions of the Devon Association* **117**, 1985, 101–27.

38. Turbary – taking turf (peat) for domestic fuel; estovers – underwood or branches for fuel or repairs; pannage – acorns and beech mast for pigs; sand, gravel and stone for use on the commoner's holding.

39. M. Turner *et al.*, *The State of Farming on Dartmoor 2002*, Exeter: University of Exeter Centre for Rural Research, 2002.

40. The information comes from numerous documents posted by the Dartmoor National Park Authority on their website www.dartmoor-npa.gov.uk and accessed in April 2001. There is no discussion in the present work of the major role of the Authority in development control.

41. See the comprehensive *Nature of Dartmoor*, published jointly by English Nature and DNPA (second edition, 2001).

42. D. K. Leighton. *Mynydd Du and Fforest Fawr. The Evolution of an Upland Landscape in South Wales*, Aberystwyth: RCAHM Wales, 1997.

43. F. Lynch, S. Aldhouse-Green and J. L. Davies, *Prehistoric Wales*, Stroud: Sutton Publishing, 2000; D. K. Leighton, 1997 (Note 42).

44. C. Musson (ed.), *Wales from the Air. Patterns of Past and Present*, Aberystwyth: RCAHM Wales, nd; RCAHM Wales, *An Inventory of the Ancient Monuments in Brecknock (Brycheinog): the Prehistoric and Roman Monuments*, Aberystwyth: RCAHM, 2 vols, 1986; F. Olding, *The Prehistoric Landscapes of the Black Mountains*, Oxford: Archaeopress, British Archaeological Reports, British Series **297**, 2000.

45. J. Davies, *The Making of Wales*, Cardiff: Cadw/Stroud: Alan Sutton Publishing, 1996.

46. D. H. Williams, *Atlas of Cistercian Lands in Wales*, Cardiff: University of Wales Press, 1990.

47. A. Ward, 'Transhumance and settlement on the Welsh uplands: a view from the Black Mountain', in N. Edwards (ed.), *Landscape and Settlement in Medieval Wales*, Oxford: Oxbow Books, 1997, 97–111.

48. There is an immense level of detail in S. Hughes, *The Brecon Forest Tramroads. The Archaeology of an Early Railway System*, Aberystwyth: RCAHM, 1990. The illustrative matter of maps, aerial and ground photographs and ink-drawing reconstructions is outstanding. Recommended walks are also given but the weight of the book is a bit inimical to its use outdoors. Christie also built a wall around the part of the Great Forest of Brecon that he owned but altogether it was too much for him financially.

49. R. M. Whyte, *Brecon*, London: Geographical Publications Ltd. *The Land of Britain*, Part 37, 1943; A. Rhys-Clarke, *Monmouth*, London: Geographical Publications Ltd. *The Land of Britain*, Part 38, 1943; M. L. Parry, A. Bruce and C. E. Harkness, *Changes in the Extent of Moorland and Roughland in the Brecon Beacons National Park*, Birmingham: Department of Geography, University of Birmingham, 1982. (Surveys of Moorland and Roughland no. 7.)

50. R. M. Whyte, 1943 (Note 49) p. 415. The named farm is not shown on the recent 1:25,000 maps.

51. If ruins pall, then part of the Abbey structure is a hotel with an atmospheric undercroft bar; the adjacent Court Farm has one of the nicest two-bedroom rental 'cottages' you might ask to find.

52. D. K. Leighton, 1997 (Note 42) pp. 26 and 51.

53. M. L. Parry, A. Bruce and C. E. Harkness, *Changes in the Extent of Moorland and Roughland in the Brecon Beacons National Park*, Birmingham: Department of Geography, University of Birmingham, 1982 (Surveys of Moorland and

Roughland no. 7.); J. C. Taylor, *Landscape Change in the National Parks of England and Wales, Final Report Vol. III The Brecon Beacons*, Silsoe: Cranfield Institute of Technology, 1991, Table 5.5.

54. D. H. Williams, 1990 (Note 46).

55. Reproduced in C. Musson (Note 44) p. 43: RCAHM 935047–63.

56. An Aerofilms photograph from their book *Wales from the Air* (foreword by Jan Morris), London: Barrie and Jenkins, 1990, 126. (No identification number nor date is given.)

57. For example, David Bellamy, *Images of the South Wales Mines*, Stroud: Alan Sutton Publishing, 1997.

58. An investigation into common land management on Fforest Fawr in 1995–8 showed that, for instance, 39 per cent of graziers did not know who would succeed them on the farm and that over two-thirds of them placed over three-quarters of their stock on unenclosed commons; see Brecon Beacons National Park, *Meithrin Mynydd*, Brecon: National Park Authority 1999 [a bilingual booklet].

CHAPTER SEVEN

Representations from the imagination

The story so far has been one in which the information provided by the disciplines of the natural sciences and of history have been strongest. Their public stance is to provide information and explanation which is 'objective' or at least 'intersubjective'. Different people presented with the same set of data (whether numbers of insects per square metre of moor or documents about eighteenth century intaking) should then provide the same set of explanations of process. Most scholars are now agreed that such ideals are rarely achieved and that both scientists and historians may well have other agendas but nevertheless there is often an agreed body of 'fact' about phenomena like moorlands and their changes through time.

In addition to this received record of the nature of the moors and their history there is another set of stories: those which are born from the human imagination and thus lodge in all kinds of media a construction (perhaps we might call it 'image' although that word has very strong visual connotations that are a little narrow) of what the moorlands are like. This interpretation may or may not be consonant with the scientific-historical version, and part of the ending of this book will be to compare the two. Conventional science and history find their way into the popular imagination and usage but notions of these (since they change) are also in the public mind. The concept of 'naturalness' in the 1949 National Parks Act is one example. (The whole thrust of the main chapters of this book is to show how strong the human influence has been.) Imaginative expositions of all kinds can of course stay static and exert as much influence in most spheres as the current 'official' view; it is possible to imagine (though not always to document) a popular notion running ahead of official policy. What this all adds up to is that myth-making is something to which the images of England and Wales (both urban and rural) are prone and the uplands are no exception. In this, the role of emotion can never be discounted, not least because the connections it makes are conceptual rather than causal or sequential.

There is then a set of constructions of the moorland environment which derive from the imagination of writers, adepts of visual media, and producers of sound. To be repetitive, these may have two lives: they may stand alone, without reference to the scientific/historical narrative, or they may feed off it and in turn inform its course. In another framework, imaginative constructions are part of the way in which rural environments are formulated as places, which in turn affects the management of their localities. So while pictures, prose and poetry may not directly affect the way the uplands are grazed or trodden (for example) they contribute to a national consciousness about these environments and help us to learn how they 'ought' to be used. We can look at some selected examples of these imaginative productions to see where they belong in the environmental scene.

WRITTEN EVOCATIONS

No corner of the landscapes of England and Wales has failed to find its topographer who will overtly or unconsciously describe the underlying environmental systems. This is true of the uplands in considerable measure and we have to acknowledge that in addition to the obvious categories of fiction and poetry there are travel accounts, guides, interpretive material and the focussed material of a utilitarian nature, like the agricultural writers of the late eighteenth century and early nineteenth century. Yet before the great surge of the Romantic movement in the late eighteenth to nineteenth centuries, the uplands of most nations were regarded with awe and distaste. They were waste places and better avoided, not least because of the kinds of people who might be found there. In general, the coming of an industrial economy changed perceptions so as to re-value these uplands, and in England and Wales most of them eventually became National Parks under the 1949 Act. So much of the writing has contributed to those metamorphoses of value.

Medieval appraisals

Some of the writing, used with due caution, tells us some of the things that people thought, before current categories like poetry, fiction, and administration become relevant. Even before the Conquest, which produced a flood of bureaucracy, the anonymous author of *Beowulf* was unhappy with the wild places where the monster Grendel's mother lives:

> . . . over the sombre moors . . . along meagre tracks, narrow, forbidding bridle-paths, uncertain ways, and beetling crags, past holes of the water-demons.[1]

We have to be careful of what 'moors' may mean in this context, though it sounds like the right sort of place. (Grendel is described as a *mære morstaþa*, a notorious moor-stalker.) Anglo-Saxon poetry does not often mention specific environments except the sea, though species may be enumerated. There is no word for 'nature' (meaning the non-human but excluding the supernatural) in Anglo-Saxon: there was apparently no conception of the natural world that excluded the supernatural.[2] Elegy, gnomic and seasonal poems from Welsh of the ninth to twelfth centuries contain many references to nature but in most cases are simply descriptive or act as foil for human qualities.[3] None of them in any case are very specifically upland, though there is a reference to a bog and a hillside, which we might imaginatively read as topographically relevant in part at least:

> Medal migned, kalet ridd,
> rac carn cann tal glann a vridd;
> edewit ny wnelher nydidd.

Soft are the bogs, hard the slope, the edge of the bank breaks under the white horse; a promise not performed is none.

In the Domesday Book, the moors tend to show up as gaps and in any case that record does not exist for the northern counties like Durham and Northumberland. In Yorkshire, many of the manors of all kinds were denoted as *wasta est*, which is interpreted by most scholars as referring to the results of depredations by King William's army in reprisals for the rising of the North in 1069. Further, there had been crop failures and famine in 1070 and 1087 and livestock pestilence in 1086.

Any desolation in the moors and their fringes was in tune with the revival of the Benedictine interest in a life of ascetic poverty. The founder of that monastic order had himself rhapsodised,

> ... for in those days the mountains distil sweetness and the hills flow with milk and honey, the valleys are covered over with corn, honey is sucked out of the rock and oil out of the flinty stone ...

When the Cistercian revival of the Benedictines came to Britain, then one of the favoured areas for monastic foundations were the valleys that ran up into the moors, in manors that were *wasta* or where the population could easily be moved away from the sacred precincts. In 1132, for example, both Fountains (Yorkshire Dales) and Rievaulx (North York Moors) were established. Far from flowing with milk and honey, the early Benedictines in the North York Moors thought of them as

> ... a place of horror and vast solitude ... uninhabited for all the centuries back, thick set with thorns, and fit rather to be the lair of wild beasts than the home of human beings.[4]

And this post-Conquest revival took such evaluations as signs they were in the right places. An edict of 1134 required the Cistercian order to reside 'far away from the talk of men', in pursuit of which Yorkshire had seventy monasteries by about 1200. Alongside their material production (wool, iron, lead), however, there was the spiritual myth: their life took place in an allegory of the Temptation in the Desert, in an environment that was as much literary convention and monastic imagination as it was physical hardship. Even so, there might be another view for in the time of Henry I (r. 1100–35), Rievaulx was described as 'providing for the monks a kind of second paradise of wooded delight'. The difference between the Desert and the Delight was of course the result of fruitful labour, an important element in Cistercian ideology.[5]

Later poetry

Whereas in fiction the moorland environment can be used simply as background for a stormy or indeed frightening story (with parallels in painting, we might note), the poet generally focuses directly on the moorland itself and, usually, its people. Indeed unless the place is woven immediately and consciously into the fabric of the poem, there is not much point in mentioning it at all.

Before the nineteenth century, the uplands were in general not much favoured, nor indeed noticed. Searches of on-line poetry collections generally produce little material. The nineteenth century is rather different, with various references to the uplands, in which adjectives like 'sullen', 'gloomy', 'misty' occur frequently. The well-known Welsh poet of the fourteenth century Davydd ap Gwilym is known for his nature poetry but he seems to be a creature of the hillside woods rather than the moors above though 'The Mist' is eloquent on the topic of hill-fog, 'Like a cassock of the grey-black air, a very sheet without an end, the blanket of lowering rain, a black weft from afar, hiding the world'.[6]

One remarkable source comes from the later fourteenth century when *Sir Gawain and the Green Knight* was written and which has detailed descriptions of hunting of deer, wild boar and fox in an upland environment. Some uplands were designated as Forests and Chases, which in both lowland and upland Britain were quite closely managed, with many residents owing duties of service in the forest; some were permanent full-time employees. Their environmental effects were first of all to make sure that no land uses forbidden by forest law encroached upon the feeding and breeding grounds of the deer. This was not usually a problem in the uplands except when cultivation limits were creeping up the hillside as in the twelfth and thirteenth centuries. Their second task was to prevent poaching, see that the habitat was maintained (especially to see that areas for the deer to lay up were present), and to assail predators, especially the wolf. Then the proper hunt formalities could be observed, with the King or his nominee taking the stags and lesser nobility undertaking such tasks as the killing of excess hinds, a process chronicled for Holy Innocents (between Christmas and the New Year) by the author of *Sir Gawain and the Green Knight*, and quoted on page 71. The same literary source details a fox-hunt (almost uniquely for pre-modern times, compared with deer and boar), with Reynard at the last:

... reft of his skin of red

a sight still to be seen on many days in the English winter months.

It would be wrong baldly to oppose lay and monastic views of the upland environment: the latter almost certainly influenced the former and in any case those who compiled any written document were likely producing literature rather than social science. We might though infer that the moors were less favoured than the woods even though they yielded meat, milk, wood, hides, peat for fuel, furze, lead, iron and even a little coal. Somehow the negative strand became dominant so that by Early Modern times, most of the evaluation of the moors was negative.[7] Even as late as 1794, when the re-appraisal of the non-material value of the uplands was under way, William Blake clung to the old views in his *Book of Urizen*. The eponymous subject is of the Dark Power, though it is not very clear whether he is synonymous with it. His habitat, however, is clearly delineated:

> Urizen explored his dens,
> Mountain, moor and wilderness,
> With a globe of fire lighting his journey –
> A fearful journey, annoy'd
> By cruel enormities, forms
> Of life on his forsaken mountains.

Moorlands can reflect both sadness and delight quite readily. Within the one poem of lament at the death of the Welsh bard Hedd Wynn in Flanders in 1917, his elegist R. Williams Parry writes[8] in one verse of a juxtaposition of a Welsh upland and the battlefield grave of his hero:

> Tyner yw'r lleuad heno tros fawnog   Gently the moon climbs above the peat
> Trawsfynydd yn dringo: bogs of Trawsfynydd
> Tithau'n drist a than dy ro,   You are sad beneath the gravel
> Ger y Ffos ddu'n gorffwyso   Near the black trenches lying

This contrasts with the moorland, mountain streams, pastures and 'woodland green' of his homeland. As a pretty sweeping assertion, we might argue that no writer in this century has gone more deeply to the heart of the moorland landscape and environment than the Welsh poet R. S. Thomas, who died in 2000 and who wrote in English. Priest in a succession of rural parishes in Wales, he fixed the bleakness and the unremitting qualities of the scene and its people as well as their more easily admirable qualities. 'The Moor' (1966) puts the positive side of his *poesis*:

> It was like a church to me.
> I entered it on soft foot,
> Breath held like a cap in the hand.[9]

In the second stanza, the appeal becomes emotional enough for tears, stillness, '. . . and the mind's cession/Of its kingdom' But 'there were no prayers said'

and both before and after this poem, Thomas adds the other side: the low productivity, the impoverished life, the afforestation and the emigration, all of
which confirm his view that Wales was being eviscerated by the English. The
agriculture is typified by:

> . . . the lean patch of land,
> Pinned to the hilltop, and the cloudy acres,
> Kept as a sheep-walk.[10]

and afforestation means, '. . . A world that has gone sour/With spruce'.[11] So it
is perhaps no surprise that Thomas comes later (1975) not only to a theolog-
ical aridity but to rejecting the present

> It is the machine wins:
> the land suffers the formication
> of its presence.[12]

His biographer Justin Wintle however detects a third phase, of greater acceptance of the here-and-now in the collections after about 1980.[13] But these are
less frequently land-related than those of the early delight and its accompanying darker side. One rather stark exception is 'Moorland' where he talks of the
hen harrier, which 'quartering the bare earth' is present where the air is rare-
fied 'as in the interior of a cathedral' but where the bird may be '. . . hovering
over the incipient/scream, here a moment, then/not here, like my belief in
God.'[14] which rather painfully internalises for Thomas the two sides of the
moorland environment and landscape.

The other great poet of recent times in this context is Ted Hughes (1930–98)
who is often cosmically distanced in his accounts:

> Heather is listening
> Past hikers, gunshots, picnickers
> For the star-drift of the returning ice[15]

So he is for the long term, for the major process in which environmental forces
can be perceived as:

> To your feet and surf upwards
> In a still, fiery air, hauling the imagination,
> Carrying the larks upward.[16]

So when we come to humans, as in *Remains of Elmet* (1979), the desolation of
an emigrating rural population, as well as industrial decline in the textile
industries of the valleys, leads to a disintegration of a former unity. Now only
the recreational use is worth noticing:

> Time sweetens
> The melting corpses of farms
> The hills' skulls peeled by the dragging climate –
> The arthritic remains
> Of what had been a single strength
> Tumbled apart, forgetting each other –[17]

Hughes reminds us that anything may eventually be submerged in a return of the ice age: we may be living in an interglacial. Nested inside this long perspective, Thomas's contribution insists that if we harbour a romantic view of these hill-lands as places of freedom and of difference from the city, then we are part of a probably ephemeral set of attitudes. Both, however, recall an essential marginality in both ecology and emotion.

To suggest that both Thomas and Hughes have realised the qualities of the moorland environment better than other poets is perhaps a risky claim in view of the need to think of W. H. Auden. He spent boyhood days and holidays in the North Pennines and in 1940 wrote:[18]

> As a locality I love
> Those limestone moors that stretch from BROUGH
> To HEXHAM and the ROMAN WALL,
> There is my symbol of us all

Considering that the poem and the region contain the derelict lead smelt-mill on Bolt's Law near Rookhope (Weardale) where he was

> . . . first aware
> Of self and not-self, Death and Dread . . .

there is a strong case for a major poet at work. But while Auden's early life and work contains many references to the region[19] and while it clearly retained a considerable place in his affections, he lived nowhere near the north of England and the strength of the place appears to fade in the later poetry. One critic at least, though, makes the case that his childhood interest in mining made him place limits on his imagination: an imaginary mine might be drained by an adit or a pump but not by magic. Hence his poetry has a sense of obligation to the real world, a theory of imagination with roots in actuality.[20] In this case, we cannot talk of direct representation of the moorland environment any more, but a sense of relationships deriving from it is suffused through most if not all of his mature work. Such a transformation is, we might argue, demonstrative of Auden's creative stature *vis-a-vis* other poets of the uplands, not that there is any point in league tables. We might argue that Auden comes closest to Bate's Heideggerian interpretation of the biologist, geographer and green activist having narratives of dwelling but that poetry should be a revelation of what it means to dwell on the earth.[21]

There is, of course, poetry after Auden and Hughes. A collection of much upland-related writing in a collection called *Poetry in the Parks*[22] contains moorlands as well as mountains (the parks are the national ones) and many less familiar authors as well as names like Betjeman, Plath and Coventry Patmore. John Ward's 'Roots in Lastingham' ties the present in with his sheep-farming ancestors on Spaunton Moor and echoes many attitudes to the hill lands when his earliest forebear (who died in 1759) lies in the churchyard, 'safe, immune from change'. Near Gunnerside in the Yorkshire Dales, Rodney Pybus in 'Mill Stones' is concerned with the aftermath of lead mining:

> the wind keening over fell tops
> the colour of pewter
> bruised with heather,
> and spiked with young bracken
> that's returning to ground made bald
> and sickened from the flues.

Perhaps the most ambitious poem of recent times however is Alice Oswald's *Dart* which is a long poem tracing the interaction of humans and the River Dart from its boggy source on Dartmoor to the sea, in a way which has led to comparisons with Joyce for the analogy with Anna Livia Plurabelle.[23] There are walkers, bailiffs, poachers, woodsmen, a chambermaid and a river pilot, sewage works and seals. It is Protean and so almost impossible to quote sensibly, noting for instance the abrupt changes of metre which mirror the pool and riffle structure of an upland river; after a lark is said to be spinning one note,

> splitting
> and
> mending
> it
>
> and I find you in the reeds, a trickle coming out of a bank, a foal of a river
>
> one step-width water
> of linked stones
> trills in the stones
> glides in the trills

This level of exploration of human-nature relationships is perhaps taking the role of poetry to a different level altogether as a means of understanding its outer complexity and what Gerard Manley Hopkins called 'inscape'.

Fiction

In the nineteenth century, the most usual statements are of the type found in Emily Brontë in *Wuthering Heights* (1847),

... lying from morning till evening on a bank of heath in the middle of the moors with the bees humming dreamily about among the bloom, and the larks singing high up over head ... and the moors seen at a distance, broken into cool, dusky dells; but close by great swells of long grass undulating in waves to the breeze; and woods and sounding water, and the whole world awake and wild with joy.

and although critics often make a connection between the author's identification with place, the roughness of the story and the harshness of the environment (there are not many days of the quoted type), the abiding impression is of a setting, a piece of scenery, with some pretty marginal characters who have perhaps not made it in the lusher but more competitive lowlands.[24] In Elizabeth Gaskell's *The Moorland Cottage* (1850), premature widowhood forces Mrs Browne and family into a secluded basin surrounded by grassy hills, which is neither cottage nor house nor farm, though there is a cow, a pig and some poultry. To get there, you have to cross a 'common' of unimproved land and see 'the swelling waves' of the horizon: it is a retreat in more than one sense, from which the family's emergence is not always happy. The moorland, its economy, and indeed its natural history are reduced to rather insensitive scenery in both novels. There is some contrast in the story which if searched electronically uses the word 'moorland' the most, R. D. Blackmore's *Lorna Doone. A Romance of Exmoor* (1869, though set in the seventeenth century). Here at least there is some detail about the moor:

But the nights were wonderfully dark, as though with no stars in the heaven; and all day long the mists were rolling upon the hills and down them, as if the whole land were a wash-house. The moorland was full of snipes and teal, and curlews flying and crying, and lapwings flapping heavily, and ravens hovering round dead sheep; yet no redshanks nor dottrell, and scarce any golden plovers (of which we have great store generally) but vast lonely birds, that cried at night, and moved the whole air with their pinions; yet no man ever saw them. It was dismal as well as dangerous now for any man to go fowling (which of late I loved much in the winter) because the fog would come down so thick that the pan of the gun was reeking, and the fowl out of sight ere the powder kindled, and then the sound of the piece was so dead, that the shooter feared harm, and glanced over his shoulder. But the danger of course was far less in this than in losing of the track, and falling into the mires, or over the brim of a precipice. (chapter 12)

The novel is notably linked to many other moorland themes by the way in which the lair of the Doones is hidden by pits, a waterslide and the 'folded shadows' of woodland. Altogether, then, the moorland fastness is linked to the grim and the frightening, although such settings as the Doones' residence owe little to actual topography and more to making a proper setting for 'giants and evil-doers'.[25]

It may simply be coincidence that these three novels were published within two decades of each other (and there may have been many others that have not lasted to come within sight of a superficial scour through the genre) but it does perhaps suggest that, allowing some time for gestation, the period after 1820 was one in which some creative writers of fiction felt that the moorlands were suitable places to be used either as symbolic backgrounds for what they had to say or indeed as dominating settings for a 'Romance'. For Dartmoor, the novelist Eden Philpotts established himself as the regional writer, though for him the moor tended to be a point of vantage or a scene for '. . . set pieces, which can be detached very readily from the body of the narrative', as a 1930s commentator put it.[26] However, Philpotts sets his novel *The Forest on the Hill* around Yarner Wood, an acid oakwood on the edge of Dartmoor and for 2002 English Nature commissioned a dramatisation of the book as a celebration of fifty years of the wood as an NNR.[27] Nonetheless, the context is that other environments, like the cities, the lowlands and the Highlands of Scotland, proved far more attractive to writers like Scott, Hardy, Dickens and Trollope. Thus the writer who was best at getting into the spirit of the land and its people on Dartmoor, Sabine Baring-Gould, is remembered very little for this creative contribution, more for his work as an antiquary, collector of folk-music and natural historian, and perhaps best of all as the author of the hymn *Onward Christian Soldiers*. But he wrote vivid descriptions of moorland topography which underlay a mission which 'avoids the territory of abstraction to reach the territory underfoot, to meet the people on the cliff and on the moor'.[28]

Baring-Gould knew all about the rough side of Dartmoor, and the remnants of a negative view of the hill lands are found even in the twentieth century. Arthur Conan Doyle revived one older assessment in his Sherlock Holmes novel, *The Hound of the Baskervilles* (1912),

> In the distance a grey melancholy hill, with a strange jagged summit, dim and vague in the distance, like some fantastic landscape in a dream . . .
>
> In front of us rose the huge expanse of the moor [with its] grim suggestiveness of the barren waste, the chilly wind and the darkling sky.

Literary evocations of that type (just look again at the adjectives) remind us of assessments of moorland as places to be shunned or at the least thought of as 'sublime' as in a Salvator Rosa oil painting. Perhaps they become rarer as the movements to open these land to more public access get more strident.

To confound any such generalities however, we can point to some recent writing which presents the moorland core of central Wales as part of a dystopia shot through (not an inappropriate verb) with alcohol and other drugs.

Niall Griffiths's novel *Sheepshagger*[29] focuses on Ianto, who comes from Llangurig, which is inland and, 'just farms and mountains'. This neither prevents him from having an extraordinary affinity and identification with nature when he is a small boy, nor keeps him from some very bad experiences which result in premature death. The very rawness of the story and its writing (inevitably it has been seen as a rural *Trainspotting*) do go into the same intellectual terrain as R. S. Thomas, not that he would have been an avid reader of such a novel with such a contemporary vocabulary even though part of Psalm 58 features as one of the epigraphs.

At a more popular level, one moorland area has acted as the focus for a series of popular novels, by the Welsh author Alexander Cordell (1914–97). He came to fame with *Rape of the Fair Country* in 1959 and followed it with several more novels set in the upper parts of the South Wales valleys during the period of industrialisation and the Chartist movement.[30] They are germane in this book in so far as they deal with the area around Nantyglo and Blaenavon: so here is a World Heritage Site with its 'own' novelist. There is little landscape description outside the industrialised areas: the country may be raped but we largely see the industrial end-product. There is a lot of action at night so that there are many vivid glows (people and furnaces) but the effects on land, water and air are not part of the dramatic fabric. The 'prequel' book *A Proud and Savage Land* (1969) has some far background in talking of the land as once 'a bedspread of forests' cleared in the Iron Age and kept clear by great herds of goats and wandering cattle, leaving the mountains bald where 'even the lovely Blorenge looked like a man after a pudding-basin haircut', though later it is 'decorated to the waist in purple heather'. The outcome seems to be that these moorlands lack the variety of lower countryside to evoke extended descriptions within works of fiction.

Topography and travel

One of the 'classical' travel writers of Great Britain was Celia Fiennes, who brushed the moorlands in her journeys on horseback of 1697 and 1698. The Peak District, the Yorkshire Dales and the area along Hadrian's Wall were all visited, as were Bodmin Moor and the fringes of Dartmoor. At Okehampton, she noticed how rapidly the rivers rose after 'one night and dayes raine' and came up to the arches of the stone bridges; not a lot of change there. She was not in general very happy with any hilly areas:

> All Derbyshire is full of steep hills, and nothing but the peakes of hills as thick one by another is seen in most of the County which are very steepe which makes travelling tedious, and the miles long, you see neither hedge nor tree but only low drye stone walls round some ground, else its only hills and dales as thick as you can imagine, but tho' the surface of the earth looks barren yet these hills are impregnated with rich Marbles Stones Metals Iron and Copper and Coale mines in their bowells . . .

Most of her encounters with upland country evoke a negative opinion or two, though nothing really strong: she was more temperate than some of the pre-Romantic writers.[31] An early seventeenth century visitor to Marsden in the Pennines of West Yorkshire found its 'rude mountains' were 'vast, stonie, moorish and barren' and the land was 'weet, sobbed and rotten' where 'sobbed' meant soggy or soaked.[32]

In the eighteenth century, only a few writers remarked on moorland: Walter White, walking on Mickle Fell in the North Pennines lay down to enjoy the solitude but found, 'the silence is oppressive – almost awful.' But the ability to commune with the outdoors and nature far outweighed artificial pleasures, not least perhaps because they '... could be had for nothing ...'.[33] Travelling from Carlisle to Newcastle along the line of Hadrian's Wall, John Tracy Atkyns was not happy:

> For fifteen miles together we travell'd through one of the most barren places in England, which they call the Fells, not a tree, nor even a Stump to be seen, now & then a straggling sheep appears, that with the utmost difficulty keeps itself from starving, perhaps once in a Year an unthinking Crow or Two fly over it ... no Modern would ever think of taking up his Habitation in such a Desart ...[34]

Nevertheless, change was on the way. By 1770, tours of picturesque districts were all the rage and infected even the detached appraisers of the economy. Young's *Tour* of 1771 divides some of its pages into an agricultural and accounting narrative with an extensive asterisked footnote (taking half of ten pages or so) of purely aesthetic judgement on the qualities of the scenery, some of which near Barnard Castle was, '... most exquisitely picturesque.'[35] In the south-west, Marshall took delight in the wildness of Dartmoor, calling it a 'grand' composition and Vancouver likewise talked of its 'bold majestic grandeur' but they were on the hinge of new assessments, for in 1795, a Parliamentary Select Committee thought that having lands in common was to be 'derived from that barbarous State of Society when Men ... had only just tasted the Advantages to be reaped from the Cultivation of the Earth', a remark still to be found *mutatis mutandis* in some archaeology textbooks today.[36] In general, lands which became profit-making attracted more interest, just as better travel conditions allowed an aesthetic interest which could easily be abandoned when the rain and wind got too pervasive.

Wales, not surprisingly, was a prime candidate for re-assessment of its landscape. Until the 1770s, the rare visitors expected – and hence saw – topographic perils and climatic hazards. Jane Zaring points out that for most of the eighteenth century Wales featured little in the travel accounts printed in the *Gentleman's Magazine*, whereas Africa and the Americas were staple fare.[37] When lowland England was the norm for beauty, then upland Wales held no attractions: Defoe was happy to leave the barren mountains for areas like the

Severn valley in Montgomeryshire, which he considered the most attractive part of that county. This changed in the 1780s and 1790s when Wales featured in most issues of the magazine since readers were now urged to contemplate ideas of an elevating, awful and magnificent kind. Thus, for example, the waterfalls of the Principality provoked a cascade of topographical description, and received an awesome accolade of approval from William Wordsworth.

Though the moors of the Berwyn Mountains might be deemed to be full of 'terrific chaos', it becomes clear that it was mostly the mountain areas with their crags and water chutes that were emphasised and that moorland topography did not fit too well into the model; the great expanses of the rounded hill and the smooth skyline did not attract attention. In that category, they accompanied many other phenomena that the travellers ignored, such as the progress of Welsh agriculture and the onset of industry. Romantic charcoal burners might be mentioned but mines, quarrying and newly enclosed pastures with commercial varieties of grass and carefully chosen breeds of cattle and sheep were not; likewise the poor rate recipients who grew in number by about twenty-five times during the years of most rapid enclosure (1760–90) in the upland parishes were seen through spectacles that were as distorting as any Claude glass. In this, there is a reflection of a widespread contrast between the social reality of the countryside and its idealised representation.[38] Little of this however is carried forward into the growing mythology which claimed that rural England might represent the whole nation; in large measure it was the land which was obviously worked by its inhabitants and marked by them that undertook this role. Equally obviously, this was likely to exclude the apparently natural moorlands.[39]

One aspect of the nineteenth century moorland scene was the presence of industry within the moorland regions, which is often overlooked or deliberately omitted by some of the more rural-romantic writers of the period. It was the great realist Elizabeth Gaskell when visiting Haworth to work on her *Life of Charlotte Brontë* who noted that the valleys nearby were lightless with the smoke from the factories and that the water-courses provided power to industry. She noted, too, the patches of oats which were the chief cereal at altitude, and then remarked that the moors themselves were either

> . . . grand, from the ideas of solitude and loneliness which they suggest, or oppressive from the feeling . . . of being pent-up by some monotonous and illimitable barrier, according to the mood of mind in which the spectator may be.

She must, too, have taken note of Charlotte's reminder that Haworth was ringed with mills as well as moors and was also impressed, apparently, with Charlotte's ability to read the signs of the sky.[40] So here at any rate some of the reality of the nineteenth century scene was being absorbed, including the ill-health and poverty of many of the inhabitants of this area marginal to a rapidly industrialising region. The later history of the Lancashire and Yorkshire Pennines was, to perhaps its most famous son, all about the connection between the land ('miles and miles of ling and bog and black rock') and cultural identity in terms of the cotton and wool trades, Rugby League football and performances of Handel's *Messiah*. Who else but J. B. Priestley?[41]

One interesting traveller in the Pennines was A. Wainwright, best known for his guides to the Lakeland fells. In 1938 he walked from Settle to the Roman Wall on the east flank and then back again down the west, staying at B & B and inns along the way.[42] The weather was mostly bad, he had a cold (and seemingly a very small number of handkerchiefs) and was in a bad state of sexual frustration. He managed, nevertheless, to chronicle most of the standard attitudes to moorland, whereby they are at once paradisiacal: on a green road in a limestone country, the walker joins the sheep in Elysium. They are, as well, re-creational:

> The hills have a power to soothe and heal which is their very own. No man ever sat alone on the top of a hill and planned a murder or a robbery, and no man ever came down from the hills without feeling in some way refreshed, and the better for his experience.

Yet at Rookhope,

> ... the bleak moors crowd round to witness the birth of the stream, and shut out the light of day. These gloomy hills are trackless, and only the cry of the plover disturbs the silence.

But 'the sight of heather to me is like the sound of a band to marching men: it encourages and stimulates energy.' A brave simile, perhaps influenced by the newspapers and radio reports of the summer of 1938, but clearly a lasting one given the attachment to heather moorland still vibrant at the end of the century. Even moors with the 'desolation of death' were teeming with life if you laid down and listened to the insects. Wainwright seemed to find the Pennines in no way inferior to his beloved Lake District and has left us with a series of verbal sketches of places (not very well done) together with an accurate account of his interior attitudes to the moorland environments.

Examination of travel writing about Wales during the period between 1918 and 1950 suggests that nature is somehow subordinate to the construction of a Celtic cultural identity. This not only has historic resonances in informing the visitor about what can be seen and experienced but also can be taken up into today's construction of Welsh nationhood.[43] Environment does not seem to be particularly important in this construction, though for some writers and scholars of those years, remote upland Wales might be a refuge and indeed reservoir of old ways which were superior to the Anglicisation that had been enforced: another example perhaps of a moral geography residing in a set of environments, no matter what their ecology or their environmental history.

THE VISUAL IMAGINATION

All the usual visual forms of expression have been attracted by these land-scapes: drawings, paintings in oils and other media and photography are found in a shop in almost every village in the moorland areas, though the canon of paintings in the local city art gallery is less likely to contain examples. The landscape type seems not to have attracted a 'top name' in the way the sea, for example, is closely associated with J. M. W. Turner.

Painting and drawing

In the seventeenth century, interest in landscape depiction was growing. Treatises were written on painting and limning and tastes were beginning to crystallise out. There were various types of ideal landscape, of which the 'prospect' was a relevant category, as was the sporting scene. But even the prospects were often of cities in a rural setting: the wild itself was only a suitable subject if Italy or the Alps were in view. Most of them, too, were done on the Continent or followed Continental models and it was only in the succeeding century that a distinctive English (sic) taste was developed. The typical artist might be Jan Siberechts, who came to England in 1672 and stayed for the rest of his life, dying in, probably, 1703. He appealed to the newly prosperous merchant class as well as to established landowners but does not exhibit the Christian optimism about the land which made some of his contemporaries align themselves with a mood of the time.[44]

Changes in attitude in the eighteenth century were most obviously manifest in the way upland areas became subjects for depiction in paintings and engravings. Many of the latter formed the illustrations to books which chronicled travels or were catalogues of antiquities and so writing about these areas became enmeshed in the new artistic consciousness. The most basic reasons for this new awareness lie outside the present discussion[45] but come to us in the search for landscapes which might be seen to conform to models of either the 'beautiful' of which the model was Claude Lorraine (1600–82)[46] or the 'sublime' which followed Salvator Rosa (1615–73).[47] A third category, the 'picturesque' was added, given impetus by the advocacy of William Gilpin (1724–1804). The gentry were great consumers of these forms of art, and perhaps especially of the picturesque. Since all art is only in a limited sense mimetic and suspension of disbelief is never complete, there always exists the possibility that the assertions of writers and patterns of painters were to some extent ironic: the creators knew that these patterns were fictive and that they were thus alienated from their true ecological linkages. In that sense they were just versions of the pastoral rather than deeply connected to the land whose story they told: a charge somewhat surprisingly directed at William

Wordsworth by some critics.[48] This is to some extent confirmed by analyses that assert the primacy of learning how to 'understand' landscape scenes so as to make them amenable to shared verbalised reaction: the end-point of landscape art (say several commentators) is words.[49] The eighteenth century can be seen as a time when this ground was being prepared with the development of a cult of the wild and the variety of nature, the expansion of travel, the growth of natural science (especially geology, botany and zoology) and improvement movements in the use of land. In about 1748, for example, a Captain Burt tried to scale Ben Nevis as part of a discovery of the hills under the nominal command of General Wade, and in 1726, Daniel Defoe had gone up the Cheviot, though he went on horseback except for the last 100 feet. The second half of the eighteenth century saw a flood of re-appraisal of all the hills, albeit it led by the appreciation of the Lake District rather than a moorland area.[50] In all this flurry of new visions, the connections to Lorraine and Rosa ensured that landscape depiction was usually something that looked back rather than to the present, a characteristic that endured at least until the Impressionists but in some cases well beyond them.

The relevance is partly simple: the uplands were likely to produce scenes which conformed to the fashionable modes of art or expression, even though pastoral lowlands were a favourite subject. Within the uplands, however, mountains, waterfalls and craggy ridges were much more likely to appeal than the moorlands. Inspection of many of the books and pictures of the period 1750–1820, for instance, reveals relatively few instances of moors as such. One contemporary of Wordsworth and Turner, James Plumtre, admitted that near Settle 'the lines of ye country [are] very fine' but in general disliked open landscapes without trees unless the terrain was actually mountainous.[51] He was just one of a class of travellers in Yorkshire and Co Durham, for example, who wanted mountains, waterfalls and surging rivers like the Tees (so High Force was very popular) but who said and drew little to do with the great stretches of moorland.[52] Nevertheless, the energy of the Picturesque was capable of penetrating Romanticism to provide a widespread and popular aesthetic appeal which is seen by some interpreters as being at heart a form of consumerism but one which has given us what we now call fine art.[53] The appeal of the uplands in particular falls into the type of landscape politics discussed by John Barrell when he suggests that in the eighteenth century the lower classes could observe only the objects in front of them, but the upper classes had a wider view and hence the ability to abstract generalities. This quality made them fit to govern since they could take the wider and more disinterested view. This clearly applied with special force to the owners of landed property who were both the detached and engaged leaders of the nation, as well as being allowed to shoot game if they had an income of over £100 a year from a freehold estate.[54]

Peter Howard's quantitative exploration of which landscapes were actually painted suggests that moorland was 'the great discovery' of what he labels the

'Heroic Period' of 1870–1910. It was a taste, he thinks, which developed very rapidly in the last quarter of the nineteenth century and was distinct from the previous fondness for mountains.[55] On Dartmoor, for example, a scene described as dreary in 1800 was reviewed in adulatory terms in 1885 and the local artist William Widgery worked his way up the River Lyd as the fashions changed, getting to the moor by about 1860. Dartmoor later became an exemplar of the idea that watercolour was especially good for moorland: books of travel and topography might have photographs of lowland Devon but move into watercolour for the moor.[56] Between 1950 and 1980, of eighty pictures of Yorkshire landscapes, twenty-seven were of dales, fells and moors, thus replacing the nineteenth century concentration on the Tees, and on Barnard Castle in Co Durham. One nineteenth century attraction of the moors was their dreariness: there was a moral to the landscape even if there were no people in view, but in 1919–50 the bleakness was often replaced by the brighter colours of heather and gorse. This might be read as seeing the moorlands echoing the dull round of industrialism in the first instance, but contributing to the pressure for landscape protection in later years.

In the nineteenth century much of the landscape art was journeyman work in the sense that it was commissioned by a landowner, or was designed to be turned into engravings to accompany a volume of topography, or simply to sell to an emerging middle class. Names now reverenced, like J. M. W. Turner (1775–1851) or John Sell Cotman (1782–1842), were largely engaged in making a living from a talent rather than setting aesthetic standards for the ages. Thus it was the demanded scene that occupied their time, with less obvious subjects confined to sketches which might or might not be worked up into a full watercolour or an oil. One result is that for many artists of the period, moorlands were most likely to be background rather than the subject itself. This latter was often dramatic, as in the eighteenth century, but the softer lines of moorlands would sometimes lend themselves to providing a contrast to a crag (preferably beetling) or a waterfall or a ruin (Plate 7.1). In this respect, the tradition was long, for landscape in art has often been an ornamental extra or *parergon*,[57] and, further, a code of looking has to be developed in which moorland has the right 'place' if it is to excite the correct responses, including the movements of the wallet.

Though never ignoring these contextual considerations, it is interesting to look at some pictures themselves.* Many books of British landscape art have been published and most native galleries have some work on the British landscape. Perhaps the first striking feature is that not many of them are 'about' moorlands, as we might expect from the above discussion, but our examples will at the very least have it as a prominent *parergon*.

To start with an early example: Jan Siberechts' '*View of Beeley, near*

---

* It is too expensive to reproduce many paintings, so the discussion will have to be conducted via a few highly selected examples, with all the dangers of so small a sample.

PLATE 7.1 *'Gibside from the South', by J. M. W. Turner. A watercolour of 1817 set in Co Durham, with the estate of church, house and column visible on the mid-ground plateau. Even those grounds seem barely demarcated and elsewhere there is an absence of enclosures. This may be an attempt to make a wilder and more mountain-like scene, as happens in some of Turner's other pictures in moorland settings.*
Source: Reproduced by permission of the Bowes Museum Company, Ltd.

Chatsworth' in 1694. This was part of a set of studies which were to result in a view of Chatsworth House but in this case was of land not yet acquired by the Devonshires. and not fully enclosed until 1711. Because of the bird's eye view of the scene (one of his trade-marks) the topography has a highly realistic air – just possibly a little mountainous – in the way of his forebears from the Low Countries. The landscape units are entirely recognisable: the enclosed lands, including some rectangular fields that are clearly the work of Enclosure Commissioners, a deciduous wood on a shadowy slope, the remains of wood-land on the opposite slope, an open hilltop and the prospect of wilder terrain beyond. The watercolour is a very faithful, almost photographic, depiction, right down to the stooks in the harvested fields. There is a similar attention to realistic detail in J. M. W. Turner's *'Grouse shooting: Beamsley Beacon'* of 1816 (Plate 4.7). At that time, grouse were shot from behind after being flushed by dogs and so the gathering is small: ponies to bring the men up from the valleys, a couple of guns and their pointers, all subordinated to the moorland except for one man's head.[58] The moor itself is equally realistic in the sense that the pool, the small outcrop of rock and the heather could come from a photo-graph. We can see that the heather is long but not 'leggy', for instance, and that it dominates the middle ground as well as the foreground. The far back-ground, as in so many nineteenth century paintings of uplands tends to be peaky, in contrast to the enclosing swells of the writers. In several lineaments, it is entirely consonant with Peter de Wint's watercolour of c. 1812, '*Yorkshire Fells*'. The contrast between grass and heather areas, the texture of the sheep-trod, and the sheep themselves, and the boulders, are as it were direct from the scene. Even the smoke hints at a more intensive land use in the very sheltered valley: that of settlement or perhaps a small industrial plant. It is very largely an image of stillness. It harks back to an 1802 picture of Wharfedale ('*Storiths Heights, Wharfedale*') by Thomas Girtin in which horizontal summit-lines are dominant, and which are broken only by the bulk of some tree-canopies but reinforced by the field boundaries lower down the hillsides.[59] The contrast then comes with those pictures in which the artist has conformed to the taste for the sublime in upland terrain and produced mountains out of moor hills. In the case of Turner, this is presumably affected by the intended use of some of the work as engravings to illustrate books of travels; with the absence of colour, then amplitude of relief is needed to maintain a dramatic effect. A good case is his watercolour of Llanthony Abbey, where the standard element of a ruined abbey (rather less dilapidated than it became) is one focus but in which the virtually horizontal skyline of the hill now carrying the Offa's Dyke Path (which does not have a modern name but of which Loxidge Tump is part) becomes something virtually Alpine. The same can be seen in his rendering of Semer Water in Wensleydale, though his '*Crook of Lune*' has flatter skylines. Not to be outdone, John Sell Cotman turned Gormire (between Thirsk and Helmsley) into a sort of mountain lake in 1802 ('*Gormire Lake, Yorkshire*') by exaggerating the vertical immensely and generally stretching out the land-scape so that a small rather intimate little lake behind a landslip is instead something of a Rydal Water size. The granddaddy of all these inflations is John Martin's oil '*The Plains of Heaven*' (1851–3), in which (so it is said) the view ascending from his native Haydon Bridge in the Tyne Valley to Allendale in the North Pennines becomes not only a beatific vision but a highly mountain-ous one as well (Plate 7.2). There are no 'real' people: it looks as if his rainy picture of '*The Deluge*' (1834) has borne them all away to the other place. Wilton says of Turner that his sketchbooks testify to his own lifelong attention to the literal truth of nature as the foundation of his art;[60] the meaning of 'literal' here is perhaps more post-modern than is commonly accepted. The Turner-Cotman tradition has continued unabated, with examples of moor-land depictions in watercolours present in many galleries' regional collections. The Mercer Gallery in Harrogate, for example, has examples of the work of David Thomas Rose (1871–1964), an engineer on the Scar House Dam in the Yorkshire Dales who produced moorland views from that region and both in the late nineteenth century and in the 1940s Beamsley Beacon became, as with Turner, a suitable subject. In a similar way, the Cartwright Hall Gallery in Bradford has work from the 1970s by Joe Pighills which continues the bleak-ness tradition.

A post-WWII landscape artist of distinct style is found in Edward Burra (1902–76). His primary interest was human energy and character but after

PLATE 7.2 *'The Plains of Heaven' by John Martin, an oil of 1851–3. Martin was from the Tyne Valley and it is said that this picture took its topography from the view towards Allendale from Haydon Bridge. If so, then it is an extreme example of the 'mountainising' tendency of nineteenth-century painters in moorland environments. There are no lakes, either.*
Source: © Tate, London, 2003.

1947 he produced a large number of watercolours of landscapes. Some were of lowland England and some of generic scenes of main roads, motorways and other industrial developments. But following in Romantic tradition, he also looked for sparsely inhabited places, with people in them only if they were workers. The focus is often on the remote at the expense of the foreground and is, says Andrew Causey in his *Complete Catalogue*,[61] 'commonly sunless and uningratiating.' The landscapes are certainly not 'picturesque' and the humanised world in the uplands can be seen as a series of enclaves in the middle of wide 'natural' horizons. For Burra, a landscape might be '. . . a place of last resort for the disenchanted' and the moorland pictures distil that essence.

His watercolour of the Hole of Horcum[62] typifies these appraisals. The deep bowl is seen as a huge bite out of an apple. The apple is very dark-skinned, as an allusion to the heather moorland and captures exactly the flat plateau surfaces of the North York Moors. The sides and floor of the bowl are mostly terracotta, with a few isolated shrubs picked out in white; the dominant colour reminds us perhaps of the bracken in autumn or even of the rusty colours of some podsols. Notably, all his moorland landscapes are faithful to the outlines of the topography.

Nobody argues that it is the duty of a painting to be 'realistic' in the sense of being like a photograph, though it is interesting that for example the Turner paintings of Rievaulx, Gibside (Co Durham) and Crook of Lune suggest that the amount of enclosed land was very small compared with today's views. They do show quite vividly their indebtedness to earlier models which were in the traditions of the sublime and the picturesque, with some emphasis on the former in the characterisation of upland Britain. Thus in order to be understood, evaluated (and, presumably, sold) they have to be placed within a school or tradition. The attachment of a label ('Norwich School', 'Impressionist') is a great aid to situating a painting, which in turn helps not only its sales but the verbal **discourses** of which it has to form a part if it is to communicate to an audience at all. The transformations of some of the dull planes which Gilpin thought so worthless into mountains may be part of the cultural input into today's evaluations of moorland environments and landscapes, just as the more sober delineations (Plate 7.3) constitute a contrast and hence create a tension.

Sound into vision

It may seem perverse to include music here, but most 'moorland' music is quite visual in its referents. No account could fail to discuss the Yorkshire

PLATE 7.3 *'Cement Works, No 2' by Eric Ravilous, watercolour and pencil, 1935. A modest-sized limestone operation in a moorland setting, probably the Peak District. Not only is a local tramway part of the plant but the key nature of the railway connection (coal in, cement out) can be glimpsed.*
Source: © Estate of Eric Ravilous 2003. All rights reserved, DACS. Reproduced by permission of Sheffield Galleries and Museums Trust.

song, 'On Ilkely Moor ba'tat': the Number One anthem of walkers and hikers in upland Britain until perhaps the advent of Rock and Roll. Its basic environmental message is one of the recycling of animal tissues (via pneumonia, death, worms and ducks) and so highly appropriate to the bracing weather of upland England. Concert music with a moorland theme is not very common. Gustav Holst wrote 'A Moorside Suite' for the National Brass Band Championships in 1928 and it perhaps reflects that purpose rather more than the evocation of landscape. Frederick Delius was born in Bradford and as a boy rode on ponies over the Ilkley. One result is 'North Country Sketches' for orchestra, the last movement of which is 'The March of Spring. Woodlands, Meadows and Silent Moors', which latter suggests memory (he lived in France) rather than wind and sheep. In general, moorlands seem to have had a restricted attraction for composers of concert music and that general assessment continues today.[63]

Photography

In 1870, it was difficult to 'take' a landscape since the equipment was so bulky and expensive. After 1888, hand-held cameras were mass produced and new meanings of landscape were made possible by the new mobilities of technology and social class. Pictures were to save the rural Eden of England from encroaching suburbia and industrial growth and in 1914–18 there was an identification of rural landscape with patriotism. Yet the uplands seem marginal to this effort: a 1915 Batsford book gets as far as 'the sheepfold on the hill' but it seems more like Sussex than Westmorland.[64]

In trying to go beyond the simple affective portrait, the apogee of moorland photography so far has been reached by Fay Godwin. Her book *Land* has only a few pictures relevant to the present work compared with *Remains of Elmet*[65] but for drama these two sources are incomparable. As the introduction to *Land* (by Ian Jeffrey, at p. xxiii) says,

> Nature is wild, and even belligerent; and human making, which tends towards the fastidious, looks small and frail by comparison. Fay Godwin's version of humanity in its setting emphasises the awkwardness of the relationship and its one-sidedness: humanity struggles to keep that unmanageable other at bay, and sometimes its efforts are no more than amateurish.

In one such photograph a human figure and his dog in front of the broken-down wall face a virtual infinity of open moor in which the light brings out a series of striations which look as if a glacier had only just retreated. The foreground wall and the enclosure in the left of the middle ground signify a human ambition to 'reclaim' the wilder terrain but it is clear that success may not be permanent. The ground cover inside and outside the enclosure, for instance, looks very similar and we imagine that the sheep (the sole evidence of current economic activity) have equal access to both. In the accompanying poem, Ted Hughes's opening words seem to reverberate from the picture:

> Wind-shepherds
> Play the reeds of desolation

This land is positively humanised compared with Godwin's pictures of upland Wales in *Land*: '*Maenserth standing stone*' (p. 67) for example, is dominated by unenclosed and more or less uniform *Nardus* grassland with randomly scattered sheep. A large cumulus cloud has more to offer by way of interest. '*Marker stone, Old Harlech to London Road*' (p. 69) lacks even the cloudscape. So the pre-romantic vision of marginality and even the desolation of the pre-eighteenth-century travellers is resurrected. It is perhaps rather stronger than the 'warning melancholy' of which the Introduction speaks (p. xiii) and is greatly emphasised by her comments on property rights in the photograph of a 'private' sign in front of part of the Forest of Bowland.

As a footnote, we might mention aerial photography as a way of seeing. In the lowlands at least, it has been said that the aerial view,

> heightened the sense of the distinction being made between the clear and synoptic vision of the preservationists and the jumbled, haphazard, narrow and short-sighted view of those building 'casually' over the countryside.

The photograph from the aircraft thus

> . . . best conveys the preservationist viewpoint. The preservationists take neither a specifically rural nor a specifically urban viewpoint, but rather seek a position of overview and detached, clear, 'enlightened' judgement.

This was in the inter-war period and about the lowland south.[66] Does the same apply to uplands and to today's changes? The windfarm might just reverse the judgements.

MIXED MEDIA: MIXED MESSAGES

The twentieth century saw the growth of new media such as film and television, in which 'fact' and fiction (i.e., documentary and drama) are both present, with the border between the two often blurred. There is no way in which the importance of television can be understated since for many people it is their main source of knowledge other than other people, whereas the feature film shown first in the cinema reaches a different and smaller audience. The documentary film is now confined to TV and video: its place between the newsreel and the main feature, remembered by a few of us, has been lost for ever.

Television

Moorland imagery is often seen on television. In the advertisements, it sometimes figures in nostalgia-laden plugs for 'real' nutrition, such as 'traditional' brown bread. More often, it is a backdrop for a motor car. Though such places might seem the natural habitat of the new breed of 4×4 owners, a mainstream saloon is more likely to be seen eating up the miles, effortlessly sailing round the curves and up the sides of the valleys. Wild the landscape may be, but this enclosed environment (not infrequently with air-conditioning) enables the driver to meet a lover or at least to get somewhere else, fast. The message seems to be that moors are wonderful places to have been in. In the programmes set in such landscapes, the predominant theme of the most popular is nostalgia, where these regions are seen as a repository of 'characters' who are that bit behind even the times portrayed in the series. Those set in the present day share some of those characteristics, especially the warm glow of a particular type of place, even if this is an interior like that of the Tan Hill Inn (526 m ASL) made more attractive by double glazing. It is in both a kind of *paysage moralisée*: in this case the moorland landscape and its valley inhabitants are representatives of an enviable way of life. Harder than the cities, of course, but more gathered: more *gemeinschaft* than *gesellschaft*.[67] Yet, the older views sometimes emerge like nunataks: the 'Moors Murderer' whose fate had regularly exercised the judiciary, the Home Secretary and the tabloid press would surely have a lower profile if the bodies of the children had been buried on farmland or in sand-dunes. The projection of an absolute and irredeemable evil has surely been intensified by the pictures of the bleak moorlands where searches have been carried out.

Cinema

A relevant example is of the translation of a novel into film in *On the Black Hill*, from the book by Bruce Chatwin (1982).[68] This is a four-generations story of a farm in a valley in the Black Mountains of Monmouthshire, though one which has plough land as well as hill sheep. There are scenes of upland sheep-herding but in general the focus is on the farm and not its enframing landscape. There is a pointer to a reading of the film in the opening shots of an open and arid-looking moor over which is superimposed WALES which then pans eastwards to a lush valley with a clustered village and church, from which the hymn that is heard is defiantly Welsh. This scene however is superimposed with ENGLAND. De-saturated colour is mostly used to enhance the 'distance' effect. There seems to be a case for saying that the moorlands are simply background and little woven directly into the tale, a feature of some of the novels discussed but not the case with poetry. A minor point that might bear a more thorough examination.

HEAVY FALLOUT?

The idea of the moorlands thus rains out upon the rest of the nation. What they contribute is first of all an image of wildness and of dominance by nature. A review of the National Parks in 1991, dealing mostly with uplands (but both mountain and moorland) talked of the essence of such places as lying '. . . in the striking quality and remoteness of much of this scenery, the harmony between man and nature that it displays, and the opportunities that it offers for suitable forms of recreation.'[69] To some this even merits the (undefined) term 'wilderness' as an extreme contrast to the cities and the 'new prairies' of East Anglian cereal growing. The main thrust of protection for the moorland landscape has come from those to whom the term 'wilderness' has a meaning in the language of spiritual cocoons which keep out the works of humans or which are a context for individual liberty away from 'the herd'. The landscape has therefore only to look 'natural', for it is a kind of blank canvas which is being sought. Hence, wildness, openness, asymmetry and homogeneity, height, the freedom to wander at will and the apparent absence of human handiwork are all central to the experience.[70] A long quote in Carolyn Harrison's book on recreation sums this up very well and just one piece will have to suffice here. 'Elaine' is talking about moorland:

> . . . you feel free that you have access to that. And um, that you can actu-ally just go – you can go right into it! It's like going into the picture . . . you can walk into the moorland. You can actually go in and you can possess it.[71]

Their second endowment is that of image: of a real (or even ideal) commu-nity. Recent, if ephemeral-seeming, instances were mentioned in the televi-sion commentary, above. Some very serious people, however, have elevated the hill lands to a virtually mystic status. This was especially the case of Wales, where Sir George Stapledon (1882–1960) championed the cause of more rational land use in the uplands. He wanted to see greater diversity of produc-tion systems and in particular the replacement of most bracken land with rye grass and clover mixtures. He opposed afforestation with conifers, chauffeurs and examination papers. But beyond these workaday concerns, he was an advocate of a holism in which the land was a source of spiritual qualities and the sustenance of the farm distilled this. He was a passionate champion of rural life as it might be, even to the point of wanting to infuse urban dwellers with some good country genes. Moore-Colyer argues that the survival, '. . . into the 1990s of Welsh upland communities with their distinctive traditions and culture . . .' owes much to the sagacity and vision of Stapledon, so that those unenamoured with such a situation (like R. S Thomas, perhaps) would rather he had been less successful.[72] The geographer H. J. Fleure (1877–1969) who also worked at Aberystwyth, was likewise convinced that Welsh people exhibited genetic continuities with the past, though he rejected any of the inter-war conclusions of a racist type that were then available. His concern for the land and the Welsh way of life, though, can be seen as rooted in a moral conviction that the rural districts were a necessary challenge to materialism and industrial values. Nevertheless, Fleure hoped that afforestation, fisheries and hydro-power might transform rural Wales into a thriving set of communities where an older moral order is re-founded on new technology.[73] A kind of echo is heard in today's espousal of ICT as a new saviour of remoter places, though we need to ask whether it is the locally-raised young who are keen to adopt e-work, or whether it is the refugees from the towns.

The type of imaginative appraisals which have been briefly mentioned here are not separate from the more detached findings of science and history which have dominated this volume. They feed into, for example, National Park policies, decision-making after the Foot and Mouth Disease outbreak in 2001 and the views of organisations like the Ramblers' Association when in pressurising mood. Thus any final part of the book must contain both even though the use of detached knowledge has been the commoner and indeed the stronger.

NOTES AND REFERENCES

1. D. Wright (ed.), *Beowulf,* Harmondsworth: Penguin Books, 1957, 60.
2. J. Neville, *Representations of the Natural World in Old English Poetry*, Cambridge: Cambridge University Press, 1999, Cambridge Studies in Anglo-Saxon England no. 27.
3. K. Jackson, *Studies in Early Celtic Nature Poetry,* Felinfach, Lampeter: Llanerch Publishers, 1995. (Facsimile reprint of original from Cambridge University Press, 1935)
4. C. Platt, *The Abbeys and Priories of Medieval England,* London: Secker & Warburg, 1984.
5. For more examples of rhetoric surrounding the Cistercians in Yorkshire, see N. J. Menuge, 'The foundation myth: some Yorkshire monasteries and the landscape agenda'. *Landscapes* 1, 2000, 22 37.
6. Adapted from the translation by K. J. Jackson, *A Celtic Miscellany,* Harmondsworth: Penguin Books, 1951, 78.
7. Eccentrics always evinced counter-trends, naturally, as with Father Ignatius (Revd J. L. Lyne 1837–1908) who built a monastery at Capel-y-Ffyn in the Black Mountains, of which Francis Kilvert recorded in his famous diary:

   > The monks as usual had chosen a pretty and pleasant place on a fine slope at the foot of the mountain where there was good soil and plenty of good water, a trout stream and sand for mortar.

8. R. Williams Parry, 'Hedd Wyn' in *Cerddi'r Bugail* [*The Shepherd's Songs*], 1931. I am grateful to Dr Lewis Davies for the translation from the Welsh. He notes that the repetitive 'gadael' in the second quoted verse could also be translated as 'quitting' or 'departed from' to give the extra emphasis. The verse form is that of the *englyn,* with strict rules as to metre, rhyme and alliteration.
9. 'The moor' in *Pietà*, London: Rupert Hart-Davis, 1966.

10. 'The airy tomb', in *Song at the Year's Turning*, London: Hart-Davis, 1955.
11. 'Afforestation', in *The Bread of Truth*, London: Hart-Davis, 1963.
12. 'Welsh summer', in *Laboratories of the Spirit*, London: Macmillan, 1975. 'Formication' must not be mis-read: 'a sensation as of ants crawling over the skin'. It seems not to go quite deep enough.
13. J. Wintle, *Furious Interiors. Wales, R. S. Thomas and God*, London: Flamingo, 1997.
14. 'Moorland', in *Experimenting with an Amen*, London: Macmillan, 1986.
15. 'Heather', in *Remains of Elmet* (with photographs by Fay Godwin), London and Boston: Faber and Faber, 1979.
16. 'Pennines in April', in *Lupercal*, London: Faber and Faber, 1960.
17. 'The sheep went on being dead', in *Remains of Elmet*, London and Boston: Faber and Faber, 1979. Photographs by Fay Godwin.
18. In *New Year Letter*, London: Faber and Faber, 1941.
19. A. Myers and R. Forsythe, *W. H. Auden. Pennine Poet*. Nenthead: North Pennines Heritage Trust Publication no. 7, 1999.
20. E. Callan, *Auden: A Carnival of Intellect*, Oxford and New York: Oxford University Press, 1983, ch. II.
21. J. Bate, *The Song of the Earth*, London: Picador Books, 2000.
22. W. Bardsley (ed.), *Poetry in the Parks. A Celebration of the National Parks of England and Wales in Poems and Photographs*, Wilmslow: Sigma Books, 2000.
23. A. Oswald, *Dart*, London: Faber, 2002.
24. E. Helsingor, *Rural Scenes and National Representation. Britain, 1815–1850*, Princeton, NJ: Princeton University Press, 1977, ch. 6.
25. S. D. Trezise, *The West Country as a Literary Invention*, Exeter: University of Exeter Press, 2000, p. 124. He thinks that Lorna's bower is 'womb-like' and owes a lot to 'erotic fantasy'. It is not very often that sex gets an airing in the context of moorland lit. crit.
26. G. Sheldon, 'Devonshire scenery as depicted in English prose literature', *Transactions of the Devon Association* 63, 1931, 283–91.
27. 'A lesson in social, cultural and natural history', *English Nature Magazine* no. 61, May 2002, 10–11.
28. Trezise, p. 207 (Note 25). The whole chapter (ch. 6) is worth reading for a warm account of Baring-Gould (1834–1924) which draws attention to the many other regional novels which he wrote and also to the regrettable fact that he censored many of the folk-songs that he collected. He did not have to bowdlerise the most famous of them, which is *Widdecombe Fair*.
29. London: Jonathan Cape, 2001.
30. The period of the novels is roughly 1800–50. There are numerous editions and recent paperback versions are available from Blorenge Books. Chartism was a working-class reform movement started in 1838 which scared landowners and government alike. It was not violent, though the Newport Rising of 4 November 1839 resulted in the deaths of twenty-four demonstrators. An association with moorlands might be asserted since there were Chartist meetings on Kersal Moor near Manchester and Hartshead Moor near Leeds in 1839. Chartism as an organised movement was defunct after 1842.
31. There are numerous editions of Fiennes's journals. I have used C. Morris (ed.), *The Illustrated Journeys of Celia Fiennes 1685–c.1712*, London and Sydney: Macdonald & Co/Exeter: Webb & Bower, 1982.
32. C. Pearce, 'A landscape of dissent: topography and identity in three Pennines valleys', *Landscapes* 2, 2002, 84–102.
33. W. White *A Month in Yorkshire*, London: Chapman and Hall, 1858. In finding silence oppressive he would have found friends among today's students.

34. Quoted in A. Wilton, *Turner and the Sublime*, London: British Museum Publications 1980. The Manuscript of Atkyns's journey (*Iter Boreale*) is in the Yale Centre for British Art, ms 1732. Celia Fiennes had trouble along The Wall (which she thought was Pictish) with soggy ground under the horses and dirty bedsheets at Haltwhistle.

35. A. Young, *Six Months Tour through the North of England 1771*. Facsimile edition: New York: Augustus M. Kelley, 1967, vol. II pp. 179–87 for example, mostly about Teesdale. This may be an example of the balance between scientific evaluation and the aesthetic at the end of the eighteenth century, about to be sundered into opposing discourses. See C. Klonk, note 45.

36. The material on the agricultural writers of the south-west is in S. Wilmot, 'The scientific gaze: agricultural improvers and the topography of south-west England', in M. Brayshay (ed.), *Topographical Writers in South-West England*, Exeter: Exeter University Press, 1996, 105–35. The remark about archaeologists is my own.

37. J. Zaring, 'The Romantic face of Wales', *Annals of the Association of American Geographers* **67**, 1977, 397–418.

38. A. Bermingham, *Landscape and Ideology. The English Rustic Tradition 1740-1860*. Berkeley and Los Angeles: University of California Press, 1986.

39. E. Helsingor, 1997 (Note 26) ch. 1.

40. All quoted from B. Wilkes, *The Illustrated Brontës of Haworth*, London: Tiger Books International, 1994, 15–16 and 221.

41. C. Pearce, 2002 (Note 32) pp. 86–7. He quotes *The Good Companions*, London: Heinemann, 1929.

42. A. Wainwright, *A Pennine Journey. The Story of a Long Walk in 1938*, London: Michael Joseph 1986/ Harmondsworth: Penguin Books, 1987.

43. P. Gruffudd, D. T. Herbert and A. Piccini, 'In search of Wales: travel writing and narratives of difference, 1918–1950', *Journal of Historical Geography* **26**, 2000, 589–604.

44. H. V. S. Ogden and M. S. Ogden, *English Taste in Landscape in the Seventeenth Century*, Ann Arbor: University of Michigan Press, 1955.

45. The opening pages of C. Klonk, *Science and the Perception of Nature. British Landscape Art in the Late Eighteenth and Early Nineteenth Centuries*, London and New Haven: Yale University Press, 1996, are very helpful on both the immediate and longer-term manifestations. The earlier idea that reality was mediated by a single, underlying common principle which might be a fixed mechanism was replaced by an organic idea of imperfection and emergence; the role of the individual might gain a greater significance as the source of a unity of perception.

46. Claude Gellé from Lorraine but usually known in England as Claude Lorraine or simply Lorrain.

47. The 'sublime' was meant to be awesome in a controlled kind of way, like today's horror movies. Its literary appeal seems to have lasted well, judging by the passages from Conan Doyle quoted in the Introduction: *The Hound of the Baskervilles* was published in 1912.

48. See the discussion in J. Bate, *Romantic Ecology. Wordsworth and the Environmental Tradition*, London and New York: Routledge, 1991.

49. M. Roskill, *The Languages of Landscape*, University Park, Pennsylvania: Penn State Press, 1997.

50. P. Bicknell, *British Hills and Mountains*, London: Collins, 1947.

51. I. Ousby, *James Plumtre's Britain. The Journals of a Tourist in the 1790s*, London: Hutchinson, 1993.

52. M. D. C. Rudd, 'The picturesque and landscape appreciation. The development

of tourism in the Yorkshire Dales and County Durham 1750–1860', University of Durham MA thesis, 1992. We need to remind ourselves that the moorlands were then much more extensive than now.

53. J. Whale, 'Romanticism, explorers and Picturesque travellers', in S. Copley and P. Garside (eds), *The Politics of the Picturesque. Literature, Landscape and Aesthetics since 1770*, Cambridge University Press, 1994, 175–95.

54. J. Barrell, 'The public prospect and the private view: the politics of taste in eighteenth-century Britain', in S. Pugh (ed.), *Reading Landscape. Country-City-Capital*, Manchester and New York: Manchester University Press, 1990, 3–40. On page 27 he quotes an eighteenth-century rhetorician (George Campbell in 1776) as writing, 'the prospect is continually enlarging at every moment . . . all the variety of sea and land, hill and valley, town and country, arable and desert, lies under the eyes at once.'

55. P. Howard, *Landscape. The Artists' Vision*, London and New York: Routledge, 1991.

56. There is an illustrated chronological compendium of Dartmoor artists and their work (from about 1760 until recently) in B. Le Messurier, *Dartmoor Artists*, Tiverton: Halsgrove/Dartmoor National Park Authority, 2002. There is however no critical analysis of the artists' changing perceptions in the volume.

57. M. Andrews, *Landscape and Western Art*, Oxford: Oxford University Press, 1999. The great arbiter of developing landscape taste, Prebendary William Gilpin of Salisbury thought in 1786 that in mountainous areas there must be ' . . . many irregularities, many deformities . . . which a practised eye would wish to correct'. Quoted in A. Wilton op. cit. 1980 (Note 34).

58. Wallace Collection, London. There are reproductions in J. Murray, *In Turner's Footsteps through the Hills and Dales of Northern England*, London: John Murray, 1984, 102 and E. Shanes, *Turner's England 1810–38*, London: Cassell, 1990, 74, which is a full page reproduction of a 28 × 39 cm original. Beamsley is south-east of Bolton Abbey in West Yorkshire. I am grateful to Mr S. Duffey of the Wallace Collection for allowing me to see the original watercolour.

59. In a private collection but reproduced in A. Wilton and L. Lyles, *The Great Age of British Watercolours 1750–1880*, London: Royal Academy of Arts/Munich: Prestel, 1993. His '*Kirkstall Abbey*' (1800, British Museum) is similar in that topographical respect. In his Scarborough II sketchbook (1816–18) Turner made an annotated pencil sketch, '*Skyscape over Hawksworth Moor, with Colour Notes*', which may have become '*Study of the Sky*' in the Skies sketchbook (watercolour) of 1818 which looks very like the enclosing swells of moorland skylines. Both are helpfully reproduced on the same page in J. Hamilton, *Turner and the Scientists*, London: Tate Gallery Publishing, 1998, 62.

60. Wilton op. cit. 1980 (Note 34) p. 37. Indeed the sketchbooks may, but the finished pictures are often different.

61. A. Causey, *Edward Burra, Complete Catalogue*, Oxford: Phaidon, 1985. There are about eighteen pictures of upland landscapes, only two of which are in public galleries: '*Black Mountain*' (1970) in Rye Art Gallery, and '*Valley and River, Northumberland*' (1970) in Tate Britain on Millbank.

62. In the Causey volume (Note 61), this is spelled 'Horcam' in the caption to Plate 24 and 'Harcum' in the list (cat. 381).

63. The Ilkely Moor b'a'tat song is said to have been written on a chapel choir outing to that locality in the nineteenth century. Two members disappeared for a while and came back to some extent dishevelled and the song drew attention to this. The tune had, not surprisingly, been a hymn tune (used for 'While Shepherds Watched . . .' among others) but thereafter had largely to be withdrawn since its use evoked inappropriate amusement. Recordings of the Holst and Delius pieces

are plentiful: the Holst for example is played by the Grimethorpe Colliery Band on Chandos Chan 4553; Chan 9355 has the Delius piece from the Bournemouth Symphony Orchestra and Richard Hickox.

64. J. Taylor, 'The alphabetic universe: photography and the picturesque landscape' in S. Pugh, 1990 (Note 54), 177–96.

65. F. Godwin (with an essay by John Fowles), *Land*, London: Heinemann, 1985; T. Hughes and F. Godwin, *Remains of Elmet. A Pennine Sequence*, London and Boston: Faber and Faber, 1979.

66. See D. Matless, 'Definitions of England, 1928–89. Preservation, modernism and the nature of the nation', *Built Environment* **16**, 1990, 179–91.

67. The term *paysage moralisée* means a landscape that contains a moral (usually commendable) message, for example that of community linked by 'home' place (*Gemeinschaft*) and not by occupation (*Gesellschaft*). The TV series *All Creatures Great and Small* (Yorkshire Dales 1940s–50s) and *Heartbeat* (North York Moors 1960s) are the best examples; *Peak Practice* (today, Peak District) is more valley-town; *Emmerdale* (soap-opera of today) is Yorkshire Dales Traditional with drugs and delinquency. It has much lower viewing figures than its urban rivals *Coronation Street* and *East Enders*.

68. Directed by Andrew Grieve, 1988. Connoisseur Video CR096 (VHS format).

69. In the Council for National Parks' sponsored (but with secretariat provided by the Countryside Commission) report, *Fit for the Future: Report of the National Parks Review Panel*, Cheltenham: Countryside Commission, 1991, CCP 334 [the Edwards Report], p. 5.

70. M. Shoard, 'The lure of the moors', in J. R. Gold and J. Burgess (eds), *Valued Environments*. London: Allen and Unwin, 1982, 55–73. The inter-war period was probably vital in setting the uplands as exemplars of the regions of tranquillity and peace that contrasted with the experiences of 1914–18. For a general 'English' view, see S. Miller, 'Urban dreams and rural reality: land and landscape in English culture, 1920–45', *Rural History* **6**, 1995, 89–102.

71. C. Harrison, *Countryside Recreation in a Changing Society*, London: TMS Partnership, 1991.

72. R. J. Moore-Colyer, 'Sir George Stapledon (1882–1960) and the landscape of Britain', *Environment and History* **5**, 1999, 221–36. In the present context, Stapleton is represented by *The Hills and Uplands of Britain: Development or Decay*, London: Faber, 1937. The 'survival of the past' idea for upland Wales reminds us of the Celtic heritage theme (and its parks) in recent tourism promotions. See Gruffud *et al.*, op. cit., 2000 (Note 43).

73. P. Gruffudd, 'Back to the land: historiography, rurality and the nation in interwar Wales', *Transactions of the Institute of British Geographers* NS **19**, 1994, 61–77.

CHAPTER EIGHT

'Too far for you to see'

AN ECOLOGICAL-HISTORICAL SUMMARY

The chapter title is a line from R. S. Thomas's early, and distinctly gloomy, poem in which we are deemed to be unable to see his reality of the hill country, with disease of man and beast and reversion of land 'to the bare moor'. In contrast he thinks we can only see that the sheep are grazing, 'arranged romantically in the usual manner'.[1] Most of the past is, of course, too far for us to see clearly: there is not enough information about it and there is too little evaluation of it by scholars. Many commentators on both past and present have cases to argue, sometimes based on the pinkish lenses of the Romantics or on a profit-making instrumentalism. Too far to see, also, is *the* future and we shall have to think of different types of prospect that might be possible.

Ecological history

The natural sciences demand that we should distance ourselves from the objects of our investigations, to act as it were as satellites hovering over the moors recording everything impartially. With all the data accumulated then explanations may be sought, preferably those with sufficient regularity to be called laws. This is an ideal never to be realised: the very collection of information is driven by cultural assessments of what is important: witness the density of research papers on, for example, grazing by sheep, or the diseases of grouse. Invertebrate populations get a much lower level of attention.

Nevertheless, the requirement of the sciences that there be a tested acceptance of a common set of 'facts' can to some extent be satisfied for moorland ecology over 10,000 years. There is an overall narrative that would leave few dissenters, until the question of attributing value to the various phases is raised: that will be discussed in due course. For the present, here is an outline ecologically-based environmental history for upland England and Wales:

1. As the effects of Pleistocene glaciation waned, the hills were covered with tundra and montane vegetation. Periglacial processes continued, contributing to slope formation and to the occurrence of patterned ground.

354

2. With the warming of the climate in early Holocene times, deciduous woodland covered the valleys and climbed their sides. In the case of the lower uplands, the summits were potentially covered with this forest.

3. During the mid-Holocene, the presence of Mesolithic folk inhibited some woodlands from attaining their full potential of cover by burning at the upper edge and by keeping natural openings free of regeneration. Blanket bog was a consequence on areas of low slope and high rainfall.

4. During the Mesolithic cultural period, the North Sea was finally completed, isolating England and Wales from continental Europe. Rising sea-levels abutted some moorlands and created cliff coastlines.

5. During the millennia between the coming of agriculture and the Domesday Book, uplands were occupied for crop agriculture as well as for stock-rearing, except when climatic downturns made this impossible.

6. The Romans did not eschew the uplands but the principal remains of their presence are roads and marching camps. This suggests that by the Iron Age there were large areas of open land on higher ground: that prehistoric people were the cause of deforestation on a large scale.

7. Medieval and early modern (say to AD 1700) ecology was dominated by an economy which variously combined cereal cultivation on valley sides, stock rearing with transhumance, and mining. The export of wool and metals kept up contacts with the wider world. Both had their effects on woodland type and distribution.

8. The eighteenth century was often a response to demands from urbanisation, especially for meat. Cattle were reared in the uplands and were driven long distances to growing cities. At the same time, transhumance was in decline.

9. As with everywhere else, the nineteenth century was a period of immense change. Where minerals existed they were often exploited more fully than before; contact with other regions reduced the need for self-sufficiency in cereals. Attention to moorland grazings produced many enclosures. Plantations of trees were set by some landowners.

10. On the drier moorlands, sporting interests concentrated on producing high densities of grouse in the nineteenth century. Stringent keepering shrank the numbers of many other species, especially predators. Fire was an important management tool for heather.

11. Woodland, farmland, settlements and open moor were all affected by the creation of reservoirs during industrialisation and on into the twentieth century.

12. In the twentieth century, afforestation with coniferous trees claimed large areas of moorland, in order to build up timber reserves. As the century progressed, they became more important for recreation. Their environmental effects are considerable.

13. The maintenance of sheep production became underlain by a multiplicity of subsidies. With the advent of those from the EU, by the 1980s the densities of sheep became probably the highest they had ever been. Heather moor in particular became a focus of concern for conservation bodies; soil erosion was also higher, exacerbated by path and track erosion along popular recreation routes of walkers and riders. Increased attention was paid to the maintenance and enhancement of deciduous woodland, mostly on valley slopes.

A broader environmental history

Such a narrative demands further interpretation. It can be seen, starkly, as a regression of biological productivity and diversity. The productivity builds up during the assembly of woodland but is diminished by the clearance of it and kept low by intense sheep-grazing, moor-burning and erosion. Biodiversity is very high in the tundra-montane phase of the early Holocene, falls during the woodland phase, is raised during early human occupation but is kept low thereafter by grazing, burning and drainage. Afforestation with conifers is rarely an aid to raising species diversity once the trees are beyond the pole stage.

So the end-point of such an interpretation is wholly negative. The uplands have been mined of their lead and tin and of their otherwise renewable biological resources as well. Additionally, such biodiversity as remains is threatened by climatic change that may produce enough warming to threaten the functioning of the ecosystems most characteristic of British uplands. The fact that most uplands have been designated as National Parks and AONBs looks like a piece of cultural waywardness that ignores the more basic ecology encapsulated in the acronym LFA, for Less Favoured Area. Yet the starkness of such a conclusion has the drawback that it omits to take into account the historical conditions which resulted in the transformations described (i.e., that people were doing what they saw to be reasonable in the light of the time) and because it is determinist in an ecological sense, a view which is rarely acceptable in western cultures. If we examine the first idea, then it is helpful to regard the ecological history of the uplands as a series of responses to the wider socio-economic contexts of land use, especially of food and fibre production. A framework for the whole of England and Wales is given by Joan Thirsk[2] and her alternation of (a) mainstream production (often under Malthusian pressures) of meat with (b) 'alternative' crops can be adapted here for the uplands (Table 8.1). In the Table, the main uses are set against their ecological effects as described above and there is an additional column which points forward to a critical element in moorland history, namely attitudes towards them.

Two major sets of views can be singled out. In the first, the 'reclamation' of the uplands for food production has been a first priority. This happened for example whenever mining communities needed to add to their diets, during the blockades of the Napoleonic Wars, in the face of demands for meat from

TABLE 8.1 Phases of 'alternative' uses, after Thirsk, with additions

	Upland England and Wales		
	Uses	Attitudes	Ecological effects
Mainstream: (Medieval period before the Black Death)	Charcoal, iron, beasts; hunting;	Monastic approval of wilderness	Reclamation by enclosure and assarting woodlands; coppicing of woods on slopes; wood-pasture formed
Alternative experience 1350–1500	Sheep and cattle;		Reversion as more animals produced in lowlands; monasteries run large flocks of sheep & cattle
Mainstream	Common management of grazing;	Waste places: backward and not contributing to the national wealth	Relatively steady boundary; grazing impact grows if summer settlements converted to permanent occupation
Alternative experience 1650–1750	Decline of summer transhumance;		
Mainstream	Water, quarries, mines, meat;	Romantic re-appraisal	Reclamation for miners' holdings at high altitudes; growing urban markets for beef & mutton increase grazing pressures; much division of commons
Alternative experience 1879–1939	Plus grouse, military;	Increased access via rail and bus	Reversion after wars when plough-up campaigns cease; grouse bring more regular firing and predator 'control'.
Mainstream 1940–1979	Military; water; foot-recreation;	A resource for urban popula-tions (water, recreation, scenery)	Reclamation encouraged by subsidies which keep upland farms viable; fencing and drainage encouraged
Alternative experience 1980s–2000	Grouse; sheep; military; vehicular recreation; EU Directives;	Scenery and recreation as commodity	Reversion as farms amalgamate; EU subsidies raise sheep numbers to all-time high: heather decreases, erosion prevalent; footpath erosion widespread
Foot and Mouth Disease 2001	Culling of many sheep; special concern for common land animals	'Losing the landscape of Wordsworth' attitude	(to be seen)

an urbanising nation, and as a way of absorbing many of the subsidies offered in the post-WWII period for ploughing, fencing and draining. When these perceptions were less dominant then other uses might claim the land: the drowning of large areas for water catchment, priority given to sport as in grouse moors, and the military use of the land for live firing. Under the surface of all of these utilitarian processes, lies the attitude that the uplands were wasted opportunities for 'proper' production and something of a national disgrace unless put to an obviously good use. The second attitude surfaced with the Romantics and sees the uplands as an antidote to the urban-industrial lowlands (including nowadays their rural economy): as a space for freedom of various kinds. So recreation, the protection of scenic views, nature conservation and above all the maintenance of difference become objects of desire.

Distilling these interpretations still further, we appear to have a dichotomy between (a) those who view the uplands as reservoirs of resources, some of them obviously material like grazing land but others less so, like recreation space for perhaps hang-gliding, and (b) those who see an essential stillness of land cover and use as possessing unique qualities of a moral and even spiritual kind for visitors and for some residents. Such divergencies lead, not surprisingly, to different assessments of both the present and of possible futures.

MILLENNIAL CONCERNS

All through the twentieth century the natural history movement included the uplands in its special places that needed conservation. Environmentally concerned people of many kinds try to bring to the fore the particular character of the ecology of the uplands, suggesting that it presents both problems and opportunities for management. This has been parallelled by a gamut of social interests which found one focus in the 1949 Act's desire that there should be a flourishing rural economy. Yet rural poverty has been persistent throughout England and Wales and the incomes available to hill farmers have recently been very poor even with high levels of subsidy. Add to that a map (Fig. 8.1) of the areas affected by the 2001 outbreak of Foot and Mouth Disease (FMD) and there is no shortage of matters which in one or more spheres raise some worries and inevitably demands for different pathways to be followed (Plate 8.1).

Environmental concerns

In recent years, one matter above all others has occupied conservation and environmental bodies: the effect of high densities of sheep on moorland. They are out-wintered but receive supplementary feeding on the hill, made possible by all manner of all-terrain vehicles but especially the quad bike. Selective grazing has increased the dominance of grasses with little forage to offer, such as *Nardus stricta*, at the expense of most other grass and sedge species. Most noticeably, the sheep have grazed out large areas of heather, which is replaced by grasses or by bracken. Since *Calluna* moorland is something of a British

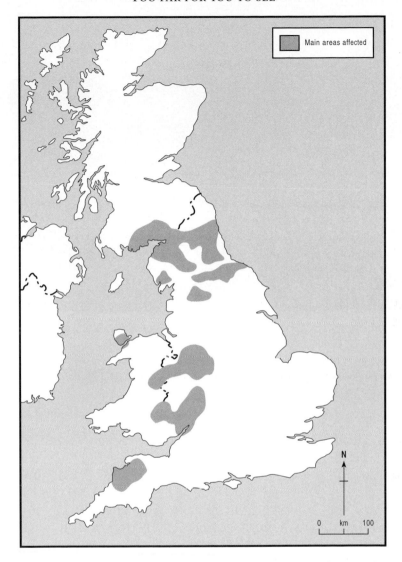

FIGURE 8.1 *A simplified map of the distribution of the outbreak of Foot and Mouth Disease in 2001. Many uplands are involved but the hill flocks were not affected totally. Northern English moorlands were the most affected as contiguous blocks and Figure 8.3 shows the area of the last restricted zone.*
Source: DEFRA Summary Map of confirmed cases at 30 November 2001. Lines have been drawn round the densest areas of symbols on the original which was accessed at www.defra.gov.uk/animalh/diseases/fmd/cases/fmdcases/map.asp on 03/12/01. Crown Copyright material is reproduced with the permission of the Controller of HMSO and the Queen's Printer for Scotland.

speciality within European plant communities, any shrinkage is viewed with concern on a EU-wide scale. Where sheep gather to be fed, soil erosion may develop and poached soils certainly act as foci for rapid and concentrated run-off rather than infiltration. Corrective action is in theory simple, but to an

PLATE 8.1 *Colonised moorland in mid-Wales. In an immediate setting of mixed wood-land but a regional coniferous plantation environment, lies the Centre for Alternative Technology at Machynlleth. What is most striking about this view from the hillside is the preponderance of post-fossil fuel energy-gathering devices: a hint of an alternative future, perhaps?*

economy already characterised by low income levels, socially and politically unacceptable.

Some farmers and landowners share with the conservation movement a feeling that the Countryside Recreation and Access Act 2000 (The 'Right to Roam' legislation, CROW) is inimical to their interests. On moorland, increased rights of access off public rights of way (no matter how many seasonal constraints are allowed) are seen as likely to disturb nesting birds, whether truly wild or sporting, and to encourage intrusion into areas formerly left to nature.

Climatic change

Superimposed on all these concerns is the possibility of climate change. Scenarios are plentiful but the nature of the work is such that certainty about future trends is impossible. If the Atlantic 'conveyor belt' were to be switched off and much colder winters were to result, then one set of conditions is likely for the uplands. On the other hand, most 'official' scenarios initially posit a more gentle progression to drier summer and wetter winters but with a rise in average temperatures year-round and more extreme events. Global models have yet to be downsized effectively and so the British Isles is covered by grids of 300–400 km in which the special characteristics of the uplands have to be inferred. The global climate models most used at the time of writing lead to a rise of average mean temperatures of 2.1°C for the 2050s in which the northern part of the UK receives higher intensities of precipitation in both winter and summer, with the most intensive events happening much more frequently, so the chances of flooding are higher. The total context nevertheless is one of more hot Augusts (with a greater potential for fires) but the likelihood of successive dry years is diminished. The greater radiation overall means a reduction in the incidence of frosts in upland areas though there would still be a large area with over thirty days' incidence. At a scenario of +2°C, there would be few areas in England and Wales with an annual temperature accumulation of less than 1,000°C and in theory uncultivable. But for most purposes, land use and social context is much more important for species survival than the amplitude of climatic change. Even though the Atlantic fringes of Europe are likely to experience the lowest manifestations of climatic change (compared with the Mediterranean or North-east Europe), small changes in climate may produce disproportionate shifts in the range of wild species and allow considerably more choice of land uses. Work in progress on the regional implications of global climate models will make the contours of possibility more apparent and the work of an Environmental Change Network which is monitoring possible indicators of climatic change will contribute detail. Data are already available which allow some comment on the proportion of annual precipitation that falls in the winter months in the uplands. There was for example an upward trend in the percentage during 1991–7 at Moor House in the North Pennines but a decrease from 1995–2000 at Sourhope in the Cheviot.[3] Modelling is made more difficult by the relative scarcity of upland precipitation measurements.[4] Nonetheless, the models predict that climate will change more quickly than at any time in the recent past and that the outcome will be unlike any climate of the past.[5]

Most of the broad habitat types would undergo some changes, though their vulnerability to rapid change is deemed to be variable. *Broadleaved woodlands,* for example may respond only slowly but their vigour and biomass would benefit from increased carbon dioxide and higher temperatures; species of southern distributions might extend into upland areas and drought might thin out some woods, allowing more understorey growth. This might alas benefit the Rhododendron. *Coniferous woodlands* are little more susceptible to rapid change, though more damage from storms might offset higher growth rates and the possibility of spread to higher altitudes; this latter is much more important in Scotland so far as native pine woods are concerned. In plantations, the storms might enforce the early harvest of crops in order to reduce windthrow losses; conservation of the red squirrel would be disrupted where this depends on a particular mixture of tree ages but species like nightjar and woodcock would benefit. *Upland grasslands* would be more productive and so supplementary feeding of livestock might be needed less; grass production could however be limited by the low nutrient status of many of the soils.

Nitrogen losses from soils might be decelerated by increased uptake into the plants.[6] Further invasion by bracken might be encouraged by higher temperatures, though not by increased carbon dioxide; modelling has suggested that the bracken stands could get denser. The category of *dwarf shrub heath* is dominated by heather moor. The conditions are likely to favour the competitive ability of heather *vis-à-vis* bracken, so the spread of the fern may be slower than on grasslands where it may well move up the hill; on the other hand, heather is very vulnerable to fire in dry spells. Planned burning may need to be extra-sensitive to soil conditions. In the few *montane* areas of England and Wales, heather may be able to reach higher altitudes. It may therefore compete with a small pool of arctic-alpine plants with nowhere to go and a small genetic diversity;[7] an 0.5°C increase in temperature might virtually eliminate montane species from Great Britain as a whole. *Blanket bog* is unlikely to be affected by warmer and wetter climate provided it is intact (i.e., is mostly *Sphagnum* moss) but erosion rates on bare peat may increase, with higher intensity of rainfall following cracking of the surface in dry summer conditions. If the peat breaks down for any reason then carbon will be released into the environment. Figure 8.2 shows that withdrawal of blanket bog to core areas of its distribution seems likely.

The likely consequences for nature conservation are more complex than first appears because of the undertakings on conservation made in Biodiversity Action Plans and in response to EU law. A redistribution of protected habitats outside the boundaries of already defined areas may undermine the significance of a site, with implications under both EU and national law. Attempts to maintain a site where climatic change is causing conservation problems may appear to be futile but necessary because of legal commitments. The whole question of 'conservation status' will need examination in a changing world, in the uplands no less than elsewhere. A net loss of habitats and species from the present situation seems likely (for every 1°C mean annual rise, the climatic space moves northwards 250–400 km) and the current emphasis on rigid site designation does not appear to favour the movement of species within or between them and habitat fragmentation is a very probable consequence of climate change.[8] A summary of some of these considerations is given as Table 8.2.

The consequences of climatic shifts for sheep are difficult to predict: is, for example, heat stress more likely? Their pastures however might be different biologically. The upper limit of bracken, for example, is currently controlled by the incidence of frost and so warmer conditions might allow the species to spread up the hill. If there were diminished densities of grazing animals then any woody species would receive additional encouragement, so there might be more early successional woodland. Protected areas like the 'radhaz' zone at Fylingdales on the North York Moors suggest however that colonisation might be slow.

The likelihood of climatic change is sufficiently great for some bodies to

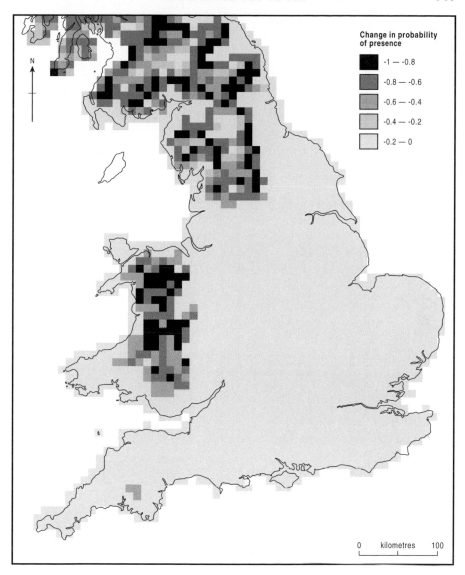

FIGURE 8.2 *The projected change in the probability of presence of blanket bog in England, Wales and southern Scotland. The baseline is 1961–90 mean monthly temperatures and the UKCIP 2050 Medium High scenario is applied. Basically, a retreat to a smaller core area of blanket bog in Wales and the Pennines is the outcome, especially where large negative change is indicated.*

Source: J. E. Hossell, B. Briggs and I. R. Hepburn, Climate Change and UK Nature Conservation, *London: DETR, 2000, Fig. 4a. Crown Copyright material is reproduced with the permission of the Controller of HMSO and the Queen's Printer for Scotland.*

TABLE 8.2 Upland semi-natural habitats and climatic change

Broad habitat	Priority habitat	Likelihood of impact	Ability to adapt naturally	Options for assisted adaptation
Broadleaved and mixed woodland	Upland oak woods	Medium, but slow and episodic evidence	Medium	Extension of habitat by planting is long-term possibility
	Upland mixed ashwoods	Medium, but slow and episodic evidence	Medium, favoured by increased rainfall in uplands	Extension of habitat by planting is long-term possibility
Dwarf shrub heath	Upland heathland (heather and crowberry)	Medium	Medium	Agri-environmental policies on grazing and fire
Wetland and bogs	*Molinia* grassland	Low	High	Agri-environmental policies on grazing and fire
	Blanket bog	Medium	Medium	Protection from degradation and erosion
Inland rock	Limestone pavements	Unknown	Probably low (very few locations) but uncertain	Unknown
Montane	Sub-alpine areas and arctic-alpine NNRs	High	Low	Low

Source: Adapted and extracted from J. E. Hossell *et al. Climate Change and UK Nature Conservation*, London: DETR 2000, Table 6, pp. 28–30. Crown Copyright material is reproduced with the permission of the Controller of HMSO and the Queen's Printer.

accept the current models and apply them to forecasts for their regions. Where the uplands are concerned then a 'common sense' modification of the likely regimes has been applied, and some implications for various environmental systems have been outlined, along with ideas for further information needs.[9] A summary of some of these explorations is given in Table 8.3. It seems worthwhile to emphasise that the economics of sheep production will be more

TABLE 8.3 Possible impact of climate change in rural uplands

'Beneficial' effects	'Detrimental' effects	Comments
Migration of new species	Loss of niches and species	'Southern' species already increasing and the corollary is also true; much may depend on availability of migration routes
Enhanced vegetation growth in a longer growing season	Erosion from more intensive rainfall	Peat growth/erosion will affect carbon uptake/loss from peats and peaty soils
Higher productivity of grasslands and hence of grazing animals	Landslides	
Greater recreational opportunities	Increased early grazing on young shoots	More recreation has impacts on vegetation, erosion and animal communities
	Vegetation may suffer from water deficits	
	Stormier: windthrow of trees; effects on wind turbines	Higher speeds, more electricity but perhaps more damage to turbines
	Fire risks	Will interact with recreation access and also with grouse moor management
	Impacts on water supply systems	Greater care in preventing losses needed; bare periods at reservoirs may increase
	Low flow periods in rivers	
	Great flashiness of streams	
	Flood risks	Interacts with sheep densities and with attempts at moorland drainage
	Heat stress of livestock	

The categories are taken from a number of documents on upland Britain starting with those in North West Regional Chamber, *Everybody has an Impact. Climate Change Impacts in the North West of England*, Manchester: Sustainability North West, 1998.

important for nature conservation in the short- to medium-term than any climatic shifts.

Social concerns

A continuing theme of this book has been the connection between the human communities and their non-human surroundings, for the practices of the one clearly have consequences for the other. So any trends in the mix of economic activities in the uplands is likely to be reflected in the ecology of the surroundings. Diversification of farming, for example, has produced soil erosion in areas where pony trekking is concentrated; being paid to grow deciduous trees will introduce diversity of flora and fauna into some upland woods. If lowland flooding, for example, in the valley of the Severn or on the Yorkshire Ouse, seems beyond the reach of local flood protection works then land managers in upland catchments may be paid to hold water in place.

Whether any of this will be sufficient to keep farmers on the hill land is an open question. What seems clear is that a younger generation, accustomed to, or certainly aspiring to, a more affluent life-style is unlikely to stay on hill farms under current conditions. The obvious trend might be towards bigger production units, with the surplus housing becoming even more often the antithesis of city life, either on a weekend basis or permanently for a few networkers and very early retirees. Whatever the grazing regime sanctioned by EU policies, there will be land that is less intensively used and therefore prone to invasion by woody plants. The military might seize the chance to diversify their training areas but little of the land would have any special attractions for recreation beyond its present allure. Quarry operators would no doubt argue that they were a source of employment in an ever-less productive economy.

There remains the position of the big landowners and especially those with largely sporting interests, which in this context means grouse. Assuming that heather moor remains a viable plant community in the face of climatic change, but that there is less impulsion to cross-subsidise sheep-farming tenants from the shooting profits, then maintenance of this land use, even in the face of its rapid water- and soil-shedding nature and even if invaded by more ramblers, seems assured. More gamekeepers might be employed; mostly (let us hope) of the ecologically-minded sort. But any major profits from shooting are likely to find their way to places other than the region where they were raised, given the multiple interests and residences of many grouse moor owners.

It is difficult to envisage circumstances which could tackle the social concerns about the rural economy of the uplands today in any way which involves massive pecuniation of this countryside in the manner of parts of East Anglia and the southern Midlands, for example. It looks as if the uplands are destined to be the home of either those whose major incomes arise elsewhere or those who are happy with cash flows well below national expectations. Not for nothing does rural Northumberland accommodate retreat houses run by Buddhist monks.

2001: Foot and Mouth Disease

Easter 2001 was marked by poor weather, which is not unusual, but also by an absence of ramblers and hill-walkers, which was virtually unprecedented. The holiday came more or less at the peak of the outbreak of Foot and Mouth Disease when notification of new cases was running at its highest level of about fifty cases per day. The outbreak was very widely spread spatially (Fig. 8.1) compared with the 1967–8 episode which was centred on the Cheshire Plain and more or less confined to the north-west Midlands and North Wales. At the peak of the earlier outbreak there were about eighty cases per day, with transmission assumed to be by wind, birds and rodents. The 2001 flare-up was different. It was initially spread by sheep movements across wide areas of the UK, and the higher densities and marketing patterns for sheep meant that from a February start, of the 1,471 infected premises by the end of April, 1,215 had sheep. It is estimated that between the disease entering the country and the ban on movement on February 23rd, some 2 million sheep were moved outside holdings. The map shows that the distribution is one of the margins of the uplands (which is where the farms are centred, not necessarily the animals): in Wales for instance there were ninety-eight cases in Monmouth and Powys and twenty in the rest of the Principality and the sporadic outbreaks towards the end of the epidemic fringed the Pennines. A true upland area, bisected by the Tyne valley, was the last area to have restrictions lifted (Fig. 8.3). In upland areas, there seems to have been transmission along valleys, as in the North York Moors and Black Mountains (Plate 8.2). In the end, if Scotland is excepted, the majority of cases were in Cumbria, Devon and North Yorkshire.

The totals of animals involved are dominated by the slaughter of sheep, though the dominant media images are of the burning and burial of cattle. By the end of 2001, nearly four million animals had been slaughtered, of which 3.2 million were sheep, 0.6 million cattle and 140,000 pigs as well as 2,000 goats and 100 deer and 1000 'other animals'. Compensatory payments were estimated at £250 million.[10]

The likely environmental consequences of the outbreak have been tabulated for the whole country by the Environment Agency.[11] They do not distinguish upland sites but a selection of their findings might highlight:

1. Monitoring of the fallout from pyres (especially for dioxin) will be an on-going task. There has to be similar monitoring of landfill disposal sites to see if they contribute to water contamination in the long-term. This is not primarily an upland problem, though there was one distinctly upland pair of sites at Mynydd Eppynt* (Powys: SN 867 348 at 390 m

* In Welsh sources the spelling is usually 'Epynt'

FIGURE 8.3 *The last restricted area of the FMD 2001 outbreak, with Hexham in Northumberland as its centre, and containing moorland areas to north and south, with the southern area containing the most moorlands.*
Source: DEFRA map for 21 September 2001, www.defra.gov.uk/footandmouth/maps/ maphexham.htm *accessed on 08/11/01. Crown Copyright material is reproduced with the permission of the Controller of HMSO and the Queen's Printer for Scotland.*

ASL for burial; SN 861373 at 340 m ASL for burning) in part of a military area. The diversity of the monitoring and reporting schemes (at Mynydd Eppynt, there have been both permanent and mobile monitoring schemes for surface and ground water, air and soil) and the number of agencies involved are an interesting reminder of the complexity of any environmental process. At the end of 2001, monitoring reports were generally positive about all potential problems. The burial site is likely to be a longer-term problem: it was lined and indeed a fissure in the pit floor was detected and dealt with early in the burial process.[12]

2. The decision by some hill farmers that this is the time to quit. Most are tenants and so the future of their lands will depend on the landowners'

PLATE 8.2 *A small notice with such immense ramifications. In two languages, a public footpath is closed because of Foot and Mouth Disease. In the Honddu valley, its presence disrupted farming and the recreation industry alike for months.*

assessments of the situation. One environmental aspect of the change will be the degree to which holdings become more or less fragmented, since the probability of Foot and Mouth Disease transmission seems to rise with the level of fragmentation within a holding: those in Cumbria were notably more broken up than the farms in Devon and Wales.[13]

3. In 2002, it was much too soon to estimate soil and ecological changes. In many areas, variations have had to be sought in the management of agri-environmental schemes since site visits have been impossible and proper mowing, grazing and manuring regimes have been subject to what the regulations call 'Force Majeure'.[14] There were fourteen major pollution incidents from disinfectant spills, slurry and carcass fluids but none seem to have had long-term effects.

4. If sheep headage payments are progressively withdrawn, there may be the creation of a zone comprising the upper enclosed pastures and the lower moorland slopes on which the early stages of succession will be established. There will be high-profile opposition to this 'scrub' from walkers and their organisations and from guardians of landscape values with a conservative outlook.[15]

There has been widespread dissatisfaction with the government's handling of the outbreak and the Minister at the time was subsequently demoted. Unlike BSE, there is to be no overall public enquiry, which has led to the setting-up of local fora as in Devon and Northumberland. These will no doubt produce suggestions for improvements in handling the disease but not address the long-term ecologies involved. It all shows that there is no shortage of concern and focus on the hill lands of England and Wales but there are distinct divergencies of outlook. Hence, the range of possible futures which flow from different outlooks and disparate basic attitudes is quite wide, even given certain constraints of physical geography.

FUTURES

There follows an attempt to depict a few possible futures for the uplands and the environmental consequences of these trajectories. They are crystallised out of more continuous spectra and so inevitably are simplifications. They also ignore one fact: that of regional difference. There is probably no reason why the future of, for example, Bodmin Moor must be the same as that of the Cheviot and the whole question of local democracy has to be examined at some later point: if the EU can have subsidiarity and if regional devolution is not inherently wrong-headed then some diversity seems inevitable. At one level, such a view reflects a microcosm of the national situation: because our environment rarely presents extremes compared with earthquake zones or tornado-ridden plains (neither is absent from Great Britain but they are minor in incidence on a global scale), then we tend to forget about it or regard it as infinitely pliable.

The uplands are less malleable though, than Essex, or the Cotswolds. They present, for example, an opportunity for the flourishing of plants and animals that are unsuited climatically to lower ground or which cannot survive in more intensively managed systems. This may prove a source of biodiversity in a warmer climate. Negatively, they are not much use to the barley barons. Neither do they appeal to the intensive pig breeders and broiler chicken units since they present higher fuel and transport costs than their lowland equivalent and their building needs rouse opposition, especially in National Parks.

Business as Usual

This scenario assumes that the farmers' and landowners' organisations persuade any government in the next ten years that the present pattern of production must at all costs be kept more or less as it is. The government then underwrites animal production from the hills and also persuades the EU that this must be done in spite of any reforms of the CAP. Given the availability of supplementary food from lowland or industrial sources, the year-round number of animals – especially cross-bred sheep – is expected to remain high. There is a further development in the breeding of animals (using techniques of genetic modification) to cope with unavoidable environmental changes, such as wetter winters. At the same time, strong encouragement is given to diversification in the hill lands, with production of other animals (e.g., red deer), recreation both active and touristic, with an emphasis on the upland scenery as a resource for these activities. There is considerable support for this scenario from the NFU, the Country Land and Business Association (the former Country Landowners' Association), the national sports associations and the Ramblers' Association. The RSPB is opposed, and English Nature is antipathetic, the Countryside Agency is unhappy but not keen to say so; their Welsh equivalents see no alternative in their particular context. Archaeologists are pleased that 'landscape archaeology' in the uplands will still be possible. Would-be windfarmers see little change in their chances. Controls are intensified from the centre with the requirement for Environmental Impact Assessments (EIA) on uncultivated and semi-natural areas.[16]

Environmental consequences are fairly obvious. Reliance on bought-in feed for the animals increases as the moor vegetation contains fewer and fewer palatable species; bracken spread adds to this. Erosion of peat increases as blanket bog is drained in order to try to add to the drier pasture land and loss of mineral soil continues; in general, runoff is quicker. The consequences are felt downstream with rivers that rise quickly to new heights and the Environment Agency is blamed for not coping well enough, although its upstream powers are quite limited. In all, the system is not very stable, as incomes are never high enough to ensure that families stay in the business; thus grouse shooting on the drier moors is well entrenched wherever grazing stints can be bought out. The results for truly wild species are very variable depending on the local attitudes of the landowners and their land managers.

Total cessation of grazing

At the opposite extreme, the structure of UK agriculture changes markedly. There is a reduced domestic demand for meat and exports of lamb fall. All the production that is needed and decreed by the EU and DEFRA can come from lowland units; the remaining farmers are bought out using the analogy of reducing the number of fishing trawlers in EU waters. The housing stock is carried up into the recreation market for second homes except in areas with notably worse climates, such as Upper Weardale. Mostly there is a brisk demand for the properties and some new owners are permanent residents who keep a few goats and like to live in a mildly Green fashion. The NFU is unhappy for a while but adapts to the new situation; conservation organisations are delighted but refrain from shouting about it; the Environment Agency in particular revises a few of its bleaker flood forecasts. The CLA's successor sees that intensively managed moors will start to look anomalous and starts to try to present shooting as environmentally desirable: a pro-hen harrier programme is introduced with fanfare but not enthusiasm. The Ramblers' Association clings to its moorland preferences and the Council for National Parks deplores the loss of the landscapes so beloved of its iconic figures like John Dower.

The environmental consequences mostly affect a belt of land which encompasses the lower slopes of the open moorland and the upper enclosures. Immediately, the lower land begins to look like reverted land in previous times, with more rushes, a greater diversity of grass species and above all, perhaps, more bracken. A few fields are kept open by the new residents and the remnants of the old system. The main change is the onset of succession towards woodland. The outcome of competition between species is not easily foreseen because most studies in the past were carried out in the presence of sheep and cattle. Moor-burning will no longer have any place as a management tool, though this may mean that its frequency will increase in an uncontrolled fashion. So although thorny shrubs (especially hawthorn, *Crataegus* spp) might seem an obvious first state in what will inevitably be called 'scrub', other species may well come to dominate this phase, depending on local sources of seed. Hazel is one obvious constituent, where it currently grows in wall and hedge-banks as in the south and west; birch seems a very likely addition to the flora. Since holly is sometimes well established by virtue of its resistance to grazing, it may well form thickets. In wet places, alder will flourish and some flushes now heavily grazed will be colonised by shrubby willows. On disturbed ground there may be more gorse and in a few localities, juniper scrub. None of this will make much progress onto the deeper peats but anything up to 25 cm depth of organic material may be colonised by species which are acid-tolerant. Perhaps the great unknown in this is the subsequent colonisation of the podsols, stagnopodsols and gley soils: will there have to be a long phase of soil amelioration by the pioneer species?[17] Not, presumably for pine, since Scots Pine can often be seen in small unfired and little-grazed patches of moorland today; likewise Sitka Spruce could well begin to look like a native tree. The diminution of close-cropped but uncultivated grassland at about 250 m ASL will close off habitat for the ring ouzel.

Total emparking

Here is another extreme case. Parliament and the Welsh Assembly agree that the uplands need no longer be a productive set of environments but should be part of 'sustainable Britain'. This means that they are to play a role in environmental protection of all kinds as their primary objective and other functions must be compatible with that design. Thus retention of water in the upper reaches of the Exe, Severn, Wye and Yorkshire Ouse, for example, is paramount; absorption of carbon by fast-growing trees is a desirable gesture even if not all that effective, compliance with EC Species' Directives and Special Areas of Conservation is a national priority and not a nuisance imposed from Brussels. Nobody much likes this scenario outside the conservation bodies and the relevant Minister's job is not strongly sought at times of election and re-shuffle.

The reintroduction of the beaver is hailed as a success and the red kite can be seen frequently; indeed there is a movement to try and stop it spreading back into the towns. In the absence of sheep the wolf seems a virtual necessity to control the abundant deer, many of which are of the smaller and more difficult to shoot species like sika and muntjac. However, runaway 'scrub' is not favoured everywhere and in some valleys there are special management programmes where 'farmers' are paid to graze cattle and sheep and thus maintain open land. The Woodland Trust has been charged with the job of encouraging the planting of millions of hectares of deciduous woodland and managing it for wildlife, underwood products and soil quality. A couple of County Wildlife Trusts have plundered their local pollen diagrams for 'climax' woodland composition and as a result are trying to accelerate the growth of a mid-Holocene native forest cover. (Filling the ecological niche of *Bos primigenius* is difficult.) Among those who perceive themselves as losers are the windfarm operators who find the new emphasis on nature-dominated landscapes rather oppressive. The whole question of outdoor recreation is still vexed. Grouse moor managers feel isolated in the new conservation-oriented regimes. Likewise, although ramblers like the increased attention paid to the environment, their right to roam is not *de facto* increased very much; highly active recreationists with machines feel they are being pushed to one side and take to more illegal use of greenways or sudden descents on glideable hills. The biodiversity of the hills is however much enhanced and they provide a refuge for all kinds of species displaced from more intensive systems at lower altitudes, a process which is reinforced by up-hill migrations of species into the warmer conditions that result from climatic change.[18] The hills become in effect a region of the nation in which solar energy is dominant and the density of the embedded energy of machines and motor transport falls. In this respect, they are held up by Greens as a portent of how the national environment ought to be managed.

Modified control for production

As a nation famed for its ability to compromise, some kind of 'middle way' is sought. The value of retaining some animal production is affirmed, along with the role of the uplands as watersheds, as tourist and recreation destinations, and as the legitimate habitats of grouse moors. The governing idea is therefore a mosaic of land cover types which makes use of the diminishing number of people who want to get most of their income from cattle and sheep. The production unit becomes larger but is underlain by a series of ESA-type arrangements rather than those tied to production. Regional bodies like the Dartmoor Commoners' Association and commoners of the Long Mynd control the animal numbers and administer the ESA conditions. Likewise, these bodies have an eye to the carrying capacity for recreation and so engage in traffic control; rather over-elaborate plans for segregation of local people and visitors by roads which they are allowed to access and by price tend to be bogged down in enforcement problems and legal challenges and so they are simplified. Even so, getting effective public road transport into sparsely inhabited regions means breaking lifetime habits. In order to emplace these arrangements, governments have had (a) to concede regional devolution of some powers to new bodies since not all uplands are National Parks or AONBs and (b) bring agriculture and forestry within the scope of the Planning Authorities.

The resulting environments do not at first seem very different since one of the fundamental ideas is that of gradual change. The valley sides show quite rapid change since the Authorities are keen to establish deciduous and native-species woodland, both by natural regeneration and nursery-based planting. This is seen as part of the water-holding strategy, though clearly the higher up the hill the woods extend the more effective the control. Above the old head-dyke level this strategy becomes more difficult for ecological reasons and successful planting and nurturing requires a lot of labour, which strains the ESA-type budgets. There is opposition, too, from weekend visitors whose expected view of the familiar prospect of bleak moorland round the sides of their Sunday paper is now bounded by trees as well as the front windscreen pillars of their car. But the scavenging of crumbs by pied flycatchers is a novel element and a few adventurous souls bring some maggot-shaped treats in the hope of attracting greater spotted woodpeckers to the parking areas. The streams may have a more even regime where their catchments are wooded and in those few uplands where limited beaver reintroduction has been tried. So there are more trout and riparian owners can lease the fishing rights for income. Enough land is open above the woodland, however, to satisfy the Ramblers' Association and the CNP that the true landscapes of these regions are being protected; any ideas harboured by Green extremists in University departments with environmental interests that Cross Fell should eventually have flower-rich open birch woodland on it again are rejected by the AONB's governing body.

Chaos not coherence?

There is a certain amount of fantasy in these scenarios. Above all, they are subject to a large number of unknowns. In the short-term, there is the question about whether government might better the rates of income foregone when environmentally sensitive measures are adopted. On moorland in the North Peak ESA in 2000, payments were £25/ha but income foregone was £42/ha and worse ratios are found for other land types.[19] Shifts to payments for area of land rather than number of animals are conjectured to result in new areas of overgrazing since farmers withdraw to core areas if they have only a small number of sheep; farmers might claim for new areas of grazing but stock at the same high levels on the original area. Nothing yet proposed seems likely to bring more cattle back onto uplands where they might reduce the prevalence of bracken.[20] More might be made of the flexibility within CAP policies for sheep production, as recommended by the post-FMD Curry report,[21] the EU Water Framework Directive (2000/60/EC) and its Rural Development regulations (e.g., Council Regulation 445/2002). Among the list of uncertainties we might find the price of energy derived from oil, the actual climatic change which occurs (as distinct from the models which we now have) and the rate of such alterations, the effect on EU policies of its enlargement, and the success of GM research into e.g., trees. (Think of a super-birch able to grow to maturity at 1,200 m ASL on wet, acid and even skeletal soils.) So there are always unforeseen developments and synergies which are, in retrospect, the stuff of chaos theory. But for all its creativity, that type of thinking cannot by definition be very detailed in its predictions, only in the outer values of its lineaments. But these are likely to change as well, as indeed might be the values which underlie our talking to each other about the environment.

DISCOURSES AND PERCEPTIONS

The German sociologist Niklas Luhmann reminded us that we do not talk directly to the natural world but that we humans talk to each other about it.[22] What many commentators have done is to point out the mediated character of those conversations (Table 8.4). There is first, to simplify a tangled web of brain-use, the ways in which we perceive the upland environment in neurophysiological terms: the dominance of visual stimuli for most people for example. Secondly, there is the cognition which involves the interfacing of those stimuli and their memory with other memories, and a vast variety of other inputs from our social world. Thus the Romantics taught us to like the open wildness of the uplands; and no doubt we could now be taught to like a lot more woodland, for example. Beyond perception, and building on cognition, there is communication. We talk to each other through language:

TABLE 8.4 Moors in/on the brain

Process	Stimuli	Cultural input (1)	Cultural input (2)
Perception	Neurophysiological	Not known	
Cognition	External continues but might be in memory more than immediate	Memory very important; reading, pictures etc	Might include 'ought' – expectations of others
Discourses	Intellectual *and* populist	Very strong: intellectual traditions, theories	Current economic pressures
'Romantic landscapes'		'This was the landscape of Wordsworth . . .'	FMD debate as a continuation
Wilderness		'Should not be in private hands . . .'	Right-to-Roam, basic rights see e.g., Shoard
National need		'The nation needs . . .'	Food and water once but now military especially
Moral reservoir		'We have much to learn from . . .'	'Characters' fictional but also Hannah Hauxwell
Stakeholders		'No moor is an island . . .'	Cannot be separated out from the rest of the economy (and ecology?)

primarily that of words but in many other idioms (of which the visual is surely primary in this instance) as the section on representation showed. The type of language used and the content of the individual discussions is usually labelled the **discourse** and it may be worth examining some of the discourses used when communication is undertaken.

Table 8.4 shows some of the types of discourse which are applicable to the discussion of moorlands, omitting for the moment that of the natural sciences.[23] They mostly run parallel with the Representations given in Chapter 7, though here it is usually the language of advocacy which is central.[24] The 'romantic landscapes' arguments present the moorlands as timeless, as areas which have undergone little change since some unspecified date. Sometimes, though, a datum will be applied: in the case of the mountains of the Lake District (not strictly speaking relevant here) during the FMD outbreak in 2001, it was lamented that the landscape of Wordsworth might be lost due to the encroachment of woody vegetation. This argument was aired much more widely when the future of hill farming was concerned: it was the preservation of today's landscape that was important. Likewise, the arguments over the new access legislation promised by the post-1997 government concentrated on the powers retained by landowners to confine and exclude walkers, especially on grouse moors, just as their predecessors had done in the 1920s.[25] These powers of ownership, argued Marion Shoard,[26] must be superseded by universal rights of access; these rights being as it were part of natural law rather than the legal systems that have accumulated at least since the Conquest. A rather similar blanket approach has characterised the hill farmers as heroes in the struggle against a hostile environment and thus a reservoir of high moral virtue. This is most obviously conducted in the realms of fiction but the way in which a lone hill dweller in Upper Teesdale, Hannah Hauxwell (b 1926), was turned into a media figure in the 1980s, shows the readiness to fan this discourse into life.[27] A more detailed examination of a particular set of ideologies was made by Rachel Woodward in her study of ways in which the military have argued their case for more live firing on moorland ranges.[28] Then, true to the age, arguments have been conducted in terms of 'who are the stakeholders in the future of the moors?' which looks like a complex set of social groups set in temporal and spatial scales, likely to show (Table 8.5) that the stakeholders are widely spread and do not exhibit any obvious ways of arranging themselves in priority order except that human concerns are always prone to push ahead of concerns for the non-human world.

Somewhere, all these perceptions feed into the role of rural Britain in forming part of the national identity. In fact, it is nearly always the image of lowland south England, the land of village greens, thatch, pub and church.[29] Whether the uplands are simply a negative area in such constructions of identity or simply that their role has not been sufficiently chronicled is at the moment a matter for the student of cultural studies rather than the historian. The social role and meanings of rural Britain as a whole are undergoing rapid

TABLE 8.5 Some 'stake-holder' groups in moorland futures

	Tenant farmers	Landowners	Local authorities	Developers of e.g., wind-power	Recreational users	Non-farming earners + retired residents	Conservation bodies
Immediate	Production from land for income	Income & sport (e.g., trees and grouse)	Some development control on site	Need suitable sites	Wear; parking; amenity	Restrictions on e.g., building materials, extensions	rarities
Local	Contribution to local economy	Forestry may add to recreation resource	Allocation of facilities; new admin bodies	Some employment	long-distance wear; contribution to economy	Contributions to economy and social life	rarities at local and regional scales; typical communities
Regional	ditto	Attracts 'outside' people		?	?		
National (UK)	Reservoir of sheep	strategic reserve of wood	Strategic views from Ministries	Into national grid	Influence on national policy	?	Nationally important biodiversity spp
Super-national (EU)	Seen as needy; Rural Development regulations coming on scene in 2000s.	conservers of heather moor; relevant companies affected by Water Directive		CO_2 control	Ideas from abroad	?	EU Directives
Global	?	?	?		?	?	Global biodiversity

NB: the question marks indicate unknowns and certainly not impossibilities.

change, though whether the uplands are a special case was not much commented upon in studies in the 1990s. Again, the FMD outbreak could cause some redirection of focus.[30]

This very attenuated discussion of the notion of discourse is not meant to detract from the value of its study for it tries to tell us the underlying meanings of what is being said during discussions and arguments. In this case, the multiplicity of discourses being conducted suggest an analogy with one of Charles Ives's pieces in which different bands march around playing different tunes, and it is difficult to hear one melody emerging from the dissonances. A more harmonious 'whole countryside' approach seems a long way off.[31]

CONCLUDING THOUGHTS

We ought never to underestimate the capacity of the British to avoid change. To have different and apparently better routeways pointed out to them is no guarantee that they will come at all close to exploring them. So whereas the history of the moorlands' environments might point to choosing a shift in direction for the management of their ecosystems (whether these be for production or protection), the weight of 'traditional' perceptions and mythologies has a strong retarding effect and indeed produces the disagreements which are usually labelled as 'contestation'.[32]

Neither, though, should we underestimate the importance of these land areas in terms of the national patrimony. At an approximate level, the uplands and their valleys comprise about five-eighths (just over 60 per cent) of the land surface of England and Wales. Of this, moorland-dominated uplands (as distinct from mountains) account for most of the area except in North Wales and the Lake District. Much of the productionist thrust for the uplands stems originally from the evaluation in the eighteenth and nineteenth centuries of how an important stretch of national territory was not contributing to the glory or the survival of the nation. That attitude has been reinforced ever since, though it has competed with the 'protection of the soul' argument ever since the industrial revolution and the changes in outlook initiated by Romantic writers and visual artists.[33] (Today's version of that mostly involves the pursuit of pleasure rather than salvation.) Specific protection of the non-human, in terms of wild species, woods and runoff rates followed at a fair number of decades. The significance to wider national considerations has also been debated in economic terms. No data for the moorlands' economy are collected separately, though hill sheep subsidies might point to some indicators of just one element of production. But what the 2001 Foot and Mouth Disease outbreak showed with some clarity is that the hills are far more important to the tourist industry than any other when money is the unit of measurement. Other protective functions such as water gathering, processing and supply, are not costed at all once the capital works of e.g., a reservoir are paid for. Hence the arguments for shifting the general direction of hill land management in favour of various kinds of protective function seem powerful. They would be much enhanced if the major lowland-upland link in this category – holding water back in the uplands in times of heavy rainfall and/or snowmelt – could be made stronger. If we look at some of the predictions from global warming science for moisture deficits in Great Britain then it is clear that water supply may be one of the great limiting factors in all kinds of socio-economic developments.

The outcome of these kinds of perceptions seems to be that a set of regions devoted 100 per cent to pleasure (i.e., as recreation areas, tourist Meccas, or just as scenery) is not in the national interest. Neither, it seems to be obvious, is full-tilt production even if its basis were to shift to lower-density, more environmentally-conscious animal-yield systems. Both these have been placed in the spotlights and gained star billing at the expense of the non-human and the non-domestic: the plant and animal communities, their energy and carbon metabolisms, the soil and peat cover of the uplands, and the water which on most visits to upland areas seems to be present in such abundance.[34]

The National Park Authorities have acknowledged (pp. 274–6) that their attention to nature conservation was poor for many years, a point repeated in the 2002 review of the role and functions of the Authorities.[35] Yet to devote large areas of land to 'nature', with production and pleasure taking lower-profile roles may seem odd, though it acknowledges immediately that it is the 'natural' systems which provide a great deal of the water used by major urban-industrial areas and is therefore part of their life-support system. Beyond that, it is possible to argue that if the UK wants to act responsibly in terms of the global environment, then there have to be shifts in material consumption patterns and in our understanding of the natural world. This latter is principally as a set of isolated phenomena, often with fences round them: urban 'farms', nature reserves, game parks, for example. There is no felt flow, no instinctive affinity with the non-human: all is *resource* of one kind or another. So there may be a renewed attention to nature everywhere via such measures as Biodiversity Action Plans and Habitat Action Plans but the uplands represent a special opportunity: the balance of trends in land use and land cover could tip towards the protective relatively readily.[36] Nevertheless, there has to be a debate over 'conservation' of nature: what exactly is 'nature' in historical terms? Is it largely what was inherited by the protectionist movements of the twentieth century, for example? Is the 'nature' of the nineteenth century (as distinct from today's or that of e.g., 1825) particularly what we wish to perpetuate? And, are the uplands today (as since perhaps the late eighteenth century and occasionally even earlier) run by values and demands that come from outside, especially the cities of the plains? If they are, is that necessarily wrong?

It is not possible to claim that if upland regions were devoted largely to the sustenance of nature then this would all change. But the presence of such large zones of a different set of systems – which were perceptually much wilder and more diverse than the present – would enhance the standing of 'nature' in our cognition of ourselves and our place in the world. The current acknowledgement of difference which underlies the designations and the popularity of the hills is a pointer that way. But follow the pointers and we come to bare and impoverished lands. Better then to follow them to a place of recognition of the ecological systems of which we are all still a part and on which we rely for life support. If some form of productive economy can be integrated with 'the wild' which acknowledges our connectivity but does not rely on, for example, large amounts of embedded energy brought from outside a given region, then this too is moving in a positive direction (Plate 8.3). If we seek a second priority after 'nature' then water management has a good case, since the uplands are not necessarily the region of surplus which we often imagine them to be, especially in summer.

Few British environmental activists and writers think much about Henry David Thoreau (1817–62), compared with the iconic status afforded to him by their equivalents in the USA. His fame relies mostly upon his journals from the time he lived alone near Walden Pond in Massachusetts. The hut was within sound of the railroad and he mentions the hooters of the locomotives as part of his sound-world, along with the bullfrogs and the birds. So he was not in what North Americans used to call a 'wilderness', where human presence is supposed to be absent. Yet one of his phrases was 'in wildness is the preservation of the world'. Since the Palaeolithic period we in Britain neither have had nor could have true wilderness in the uplands. Yet that slight difference in one word is our clue: we have an opportunity to improve the diversity and the amount of our wildness, and to hope – more is not possible – that our descendants' chances of living beyond the pure imperatives of survival are preserved.

PLATE 8.3 *Part of a hypothetical upland valley. Onto the view shown in the upper picture, is drawn a set of potential land use zones for an adjusted upland pattern. The uppermost zone (A) is still open moorland; then zone (B) unites some upper fields with existing woodland and scrub to form a deciduous forest zone. Zone (C) is devoted to agriculture, with a few patches of woodland to help keep up biodiversity. At (D), the recreation and tourism is centred and there could be a community focus there as well.*

NOTES AND REFERENCES

1. 'The Welsh Hill Country', in *Song at the Year's Turning*, London: Rupert Hart-Davis, 1955.
2. J. Thirsk. *Alternative Agriculture. A History from the Black Death to the Present Day*, Oxford: Oxford University Press, 1997.
3. The above paragraphs are based on J. E. Hossell, B. Briggs and I. R. Hepburn, *Climate Change and UK Nature Conservation*, London: DETR, 2000. Available in May 2001 on the detr.gov.uk website. See also UK Climate Impacts Programme (UKCIP) Technical Report No 1, *Climate Change Scenarios for the United Kingdom*, 1998. (A joint publication of UKCIP, The Meteorological Office and the Climatic Research Unit of the University of East Anglia); M. L. Parry, *Assessment of Potential Effects and Adaptations for Climatic Change in Europe: The Europe ACACIA Project*, Norwich: Jackson Environment Institute, 2000; N. C. Pepin, 'The use of GCM scenario output to model effects of future climatic change on the thermal climate of marginal maritime uplands' *Geografiska Annaler* 77 A, 1995, 167–85; N. C. Pepin, 'Scenarios of future climate change: effects on frost occurrence and severity in the maritime uplands of northern England', *Geografiska Annaler* 79 A, 1997, 121–37. A preliminary attempt at linking climate and land use was made by M. Hulme, J. E. Hossell and M. L. Parry, 'Future climate change and land use in the United Kingdom' *Geographical Journal* 159, 1993, 131–47. This is a very intensive research area and changes in data output and scenarios are frequent. Data from the Environmental Change Network (ECN) are available at http://www.ecn.ac.uk and other data and programmes at www.ukcip.org.uk/ukcip.html
4. R. S. Purves and N. J. R. Hulton, 'A climatic-scale precipitation model compared with the UKCIP baseline climate', *International Journal of Climatology* 20, 2000, 1809–21.
5. B. Huntley and R. Baxter, 'Climate change and wildlife conservation in the British uplands', in T. P. Burt *et al.* (eds), *The British Uplands: Dynamics of Change*. Peterborough: JNCC Report no. 319, 2002, 41–7.
6. P. Ineson *et al.*, 'Effects of climate change on nitrogen dynamics in upland soils. 1. A transplant approach', *Global Change Biology* 4, 1998, 143–52; P. Ineson *et al.*, 'Effects of climate change on nitrogen dynamics in upland soils. 2. A soil warming study', *Global Change Biology* 4, 1998, 153–61.
7. B. Huntley, 'Plant species' response to climatic change: implications for the conservation of European birds', *Ibis* 137 (Suppl.), 1995, S1127–38; B. Huntley, 'The responses of vegetation to past and future climate changes', *Ecological Studies* 124, 1997, 290–311; P. M. Berry, T. P. Dawson, P. A. Harrison and R. G. Pearson, 'Modelling potential impacts of climatic change on the bioclimatic envelope of species in Britain and Ireland', *Global Ecology and Biogeography* 11, 2002, 453–62.
8. Y. C. Collingham and B. Huntley, 'Impacts of habitat fragmentation and patch size upon migration rates', *Ecological Applications* 10, 2000, 131–44.
9. See for example North West Regional Chamber, *Everybody has an Impact. Climate Change Impacts in the North West of England*, Manchester: Sustainability North West, 1998 (www.snw.org.uk/enwweb); J. F. Farrar and P. Vaze (eds), *Wales: Changing Climate, Challenging Choices*, Cardiff: National Assembly for Wales, 2000. From 1997, the UK Climate Impacts Programme (UKCIP) has organised a number of research projects, including one on natural resources (Modelling Natural Resource Responses to Climate Change, MONARCH). Begin at www.ukcip.org.uk/ukcip.html

10. In December 2000 there were 10.8 million cattle and 27.6 million sheep in the UK, along with nearly 6 million pigs. Data from various DEFRA and NFU websites.

11. A summary of the progress of the disease and its immediate environmental effects is given by The Environment Agency, *The Environmental Impact of the Foot and Mouth Disease Outbreak: an Interim Assessment*, Bristol: The Environment Agency, 2001. Table 10 (p. 26) is especially helpful. The DEFRA reports published in 2002, *Origin of the UK Foot and Mouth Disease Epidemic in 2001*, and *Foot and Mouth Disease 2001: Lessons to be Learned Inquiry* are both interesting but not especially environmental in their findings.

12. The details were mostly posted on the websites of the Environment Agency and the Welsh Assembly. There was a great deal of local concern about air pollution from the burn (especially PM-10 and dioxin levels and the possibility of prions from BSE). In April 2001 the highest values were for PCBs about 1 km downwind from the pyre. The ash will be finally moved to a licensed landfill.

13. R. R. Kao, 'Landscape fragmentation and foot-and-mouth disease transmission', *Veterinary Record* **48**, 2001, 746–7. This finding is put into an epidemiological context in N. M. Ferguson, C. A. Donnelly and R. M. Anderson, 'Transmission intensity and impact of control policies on the foot and mouth epidemic in Great Britain', *Nature* **413**, 2001, 542–7.

14. At least a Welsh Assembly FMD Factsheet (47–25/06) uses this term in its information on how to obtain derogations from the various schemes. It could be seen at http://www.wales.gov.uk/newsflash/content/factsheets/series2/factsheet47.pdf in December 2001.

15. I. G. Simmons, 'The possible ecological consequences' [of FMD], *Landscapes* **2**, 2001, 13–18. This issue (vol. 2, no. 2) contains a mini-symposium on the outbreak.

16. In 2002, DEFRA and the Welsh Assembly were in the process of consultation and formulation of regulations.

17. The plots of pioneer species established by Geoffrey Dimbleby in the 1950s at Broxa in the North York Moors will perhaps become even more interesting.

18. B. Huntley, 1995 Note 7.

19. Data from the (then) MAFF/English Nature report of the *Task Force for the Hills*, 2001, p 74; accessible at www.defra.gov.uk/farm/hillsrep/report.pdf There were eight pages of recommendations.

20. Table 3.4, 'Upland Livestock' of Joint Nature Conservation Committee, *Environmental Effects of the Common Agricultural Policy and Possible Mitigation Measures*, Peterborough: JNCC, 2002. (This report covers all types of agriculture and not just the uplands.) Accessed on 12 December 2002 at www.defra.gov.uk/farm/sustain/envimpacts/envimpacts.pdf

21. Report of the Policy Commission on the Future of Farming and Food (Chairman: Lord Curry), *Farming and Food: a Sustainable Future*. London: DEFRA, 2002. Accessible at www.cabinet-office.gov.uk/index/CommissionReport.htm HM Government set out its response late in 2002, noting that the UK share of the EU RDR budget was low since it was based 'on our low level of agri-environment spending in the 1990s'.

22. N. Luhmann, *Ecological Communication*, Cambridge: Polity Press, 1989.

23. Even here, there is diversity. The excellent study by English Nature, *The Upland Challenge*, Peterborough: English Nature, 2001, has as its second sentence in the Summary (p. 5), 'behind the face of scenic beauty, however, the English uplands are suffering from economic crisis, social change and environmental degradation'. Read in conjunction with G. Cox, '"Reading" nature: reflections on ideo-

logical persistence and the politics of the countryside', *Landscape Research* **13**, 1988, 24–34.

24. It is often argued that the discourse of the natural sciences is not simply that of 'objective' description and explanation but that of control.

25. T. Stephenson, *Forbidden Land. The Struggle for Access to Mountain and Moorland.* Manchester and New York: Manchester University Press, 1989.

26. M. Shoard, *A Right to Roam*, Oxford: Oxford University Press, 1999.

27. In 2001, a trawl through First Search yielded 45 books and taped books, with some duplication of editions. Titles like *The Commonsense Book of a Countrywoman, Daughter of the Dales and Seasons of my Life* convey the message that was being constructed.

28. R. Woodward, 'Gunning for rural England: the politics of the promotion of military land use in the Northumberland National Park', *Journal of Rural Studies* **15**, 1999, 17–33.

29. See for example, A. Howkins, 'The discovery of rural England', in R. Colls and P. Dodd (eds), *Englishness. Politics and Culture 1880–1920*, London: Croom Helm 1986, 62–88.

30. In the 1990s, see for example, P. Cloke and N. Thrift, 'Class and change in rural Britain', in T. Marsden, P. Lowe and S. Whatmore (eds), *Rural Restructuring. Global Processes and their Responses,* London: David Fulton, 1990, 165–81; J. Murdoch and A. C. Pratt, 'From the power of topography to the topography of power', in P. Cloke and J. Little (eds), *Contested Countryside Cultures,* London and New York: Routledge, 1997, 51–69; K. Halfacree, 'Contrasting roles for the post-productivist countryside', in Cloke and Little (eds), *op.cit.,* 70–93.

31. M. Tilzey, 'Natural areas, the whole countryside approach and sustainable agriculture', *Land Use Policy* 17, 2000, 279–94.

32. There is a relevant general discussion of social and political contexts of the British countryside as a whole in T. Marsden, J. Murdoch, P. Lowe, R. Munton and A. Flynn, *Constructing the Countryside,* London: UCL Press, 1993; see also T. C. Smout, *Nature Contested: Environmental History in Scotland and Northern England since 1600,* Edinburgh: Edinburgh University Press, 2000. I raised the issue in the context of early landscape development in 'Prehistory and planning on the moorlands of England and Wales', *Landscape and Urban Planning* 17, 1989, 251–60.

33. It can be argued that philosophical developments since the seventeenth century have progressively distanced humans from nature, at any rate in the mind. To what extent this is reflected in representations of and discourses about moorlands is a subject for a full-length treatment. A humanistic introduction in general terms is provided by J. Bate, *The Song of the Earth,* London: Picador, 2000.

34. But many uplands are not now obvious sources of surplus water. The Environment Agency's strategy is on the whole upbeat about water demand and supply but certainly does not emphasise the uplands as a comparable document would have done until perhaps 25 years ago. Environment Agency, *Water Resources for the Future. A Summary of the Strategy for England and Wales,* London: Environment Agency, 2000.

35. DEFRA, *Review of National Park Authorities,* London: DEFRA, 2002. Accessed on 9 Jan 2003 at www.defra.gov.uk/wildlife-countryside/consult/natpark/report.pdf

36. D. B. A. Thompson, 'The importance of nature conservation in the British uplands: nature conservation and land use changes', in T. P. Burt *et al.* (eds), *The British Uplands: Dynamics of Change.* Peterborough: JNCC Report no. **319**, 2002, 36–40.

Locations

Most of the places in the text are given a National Grid reference here. Large areas usually have a four-figure number to the approximate centre of the area; more precise localities are given a six-figure reference. The list includes places not strictly relevant to the main theme but which are mentioned in the text; in the Conclusions chapter a couple of very specific references are given in the text.

Abbey Dore, Hereford and Worcester	SO 386304
Abbeystead, Forest of Bowland, Lancashire	SD 5654
Abercraf, South Wales	SN 8212
Aberfan, South Wales	SO 0700
Abergavenny, South Wales	SO 3014
Aberystwyth, West Wales	SN 5881
Ailsborough, Dartmoor	SX 592677
Allendale, North Pennines, Northumberland	NY 8555
Alston, Cumbria	NY 7146
Alston Moor, Cumbria	NY 7238
Anglezarke Moor, Lancashire	SD 6417
Appleby-in-Westmorland, Cumbria	NY 6820
Arkengarthdale, Yorkshire Dales	NZ 0102
Bainbridge, Yorkshire Dales	SD 9390
Barden Fell	SE 585085
Barden Moor	SE 575025
Barnard Castle, County Durham	NZ 0516
Barnstaple, Devon	SS 5533
Beamsley Beacon, West Yorkshire	SE 098524
Beaufort, South Wales	SO 1611
Beeley, Derbyshire	SK 2667
Bellever, Dartmoor	SX 6577
Ben Nevis, Highland Region, Scotland	NN 1671
Berwyn Mountains, North Wales	SJ 0634
Black Hill, North Dartmoor	SX 604846
Blacklane Brook, Dartmoor	SX 627688
Black Mountain, (*y Myndydd Du*) South Wales	SN 7618
Black Mountains, (*Mynyddoedd Duon*) Gwent and Brecon, at Capel-y-ffin	SO 255315
Black Ridge Brook, Dartmoor	SX 579842
Black Tor Copse, Dartmoor	SX 565890
Blaenavon, South Wales	SO 2509
Blaen Onnu, Brecon Beacons, South Wales	SO 155167
Blanchland, Northumberland	NY 9650
Bluewath Beck, North York Moors	NZ 7401

Bodmin Moor, Cornwall	SX 1876
Bolton Abbey, Yorkshire Dales	SE 0754
Bolt's Law, Weardale, County Durham	NY 950450
Bonfield Gill Head, North York Moors	SE 598958
Bossington Hill, Exmoor, Somerset	SS 905485
Boulby, North York Moors	NZ 7619
Bowland, Forest of, Lancashire,	SD 6554
Bowmont Valley, Cheviots, Northumberland	NT 8222
Bradfield, West Yorkshire	SK 2692
Bradford, Yorkshire	SE 1633
Braithwaite, South of Middleham North Yorkshire	NY 0624
Brampton East Moor, Derbyshire	SK 295700
Brassington, Yorkshire Dales,	SK 2354
Breckland, Norfolk	TL 8590
Brecon Beacons (*Bannau Brycheniog*), South Wales	SO 0121
Breiddin, mid-Wales	SJ 2915
Bridestowe, Devon	SX 5189
Broads, Norfolk	TG 4222
Broomhead, Pennines, North Yorkshire	SK 224951
Broomhead Moor, North Yorkshire	SK 2295
Brough, Cumbria	NY 7914
Broxa, North York Moors	SE 9495
Brynamman, South Wales	SN 7814
Bryn Mawr, South Wales	SO 1912
Buckfast Abbey, Devon,	SX 7467
Buckfastleigh, Devon,	SX 7366
Burrator, Dartmoor, Devon	SX 5568
Buxton, Peak District, Derbyshire	SK 0673
Byland Abbey, North York Moors	SE 5478
Calderdale, Yorkshire West Riding,	SD 8534
Cannock Chase, Staffordshire	SK 0017
Capel-y-ffin, Black Mountains, Gwent	SO 2531
Cardiff, South Wales	ST 1876
Carlisle, Cumbria	NY 4055
Carlton Bank, North York Moors	NZ 5203
Carnarvon, North Wales	SH 4862
Castle Bolton,Yorkshire Dales	SD 0392
Castleton, Derbyshire	SK 1583
Cauldron Snout, Upper Teesdale, County Durham	NY 8228
Cefn Gwernffrwd, mid-Wales	SN 740490
Central Marches, Shropshire and Herefordshire	SO 4872
Ceredigion, South Wales	SN 5060
Chatsworth House, Peak District, Derbyshire	SK 250701
Cheviot, The, Northumberland	NT 9020
Clargill Head mine, near Alston, Cumbria	NY 772364
Clouds, The, near Kirkby Stephen, Cumbria	SD 740000
Cloutsham, Exmoor	SS 892432
Clydach Terrace, Brynmawr, South Wales	SO 1813
Cobscar Mill, Yorkshire Dales	SE 057932
Cock Heads, North York Moors	NZ 720010
College Valley, Cheviots, Northumberland	NT 8928
Cow Green Reservoir, Upper Teesdale, County Durham	NY 8030

Cragside, Northumberland	NU 074022
Craig Ogwr, South Wales	ST 940946
Crawley Edge, Stanhope, Weardale, County Durham	NY 000396
Cribarth, South Wales	SN 830147
Cromford, Derbyshire	SK 2956
Cronkley Fell, Upper Teesdale, County Durham	NY 8427
Cronkley Pasture, Upper Teesdale	NY 855292
Crosby Ravensworth, Cumbria	NY 6214
Cross Fell, highest point of North Pennines, Cumbria	NY 6834
Cwmbrân, South Wales	ST 2995
Cwmhir Abbey, Powys, mid-Wales	SO 0571
Cwmparc, South Wales	ST 950960
Dalby Forest, North York Moors,	SE 8888
Dale Dike Reservoir, near Bradford, West Yorkshire	SK 2491
Danby, North York Moors	NZ 7008
Dark Peak, The, Derbyshire	SK 0590
Dart Estuary, Devon	SX 8750
Deepcar, West Yorkshire	SK 293982
Derby, Derbyshire	SK 3536
Devonport, Devon	SX 4554
Dolaucothi, mine, Powys, mid-Wales	SN 660400
Dolgellau, North Wales	SH7217
Downholland Moss, Lancashire	SD 320208
Dufton Moss, Upper Teesdale, County Durham	NY 872293
Dunford Bridge, Pennines, West Yorkshire	SE 137010
Durham City, County Durham	NZ 2642
East Carmarthenshire, South Wales	SN 6020
East Dart Head, North Dartmoor	SX 608854
East Friar Houses, Upper Teesdale, County Durham	NY 895285
Eastgate, Weardale, County Durham	NY 955385
East Moors, Derbyshire	SK 2868
Ebbw River, South Wales	ST 2489
Edale, Derbyshire	SK 1285
Eden, vale of, Cumbria (at Appleby-in-Westmoreland)	NY 686204
Eggesford, Devon	SS 6811
Egglestone, Teesdale, County Durham	NY 9923
Elan Valley, mid-Wales	SN 857756
Endcliffe, Peak District, Derbyshire	SK 250638
Eskdale, Cumbria	NX 1801
Eskmeals, Cumbria	SD 089925
Esk valley, North York Moors	NZ 8808
Exmoor, Somerset	SX 8035
Exmoor Forest, Somerset	SX 7642
Extwhistle Moor, Lancashire	SD 909342
Fan Frynych, South Wales	SN 955228
Featherbed Moss, Derbyshire	SK 098928
Ferndale, South Wales	ST 0097
Ffestiniog, North Wales	SH 7042
Fforest Fawr, South Wales	SN 8919
Ffos y Fran, South Wales	SO 070060
Flow area, Caithness and Sutherland	NC 8050
Forest and Frith, Upper Teesdale, County Durham	NY 8432

Hunterston, Strathclyde, Scotland	NS 1752
Ilkley Moor, West Yorkshire	SE 1146
Ingleborough, North Yorkshire	SD 7474
Isle of Man, Irish Sea	SC 3281
Isle of Wight, South England	SZ 4985
Jervaulx Abbey, North Yorkshire	SE 1785
Keld Heads mill, Swaledale, Yorkshire Dales	SE 0891
Kempswithen, North York Moors	NZ 655085
Kielder Forest, Northumberland	NY 6288
Killhope, Weardale, County Durham	NY 828431
Kilnsey, Wharfedale, Yorkshire Dales	SD 9767
Kinder Plateau, Peak District, Derbyshire	SJ 0888
Kinder Scout, Peak District, Derbyshire	SK 1487
Kirkby Stephen, Cumbria	NY 7708
Knaresborough, North Yorkshire	SE 3557
Kitty Tor, North Dartmoor	SX 567874
Lady Clough Moor, Derbyshire	SK 104921
Lake Vyrnwy, mid-Wales	SH 9921
Lammermuir Hills, South-East Scotland	NT 5762
Lanchester, County Durham	NZ 1647
Landerfel, South Wales	ST 264952
Langdon Beck, Upper Teesdale, County Durham	NY 853312
Langley Mill, Northumberland	NY 830612
Langridge, Dartmoor	SX 655805
Leeds, Yorkshire	SE 3034
Lee Moor, Dartmoor	SX 5761
Lees Cross, Peak District	SK 250638
Leicester, Leicestershire	SK 5804
Levisham Moor, North York Moors	SE 835925
Leyburn, Wensleydale, North Yorkshire	SE 1190
Lindisfarne, Northumberland	NU 1443
Littledale, Forest of Bowland (Lancs)	SD 575618
Liverpool, Merseyside	SJ 3490
Llanbedr, The Black Mountains, South Wales	SO 2420
Llantarnam, South Wales	ST 308933
Llanthony Abbey, Gwent, Wales	SO 288278
London, South-East England	TQ 3079
Longdendale, Derbyshire	SJ 0497
Long Mynd, Shropshire	SO 4193
Loxidge Tump, Honddu Valley, Gwent, South Wales	SO 288292
Low Force, Upper Teesdale, County Durham	NY 903280
Lyn y Fan Fach, South Wales	SN 803218
Malham, Yorkshire Dales	SD 9063
Malham Tarn Yorkshire Dales	SD 880660
Manchester, Greater Manchester	SJ 8598
Margam, Glamorgan, South Wales	SS 7887
Marks Tey, Essex	TL 910240
Marrick in Swaledale, North Yorkshire	SE 0798
May Moss, North York Moors	SE 875960
Meldon Hill, North Pennines, Cumbria	NY 771291
Mendip Hills, Somerset	ST 5055
Menwith Hill, North Yorkshire	SE 205575

Merthyr Tydfil, South Wales	SO 0506
Merthyr Vale, South Wales	ST 0799
Mickle Fell, North Pennines, County Durham	NY 805245
Middleham, North Yorkshire	SE 1287
Middlehope, Weardale, County Durham	NY 905395
Middleton-in-Teesdale, County Durham	NY 9425
Mill Close mine, Darley Dale Derbyshire	SK 2663
Monmouth, South Wales	SO 5113
Moor House, North Pennines, Cumbria	NY 758328
Muggleswick, County Durham	NZ 0450
Mulgrave, North York Moors	NZ 8412
Mynydd Eppynt, Powys, mid-Wales	SN 9543
Mynydd Hiraethog, North Wales	SJ 9555
Mynydd Maendy, South Wales	SS 980860
Mynydd Preseli (Prescelly Mtns), West Wales	SS 1033
Mynydd Ty'n Tyle, South Wales	SS 9897
Neath, Glamorgan, South Wales	SS 7497
Nenthead, Cumbria	NY 7843
Newbiggin, Teesdale, County Durham	NY 9127
Newcastle upon Tyne, Tyne and Wear	NZ 2464
New Forest, Hampshire	SU 2806
Nidderdale, North Yorkshire	SE 1272
North Gill, North York Moors	NZ 726007
North Tynedale, Northumberland	NY 8583
Nottingham, Nottinghamshire	SK 5741
Nun's Cross Farm, Dartmoor	SX 606698
Offa's Dyke path, (part of), Powys/Herefordshire	SO 290295
Okehampton, Devon	SX 5895
Okehampton Park, Dartmoor	SX 5894
Oldham, Lancashire	SD 9204
Otterburn, Northumberland	NY 8893
Pately Bridge, North Yorkshire	SE 1565
Peak Forest, Derbyshire	SK 1179
Pembrokeshire Coast, West Wales	SM 8520
Pendle, Lancashire	SD 8142
Pentre Ifan, ancient monument, West Wales	SN 099369
Pen-wyllt, South Wales	SN 856160
Piles Copse, Dartmoor	SX 644622
Pinkery Pond, Exmoor	SS 723423
Pinswell, Dartmoor	SX 582837
Plym Estuary, Devon	SX 5054
Plymouth, Devon	SX 4755
Pontypool, South Wales	SO 2800
Porlock, Somerset	SS 8846
Port Talbot, Glamorgan, South Wales	SS 7690
Postbridge, Dartmoor	SX 6579
Preseli, moorland in West Wales, at Mynydd Carningli	SD 056370
Princetown, Dartmoor	SX 5873
Pule Hill Base, West Yorkshire	SE 032100
Pwll Byfre, South Wales	SN 875167
Quantock Hills, Somerset	ST 1537
Radnor-Clun Forests, Wales	SO 2064

Ramshaw Rocks, Staffordshire   SK 020625
Rassau, South Wales   SO 148125
Rattlebrook Head, Dartmoor   SX 560857
Red Lake, Dartmoor   SX 645665
Reeth, Swaledale, North Yorkshire   SE 0399
Resolven, South Wales   SN 848023
Rhondda Fach, South Wales   ST 0096
Rhondda Fawr, South Wales   SS 9994
Rhosgoch Common, mid-Wales   SO 190480
Ribblehead, Yorkshire Dales   SD 760795
Riding Mill, Northumberland   NZ 0161
Rievaulx Abbey, North Yorkshire   SE 5784
River Avon, Dartmoor   SX 7157
River East Allen, Northumberland   NY 8354
River Erme, Dartmoor   SX 6355
River Greta, North Yorkshire   SD 6271
River Honddu, South Wales   SO 0236
River Irthing, Cumbria/Northumberland   NY 6670
River Lyd, Devon   SX 4683
River Ouse, South Yorkshire   SE 4959
River Taff, South Wales   ST 1777
River Usk, South Wales   SO 2615
River Walkham, Dartmoor   SX 530705
Roaches, The, Staffordshire   SK 011614
Robinson's Moss, Derbyshire   SE 045001
Rookhope, North Pennines, County Durham   NY 9442
Rosebush, Preseli Mountains, West Wales   SN 075295
Rosedale, North York Moors   SE 7396
Rosedale Moor, North York Moors   NZ 695015
Rossendale, Lancashire   SD 8525
Rothbury, Northumberland   NU 0501
Rothbury Forest, Northumberland   NU 0802
Rough Tor, Bodmin Moor   SX 1482
Rough Tor, Dartmoor   SX 6079
Rudland, North York Moors   SE 655940
Ryedale, North York Moors   SE 5785
St Austell Moor, Cornwall   SW 980535
St John's Chapel, Weardale, County Durham   NY 8838
Sallay Abbey, Lancashire   SD 776464
Scammonden, Pennines nr Huddersfield   SE 040154
Scamridge Dykes, North York Moors   TA 9085
Scarborough, North Yorkshire   TA 0488
Scar House Dam, Nidderdale, North Yorkshire   SE 0577
Scugdale, North York Moors   SE 523992
Sellafield, Cumbria   NY 0204
Semer Water, Wensleydale, North Yorkshire   SD 9287
Sennybridge, South Wales   SN 9228
Settle, North Yorkshire   SD 8163
Severn Valley, Montgomery   SJ 0233
Severn Valley, Gloucestershire   SO 8530
Shap, Cumbria   NY 5614
Shaugh Moor, Dartmoor   SX 555635

Glossary

Adit mine
A mine where access is by a near-horizontal passage driven into the hillside, usually following a mineral vein or a coal seam.

Allelopathic
Referring to the effects of chemicals produced by plants upon other organisms. Many plants, notably bracken, produce chemical compounds which inhibit the establishment and growth of competitors, either of their own or other species. Chemical toxins in plants can also deter grazing animals.

Alluvium
Usually fine grained sediment carried and deposited by water, as in river valley flood plains. Fertile alluvial soils can support rich wetland ecosystems and if well drained are good for crops and meadow. The process of the accumulation of alluvium is called alluviation.

Ancient woodland
Woodland known to have been continually in existence since AD 1600.

Assarting
From the Latin *ex* 'out of' and *sarire* 'to hoe or weed'. To develop or reclaim land for agriculture from forests or wastes. In the medieval period it indicates rising populations and a consequent demand for land, usually for food crops but also sometimes for pasture.

Biodiversity
The number of different species supported in a given area of land or water.

Biophysical Systems
Systems on the earth's surface which consist of biological and physical components. Usually employed as a shorthand for 'natural' systems without human presence or intervention.

Biota
A group of plants and animals occupying a space together.

Blanket Bog
Peat developed directly over mineral ground on a low angle of slope, normally in an upland environment where the rainfall is high and year-round. It is commonly 1–2 m thick but can get to 5–6 m.

Bloomery
A furnace for smelting iron ore into the metal.

Bole
A simple furnace for smelting lead ore.

Broadleaved or Deciduous Woodland
Woodland dominated by deciduous trees, which lose their leaves in winter. It formed the dominant ecosystem from the mid-postglacial onwards. Although none of this original **wildwood** survives there are extensive areas of broadleaved woodland in lowland Britain today, supporting a rich and diverse ecosystem. The upland summits may never have been fully colonised by this vegetation type in the past and there are few upland broadleaved woods today.

Brown Earths
Typical but not universal fertile soil of deciduous forest, also called 'forest brown earths'. A mixture of organic materials and minerals from the ground rock. The organic compound decreases with depth and the soil is usually slightly acid. It has a humic A horizon and a B horizon in which minerals are released by weathering.

Chopwood
Wood cut into short lengths and split to make it suitable for burning in a furnace.

Colluvium
Mineral sediment moved downslope by erosive processes and deposited at the foot of the slope or in a valley.

Coniferous Woodland
Woodland dominated by coniferous trees. In Britain Scots Pine (*Pinus sylvestris*) is the only native conifer although many other types have now been planted. Coniferous woodland has its own particular ecosystem and range of flora and fauna associated with it.

Corrie
In geomorphology, a large hollow with a steep back wall cut by ice. A small lake ('tarn' in English, 'llyn' in Welsh) often occupies the hollow. More common in mountains than on moorlands. (In popular parlance, a long-running soap-opera but about city life, not the uplands.)

Crown fire
A forest fire in which the crowns of the dominant trees are ablaze and not just smaller trees or shrubs. Such conflagrations are usually confined to coniferous stands. It may start with a fire on a fuel-rich forest floor which then runs up the tree-trunks (especially if aided by resin drips) into the crowns. Once established and fanned by a moderate or strong wind then it is very difficult to extinguish.

Discourse
The language and terms in which a debate is conducted or in which investigations are made, or in which understanding is sought. The rules of the natural sciences might constitute one such discourse, the terminology, assumptions and legislation of development planning another.

Doline
A dry-floored depression in limestone terrain.

Drift
Unconsolidated superficial deposits, especially those left behind by glaciers.

Dwarf Shrub Heath
Plant community dominated by ericaceous shrubs like crowberry and other shrubs of low stature, often willow. A particular stage in Late Glacial plant succession after tall herb communities and prior to the establishment of tall shrub and then tree cover. Today its most common representative on the moorlands is heather (*Calluna vulgaris*) moor, followed by stands of crowberry (*Empetrum nigrum*).

Ecotone
The narrow transitional boundary between different ecological communities, typically species rich, such as the land/water interface.

Ecumene
The entire inhabited world.

Enclosure
A farming practice in which fields are hedged, fenced or walled and owned by an individual. It usually replaces 'open' fields where the land is held in common by a community. It was common in the lowlands from late medieval times onwards and completed by Parliamentary enclosure from the mid-eighteenth century. The uplands were sometimes different since any land taken in from the moor was likely to be enclosed.

Feral
A wild population of animals descended from domesticated stock.

Flush
An area of sediment and vegetation which is normally wet year-round and where a spring brings mineral-rich water to the surface.

Fold-Yard
An enclosed space in a settlement dominated by stock rearing, clearly for the purpose of containing the animals for counting or treatment. In some cases it is thought that the fold-yards of monastic granges have become the cores of subsequent hamlets.

Forest
After the Norman conquest, this becomes a legal term for lands preserved for the King's hunting. (If granted to lesser aristocracy then it is termed a Chase.) There may or may not be trees. In text, the use of capital 'F' usually means that the legal meaning is implied.

Forestry
The modern commercial and strategic practice of growing trees in managed areas such as plantations.

Gavelkind
A tenure by which lands descended from the father to all sons (or failing sons, to all daughters) in equal portions, and not by primogeniture.

Gley Soil
A soil where the water table is near the surface and there may be seasonal waterlogging, often beside water bodies or over impervious rock. Surface humic matter overlies a mottled layer where alternating oxidation and reduction has occurred.

Gleyed
Soils affected by the processes of waterlogging, oxidation and reduction that result in gley soil.

Granges
Granges were mostly farms and livestock rearing centres created by monastic communities colonising uplands and wastes in the early Middle Ages and were staffed by lay brethren. Fountains Abbey in Yorkshire had twenty-six granges, the largest number attached to any monastic house.

Gripping
The digging of narrow channels in upland peats to drain their surfaces. The usual results are increased erosion and gulleying.

Headage payments
Subsidies paid to farmers on the basis of the number of animals which they kept.

Head-dyke
The bank-and-ditch marking the upward limit of improved land, thus demarcating the lower boundary of moorland.

Heaf
An area of moorland to which a flock of sheep have a territorial attachment.

Holocene
The present warm epoch of the Quaternary period. It commenced c.10,000 BP and in Britain is also called the Flandrian interglacial stage. We usually assume that the glacial phases of the Pleistocene are finished for ever, without any evidence.

Hypsithermal Interval
The period of highest Holocene temperatures. In temperate latitudes, it was not necessarily synchronous; in the British Isles, it was more or less over by the time of the adoption of agriculture in the Neolithic.

Improvement
An agricultural development characteristic of the lowlands in which the productive capacity of the land was raised, especially for arable crops. As a movement, it began about 1730 and made the uplands look even less unproductive and hence in need of enclosure and 'improvement'.

Interglacial
A period of warmer climate within an Ice Age, when ice-sheets or very cold temperatures are absent and forests can colonise the land. Sea-levels become high as ice-sheets melt. A brief temperate period in a generally much colder era is called an **interstadial**.

Isostatic Recovery
Vertical readjustment of the earth's crust which has been depressed by the weight of a glacial ice sheet. Land at the centre of ice sheet loading may rise by well over a hundred

metres or more after ice removal. Land nearer the ice margins will need less recovery. Rapid at first, the process takes millennia and is still taking place in Scotland.

Karst
Limestone terrain without surface drainage and with outcrops of bare rock.

Lapse rate
The fall-off of temperature with altitude.

Lazy-beds
An upland method for growing crops, especially potatoes. The seed is laid on the surface and covered with soil dug from trenches on either side, thus providing for drainage.

Lithics
The approach to stone age cultures through their tool types and numbers; microliths are very small stone tools.

Lynchet
An agricultural terrace caused by arresting downhill soil movement and creating a flat strip available to be ploughed.

Mesolithic
The middle stone age, between the end of the Upper Palaeolithic and the advent of the Neolithic. Defined culturally by the use of very small flint points called microliths, as well as a range of other diagnostic tool types. Used also to describe the hunting and gathering communities using this microlithic culture and their way of life. It was present during the postglacial transition from open terrain to wooded environments.

Montane Areas/Zones
Mountains and plateau summits, which support characteristic plant and animal communities because of their high altitude soils and climate. Few moorlands are high enough to carry this flora and fauna.

Mor
A type of conifer forest humus layer of organic material usually matted or compacted, it is very acid with little microbial content except for fungi. Mor is also found on open heath and moorland in moist cool climates. In print it is sometimes italicised.

Nature
Commonly defined as environmental conditions and forces unaffected by human agency. The realisation that both the concept and the material reality are subject to culturally contingent cognition leads some writers to use the term as 'nature'.

Neolithic
The period of prehistory in which domesticated plants and domesticated animals other than the dog were introduced to Britain, about 4000 BC

Net Primary Production
The amount of actual plant tissue produced by a living organism or set of organisms per unit area per unit time (e.g., gm per sq m per year or kilos per hectare per year).

Numinous
A sense of divinity.

Ombrogenous
Peat growth above the water table and so fed entirely by rainwater. Ombrogenous peat is very acid and nutrient poor and so supports a limited range of acid-tolerant plants, mainly *Sphagnum* mosses.

Orographic
Most frequently used in the sense of precipitation whose annual totals rise with altitude.

Orthostats
Erect standing stones.

Palaeolithic — see **Upper Palaeolithic.**

Pastoralism
An agricultural system based on the management of domesticated animals, usually in herds, as with cattle, goats and sheep. Sometimes implicit in the term is the idea of seasonal movement (see **transhumance**) so that pig pastoralism is not encountered.

Perambulations
The marking of the boundaries of estates or parishes by walking round them and then applied to documents which set out the limits formally. They were appended to legal charters from Anglo-Saxon times onwards and might contain details of land management.

Periglacial
The zone outside an ice sheet or glacier but which is subject to very cold conditions such as diurnal freeze-thaw. Material tends to move downslope and mature soils are unable to form; plants tend to be **ruderal** in habit.

Perimarine
The zone of freshwater plant communities and sedimentation adjacent to the coast above the limit of direct influence by marine conditions but where there is indirect marine influence through changes in the groundwater table or salt spray.

Pleistocene
The geological term for the older period of the Quaternary, ending about 10,000 years ago.

Podsol
Acid soils developed in areas of high rainfall and rapid drainage. The leaching effect produces an E horizon of ash grey colour due to the removal of iron and humus and a B horizon where the iron and humus accumulate sometimes forming iron pan. Podsols are characteristic of heaths and moorlands and are mostly caused by deforestation.

Production Ecology
The science of measuring the amount of living tissue produced per unit area per unit time. Usually expressed as dry weight.

Regolith
The mantle of loose weathered material which overlies solid rock.

Retrodict
To predict backwards. A hopefully more precise form of reconstruction.

Return period
An estimate of the likely number of years that will pass before an event happens again. In upland England, for example, the likely interval for a moor burn of a single location from grouse management is 15–20 years, from lightning-set fires perhaps thousands of years.

Ruderal
The life strategy of plants which flourish on open, disturbed ground and are intolerant of competition. Examples of such conditions were found in the uplands at the end of the Pleistocene under periglacial conditions and again on the spoil heaps of mines.

Shieling
A place of summer pasture and dwelling when **transhumance** is practised. The word is of Norse origin (cf Nor. *sæter*). In Welsh the equivalent word is *hafod*.

Solifluction
The slow mass movement downhill of rock waste and debris saturated with water. Most commonly found under **periglacial** conditions.

Stagnopodsol
A very poorly drained, often gleyed, podsolic soil.

Stannaries
Tin-mining districts.

Succession
A progressive set of changes in an ecosystem from initial colonisation of a bare surface or open water to a mature and self-replicating system.

Swaling
A dialect word for burning off heather or gorse, for example.

Swangs
Linear drainage channels cut by meltwater on the North York Moors during deglaciation. Now sediment-filled.

Sward
A land surface dominated by grasses.

Taphonomic
The processes by which fossil materials are transported from their place of origin (usually that of their death) to the point of deposition and incorporation in preservative sediments.

Tephra
Pyroclastic material (hot volcanic rock) which is fine enough to be transported by air. Tephra deposits can be found a long way from their volcano of origin and can be dated by potassium-argon or fission-track dating, or by radiocarbon dating of associated organic remains. Each tephra has its own chemical signature and can be referred to a particular eruption.

Till
Fine, usually unstratified glacial deposits; boulder clay.

Tillage
The cultivation of the soil, starting with ploughing or digging.

Tor
An exposure of rock in situ that forms a conspicuous outcrop on a valley side or at a summit.

Transhumance
Agricultural system involving seasonal movement of animals between upland and lowland pastures. This is normally a summer movement upwards and the products are (a) well-fed animals and (b) preservable dairy items such as cheese. In upland Britain, declining by the seventeenth century and pretty well extinct by the end of the eighteenth century.

Turbary
The right to take peat from another's land, or common land, or simply the place where peat is dug.

Upper Palaeolithic
The most recent part of the old stone age, contemporary with the latter part of the last glacial period. Hunter-gatherer communities with a tool typology based upon flint blades and in Britain occurring within the mainly open habitat environments characteristic of the glacial and late glacial periods, although conditions were probably too severe for settlement during glacial maxima.

Vaccary (pl. vaccaries)
Cattle stations established in medieval times for the utilisation of upland areas. The high open moorland was ideal for cattle grazing in summer and many cattle stations were established, notably in the Pennines, by ecclesiastical houses or individual land holders.

Waste
From medieval times onwards, unreclaimed land and generally not subject to permanent settlement. It was not unproductive, since it was a repository of common resources such as grazing, turbary and other fuels. Eventually the term meant any unenclosed land covered with wetland, moor, heath or scrub. In academic work, the time spent in unconstructive bureaucratic tasks.

Wildwood
The concept of large swathes of more or less unbroken forest, apparently unaffected by human agency. The more research is done, the less it seems likely that the large forested areas of the mid-Holocene were actually totally pristine in terms of human impact. (Most people encounter the word in *The Wind in the Willows*.)

Wood-pasture
An area of grassland with a variable amount of trees in it, but tending to the open rather than being an obviously woodland formation. Created by management from denser woodland.

Zeitgeist
The spirit of the times. A German word, it is sometimes italicised.

Acronyms and Abbreviations

Key
Italics – organs of central government
<u>Underline</u> – designations by central government, including those which emplace European Union legislation

AONB – <u>Area of Outstanding Natural Beauty</u>
ASL – Above Sea Level
BAP – Biodiversity Action Plan
BMEWS – Ballistic Missile Early Warning Station
BP – Before Present
CAP – Common Agricultural Policy of the European Union
CCW – *Countryside Council for Wales*
CLA – County Landowners Association (now the Country Land and Business Association)
CNP – Council for National Parks
CPRE – Council for the Protection of Rural England
CROW – Countryside and Rights of Way Act 2000
DEFRA – *Department of Environment, Food and Rural Affairs,* which succeeded MAFF in 2001
DETR – *Department of Transport, Environment and the Regions*
EIA – Environmental Impact Assessment
ESA – <u>Environmentally Sensitive Area</u>
EU – European Union
FMD – Foot and Mouth Disease
GCM – Global Change Model
HAP – Habitat Action Plan
JNCC – Joint Nature Conservation Committee
JSCNP – Joint Steering Committee for National Parks
LFA – Less Favoured Area
LIA – Little Ice Age
LUS – Land Utilisation Survey (conducted in the 1930s)
MAFF – *Ministry of Agriculture, Fisheries and Food,* which became DEFRA in 2001
MoD – *Ministry of Defence*
MSS – Moorland Stewardship Scheme
NA – Natural Area
NFU – National Farmers' Union
NNR – <u>National Nature Reserve</u>
NPA – National Park Authority
NPMP – National Park Management Plan
OD – Ordnance Datum (mean sea level)
RSPB – Royal Society for the Protection of Birds

SAC – <u>Special Area of Conservation</u> (an EU designation)
SPA – <u>Special Protection Area</u>
SSSI – <u>Site of Special Scientific Interest</u>
WCA – Wildlife and Countryside Act 1981
YHA – Youth Hostels Association

Select bibliography

Atherden, M. A., *Upland Britain. A Natural History*, Manchester: Manchester University Press, 1992.

Bate, J., *The Song of the Earth*, London: Picador Books, 2000.

Blunden J. and N. Curry (eds), *'A People's Charter'? Forty Years of the National Parks and Access to the Countryside Act 1949*, London: HMSO, 1990.

Brown A. G. and T. A. Quine (eds), *Fluvial Processes and Environmental Change*, Chichester: Wiley, 1999.

Charman, D., *Peatlands and Environmental Change*, Chichester: Wiley, 2002.

Clapham, A. R. (ed.), *Upper Teesdale*, London: Collins, 1978.

Collins, E. J. T., *The Economy of Upland Britain 1750–1950: an Illustrated Review*, Reading: University of Reading Centre for Agricultural Strategy, 1978.

Department of the Environment, *Report of the National Park Policies Review Committee*, London: HMSO, 1974 [the Sandford Report].

DETR, *Protection and Better Management of Common Land in England and Wales*, London: Department of Environment, Transport and the Regions, 2000.

Edwards, N. (ed.), *Landscape and Settlement in Medieval Wales*, Oxford: Oxbow Books, 1997.

English Nature, *State of Nature. The Upland Challenge*, Peterborough: English Nature, 2001.

Farrar, J. F. and P. Vaze (eds), *Wales: Changing Climate, Challenging Choices*, Cardiff: National Assembly for Wales, 2000.

Fielding, A. F. and P. F. Haworth, *Upland Habitats*, London: Routledge, 1999.

Fleming, A., *Swaledale: Valley of the Wild River*, Keele: Keele University Press, 1997.

Forestry Commission, *A New Focus for England's Woodlands*, Cambridge: Forestry Commission, 2001.

Hassan, J., *A History of Water in Modern England and Wales*, Manchester and New York: Manchester University Press, 1998.

Himsworth, K. H., *A Review of Areas of Outstanding Natural Beauty*, Cheltenham: Countryside Commission CCP **140**, 1980.

Hossell, J. E., B. Briggs and I. R. Hepburn, *Climate Change and UK Nature Conservation*, London: DETR, 2000.

Hudson, P., *Red Grouse. The Biology and Management of a Wild Gamebird*, Fordingbridge: The Game Conservancy Trust, 1986.

Hughes T. and F. Godwin, *Remains of Elmet. A Pennine Sequence,* London and Boston: Faber & Faber, 1979.

Leighton, D. K., *Mynydd Du and Fforest Fawr. The Evolution of an Upland Landscape in South Wales*, Aberystwyth: RCAHM Wales, 1997.

MacEwen A. and M. MacEwen, *National Parks, Conservation or Cosmetics?*, London: Allen and Unwin, 1982.

Miller, G. R., J. Miles and O. W. Heal, *Moorland Management: a Study of Exmoor*, Cambridge: ITE, 1984.

Musson C. (ed.), *Wales from the Air. Patterns of Past and Present*, Aberystwyth: RCAHM Wales, nd.

Myers, A. and R. Forsythe, *W. H. Auden. Pennine Poet,* Nenthead: North Pennines Heritage Trust Publication no. 7, 1999.

Orwin, C. S. and R. J. Sellick, *The Reclamation of Exmoor Forest,* Newton Abbot: David & Charles, 1970.

Pearsall, W. H., *Mountains and Moorlands,* London: Collins New Naturalist, 1950.

Raistrick A. and B. Jennings, *A History of Lead Mining in the Pennines,* Newcastle: Davis Books/Littleborough: Kelsall Publishing, 1983.

Ratcliffe, D. A., *Bird Life of Mountain and Upland,* Cambridge: Cambridge University Press, 1990.

Riley, H. and R. Wilson-North, *The Field Archaeology of Exmoor,* Swindon: English Heritage, 2001.

Sheail, J., *An Environmental History of Twentieth-Century Britain,* Basingstoke and New York: Palgrave, 2002.

Shoard, M., *A Right to Roam,* Oxford: Oxford University Press, 1999.

Simmons, I. G., *The Environmental Impact of Later Mesolithic Cultures. The Creation of Moorland Landscape in England and Wales,* Edinburgh: Edinburgh University Press, 1996.

Smith, R. T. and J. A. Taylor (eds), Bracken. *Ecology, Land Use and Control Technology,* Lancaster: Parthenon Press, 1986.

Smout, T. C., *Nature Contested. Environmental History in Scotland and Northern England since 1600,* Edinburgh: Edinburgh University Press, 2000.

Spratt D. A. and B. J. D. Harrison (eds), *The North York Moors: Landscape Heritage,* Newton Abbot & London: David & Charles, 1989.

Stephenson, T., *Forbidden Land. The Struggle for Access to Mountain and Moorland,* Manchester: Manchester University Press, 1989.

Tallis, J. H., R. Meade and P. D. Hulme (eds), *Blanket Mire Degradation. Causes, Consequences and Challenges,* Aberdeen: Macaulay Land Use Research Institute, 1997.

Tapper, S., *Game Heritage,* Fordingbridge: The Game Conservancy Trust, 1992.

The Environment Agency, *The Environmental Impact of the Foot and Mouth Disease Outbreak: an Interim Assessment,* Bristol: The Environment Agency, 2001.

Thirsk, J., *Alternative Agriculture. A History from the Black Death to the Present Day,* Oxford: Oxford University Press, 1997.

Thompson, D. B. A., A. J. Hester and M. B. Usher (eds), *Heaths and Moorland: Cultural Landscapes,* Edinburgh: HMSO, 1995.

Trezise, S. D., *The West Country as a Literary Invention,* Exeter: University of Exeter Press, 2000.

Tsouvalis, J., *A Critical Geography of Britain's State Forests,* Oxford: Oxford University Press, 2000.

Usher, M. B. and D. B. A. Thompson (eds), *Ecological Change in the Uplands,* Oxford: Blackwell Scientific Publications, Special Publications of the British Ecological Society no 7, 1988.

Vyner B. (ed.), *Moorland Monuments: Studies in the Archaeology of North-East Yorkshire in Honour of Raymond Hayes and Don Spratt,* London: CBA Research Report 101, 1995.

Wheeler, D. and J. Mayes (eds), *Regional Climatology of the British Isles,* London and New York: Routledge, 1997.

White, R., *The Yorkshire Dales. A Landscape through Time,* Ilkley: Great Northern Books, 2002.

Winchester, A. J. L., *The Harvest of the Hills. Rural Life in Northern England and the Scottish Borders 1400–1700,* Edinburgh: Edinburgh University Press, 2000.

Index

Note: animals and plants will be found under the headwords **fauna** and **flora** respectively